Assessment of Ventricular Function

Ettore Majorana International Science Series
Series Editor:
Antonino Zichichi
European Physical Society
Geneva, Switzerland

(LIFE SCIENCES)

A Continuation Order Plan is available for this series. A continuation order will bring delivery of each new volume immediately upon publication. Volumes are billed only upon actual shipment. For further information please contact the publisher.

Assessment of Ventricular Function

Edited by

Angelo Raineri
University of Palermo Medical Center
Palermo, Italy

Robert D. Leachman
Texas Heart Institute
Houston, Texas

and

Jan J. Kellermann
The Chaim Sheba Medical Center
Tel-Hashomer, Israel

Plenum Press • New York and London

Library of Congress Cataloging in Publication Data

Main entry under title:

Assessment of ventricular function.

(Ettore Majorana international science series. Life sciences; v. 21)
"Proceedings of a symposium on Assessment of Ventricular Function, held Oct.
3–9, 1983, at the Ettore Majorana Center for Scientific Culture, in Erice, Sicily, Italy"—
T.p. verso.
1. Heart function tests—Congresses. 2. Heart—Ventricles—Diseases—Diagnosis—
Congresses. I. Raineri, A. II. Leachman, R. D. III. Kellermann, Jan J. IV. Title. V. Series:
Ettore Majorana international science series. Life sciences; 21. [DNLM: 1. Heart Function Tests—congresses. 2. Heart Ventricle—physiology—congresses. W1 ET712M
v.21/WG 202 A8461 1983]

RC683.5.H4A78   1983   616.1′2   85-17007
ISBN 978-1-4684-8005-4   ISBN 978-1-4684-8003-0 (eBook)
DOI 10.1007/978-1-4684-8003-0

Proceedings of a symposium on Assessment of Ventricular Function,
held October 3–9, 1983, at the Ettore Majorana Center for Scientific Culture,
Erice, Sicily, Italy

©1985 Plenum Press, New York
Softcover reprint of the hardcover 1st edition 1985

A Division of Plenum Publishing Corporation
233 Spring Street, New York, N.Y. 10013

PREFACE

This is the 3rd Course that has been organized at the prestigious E. Majorana Centre.

The choice of the theme is not a casual one. In fact after rehabilitation and prevention which were the topics discussed in the previous courses of 1979 and 1982, this Course deals with subjects more deeply connected with the mechanisms of cardiac function and the pathological aspects that under this point of view characterize many specific diseases.

The rapid development in the application of technology to the problems of Cardiological diagnosis has made it possible to study ventricular function in the evaluation of patients with apparent or suspected heart disease.

Knowing that we are facing one of the most complex subjects in modern cardiology we wish to arouse the interest of researchers and cardiologists by comparing different experiences with the aim of giving an overall survey of the problems.

The program has been divided into 4 chapters:

1. Systolic and diastolic ventricular function.

2. Specific studies of ventricular function using invasive and non-invasive techniques.

3. Diseases with altered ventricular function.

4. Pharmacological manipulation of ventricular function.

Our aim is not only to stimulate useful discussions in the faculty but also among the participants. I hope that we can achieve this.

I should like to take this opportunity of thanking the co-directors of the Course Prof. Jan Kellermann and Prof. Robert

Leachman who have played an important part in the scientific organization.

I would also like to thank all the members of the faculty for finding time to come to Erice and giving us the benefit of their knowledge and experience. I think that the participants deserve to be thanked too for showing an interest in this Course.

I would like to wish everyone profitable work.

Angelo Raineri

Director of International
School of Cardiology
"Ettore Majorana Centre"

ACKNOWLEDGEMENTS

The success of this course was the result of the medical staff's hard work of the Cattedra di Fisiopatologia Cardiovascolare, Palermo University, and the secretarial assistance of Irena Salmeri and Claudio Schirò.

It is a great pleasure to express my thanks to my collaborators, Dr. Pasquale Assennato and Dr. Antonio Rubino, for their assistance in publication of these proceedings.

Angelo Raineri

CONTENTS

CHAPTER I
SYSTOLIC AND DIASTOLIC VENTRICULAR FUNCTION

CHAPTER II
SPECIFIC STUDIES OF VENTRICULAR FUNCTION USING
INVASIVE AND NON INVASIVE TECHNIQUES

CHAPTER IV
PHARMACOLOGICAL MANIPULATION OF VENTRICULAR FUNCTION

CHAPTER I

SYSTOLIC AND DIASTOLIC VENTRICULAR FUNCTION

CHAPTER I:

INTRODUCTORY REMARKS

R. D. Leachman

Texas Heart Institute
Houston, USA

The intention of this meeting is to review the current concepts of heart muscle function and the effects that various diseases have on that function. The physiology of the heart has long intrigued the mind and since the days of Hales, there have been multiple experiments to analyze the complex factors of heart function. Many of these experiments and conclusions have occurred simultaneously with the technological advances that provided new methods of study. Claude Bernard has studied the circulation by invasive means in the 1830's, but the kymographic recording techniques gave Frank and Starling the method to evaluate the "laws of the heart" in the 1890's and early 1900's. A. V. Hill with his thermocouples in the 1920's was able to study the energy cycle of muscle contraction, and Carl Wiggers' elegant arrangement of mirror galvanometers gave him the opportunity to describe with great accuracy the timing intervals of pulse pressures in various parts of the circulation.

The heart can be considered as a pump and the heart can be considered as a muscle. The stroke volume of the heart was related to the volume distention of the ventricle, by the Frank-Starling observations and it was evident that over distention resulted in a decreased output, and that various metabolic and electroyte changes affected the muscular capacity to pump. In the course of his investigations, Carl Wiggers described the 3 principal factors affecting the heart as a pump namely: 1) Preload, 2) Afterload and 3) Contractile state - in the 1930's.

Fenn, Glasser and A. V. Hill, between 1922-1930 had described skeletal muscle function and behavior, particularly with regard to switch tension, tension length relations, and differences in isotonic and isometric contraction with regard to velocity of contraction and

peak tension generated. These concepts of skeletal muscle function
were examined in heart muscle strip preparations, on papillary
muscles by Braunwald, Sonnenblick, Ross and other groups of investigators in the 1960's. Indeed, isolated myocardial fibers, func-
tioned very much like skeletal muscle.

Increased tension was generated by heart muscle strips when they
were optimally stretched. The velocity of shortening and "The Frank-
Starling curve" were noted to be altered by changes in the "contractile state" of the heart, and soon became clinical measures of
myocardial function.

Technology for the high speed registration of heart sounds,
pressure pulses, and precardial pulsations gave cardiologists new
non invasive ways to evaluate ventricular function, and to observe
changes in heart function produced by drugs, exercise etc.

The available technology has been expanded with echocardiography
to examine precise mechanical events in the cardiac cycle and with
nuclear imaging techniques to examine variations in contraction
patterns, ventricular volumes and ejection fractions in a wide
variety of circumstances.

Most of the above investigations of heart function have been
related to systolic function and only indirectly to diastolic
function. The great physiologic debate a century ago on the vis a
fronte or vis a tergo seems resolved, but there may be diastolic
sucking in some circumstances. A.V. Hill's thermocouples allowed him
to study the metabolic recovery of skeletal muscle and I believe the
next chapter in cardiac physiology will be devoted to diastolic
function of the heart and the factors that influence it. With new
pharmacologic agents to affect the calcium mechanism, and with new
measuring techniques the various factors involved in diastolic
relaxation will likely become better defined.

The obvious goal of the clinical cardiologist is to maintain his
patients with the best possible heart function. Indeed, preservation
of the myocardium is the ultimate objective of treatment along with
eliminating symptoms. New medical and surgical techniques force the
clinician to make choices about the best treatment at any point in
time. Study of ventricular function, we hope, will give us the
information necessary to identify the patient whose heart is in
jeopardy with pressure or volume overload, so that appropriate treatment may be designed to prevent irreversible heart damage.

We have with us here at this conference, experts in all these
areas of investigation. I am confident that all of us will leave
this meeting with new ideas and new challenges for further clinical
and basic investigation, as well as with a better understanding of
the methods for evaluation of the patients who depend upon us for
guidance in their health needs.

MYOCARDIAL CONTRACTILITY, AFTERLOAD

MISMATCH AND VENTRICULAR DYSFUNCTION

Kirk L. Peterson

University of California
San Diego, USA

Although abnormalities in right ventricular function and/or the pulmonary vascular bed can be of pathophysiologic importance, it is more commonly an alteration of one or more determinants of left ventricular function which create an adverse hemodynamic status in the cardiac disease states encountered in man. Of the four major factors which control left ventricular function, i.e. heart rate, myocardial contractility, preload or end-diastolic fiber stretch, and afterload or the force which resists shortening, the latter is now appreciated as playing a particularly critical role in the mechanisms underlying heart failure. For example, the interdependence of wall shortening and afterload, at any given preload, has allowed understanding of the favorable influence of vasodilating agents in treating congestive heart failure. Moreover, the serial application of both invasive and non-invasive diagnostic techniques has provided improved insight into the role of afterload in chronic valvular heart disease, the potential for improving left ventricular shortening by valve surgery, and the relation between postoperative changes in afterload and regression of hypertrophy.

CONSTRUCTS FOR EVALUATION OF LEFT VENTRICULAR FUNCTION

There is a now a wealth of clinical studies which validate both angiographic (contrast or radionuclide) and echocardiographic methods for measurement of left ventricular volumes and dimensions, thereby providing relevant estimates of preload and myocardial shortening[1-3]. Quantitative assessment of afterload has been more problematic and described by one of two constructs (Table 1): 1) calculation of wall stress as a function of chamber dimension, pressure, and wall thickness[4-6] and 2) computation of aortic input impedance

5

Table 1. Quantitative Descriptors of Left Ventricular
          Afterload

---

1. <u>Wall Stress(σ)</u>

$$\sigma = [P * R]/ 2 h$$

Where P = intracavitary pressure, R = chamber radius,
and h = wall thickness.

(The formula given applies to a spherical model)

2. <u>Aortic Input Impedance and Systemic Resistance</u>

    - Impedance spectrum is the ratio of pressure (P)
      harmonics to corresponding flow (F) harmonics,
      derived from a Fourier analysis of the respective
      wave forms.

    - Systemic resistance(R) equals the impedance modulus,
      or the ratio of amplitude of pressure to amplitude
      of flow, at the fundamental frequecy (Heart rate)
      or 0 Hz. In a constant flow, non-pulsatile system,
      resistance can also be calculated as the ratio of
      mean pressure to mean flow.

derived from a Fourier analysis of ascending aortic pressure and
flow[7-10]. With the latter approach, a spectral plot of the ratios
of pressure to flow for each harmonic serves to describe comprehen-
sively the input impedance moduli; however, it does not provide a
single instantaneous or mean number which quantifies afterload.
Thus, many investigators have chosen to use arterial resistence,
calculated by dividing mean flow into mean pressure, or characteristic impedance obtained by averaging impedance moduli between 2 and
12 Hz[7,9,10]. The primary theoretical argument for use of aortic
input impedance is that the physical properties of the systemic
arterial system are constant during ejection and independent of
cardiac function, which is contrary to the case with a wall stress
analysis. While theoretically useful, measurement and computation of
aortic input impedance, simultaneous with ventricular shortening, is
technically difficult; moreover, a recent animal study from our
institution would suggest that during aortic impedance changes, the
corresponding alterations in wall stress and shortening characteristics are completely described within the framework of the
traditional force-velocity-length relation[10]. Thus, in our
clinical studies we have chosen to assess afterload using <u>mean</u>
stress, i.e. the integral of the stress-time curve, during ventricular shortening.

Measurement of contractile state independent of the effects of preload and afterload has remained a difficult task. Numerous parameters derived from the isovolumic phase of systole (e.g. dP/dt, dP/dt/P) have been investigated, but none have been shown to be free of the influences of preload and afterload, and some are dependent on constants of elasticity and muscle models which likely are not appropriate to human heart disease[11,12]. Direct analysis of the left ventricular wall or cavity-endocardial interface during ejection has revealed, however, a characteristic inverse relation between the extent or velocity of shortening and the afterload (mean stress), with appropriate shifts in the curve by changes in preload or contractile state[13]. The observed responses resemble closely those of the whole isolated cardiac muscle when contracting isotonically. Force-velocity, pressure-diameter, pressure-volume, and stress-volume curves all have been generated systole and have been valuable for defining the interplay between myocardial contractility and afterload[14-16]. It is apparent that the common ejection phase indices (ejection fraction, mean velocity of circumferential fiber shortening, mean normalized ejection rate) are all sensitive to acute changes in afterload[17]. However, these indices can be quite valuable for assessing basal contractility before and after a chronic adaptation has taken place since generally such changes in ventricular geometry and size serve to maintain wall stress within a normal range[11-13,18].

In recent years, a number of investigators have observed the utility of the end-systolic pressure-volume and stress-volume relations as unique descriptors of myocardial contractility independent of the effects of preload and encompassing the effects of afterload. These studies have been accomplished in isolated heart preparations, the chronically instrumented dog, and also in man[19-24]. In order to measure this parameter of myocardial contracility in the ejecting heart of the intact organism, one must manipulate afterload and preload of the left ventricle using either pharmacologic or mechanical means. We have found angiotensin II to be useful for augmenting afterload, and either nitroprusside and/or inferior vena cava cuff occlusion for reducing preload. Phenylephrine, a relatively potent peripheral vasoconstrictive agent, appears to shift the relation to the left (increase contractitility), presumably secondary to at least some stimulation of beta-1 receptors in the myocardium[23]. Although the end-systolic indices (Emax, Vo) are promising for assessing myocardial contractility, our initial studies in the chronically instrumented dog indicate that they are more useful for detecting augmentation, rather than depression, of contractility[24].

DEFINITION OF AFTERLOAD MISMATCH

When myocardial contractility is normal and the myofibers are operating at a normal fiberstretch of approximately 1.9 microns,

there is substantial preload reserve available for maintaining stroke
volume relatively constant as afterload changes[25]. However, when
the preload reserve is fully utilized (myocardial failure, chronic
volume overload) or unavailable (serve concentric hypertrophy, inadequate venous return) and the wall forces are elevated for either a
normal or depressed level of contractility, a state of "afterload
mismatch" exists[13]. Under these circumstances, the extent of
shortening of the ventricle is inordinately depressed, and the basal
stroke volume remains inappropriately low. Conversely, an afterload
mismatch state can also occur when the afterload is inappropriately
low for a normal or depressed level of contractility; the supernormal
stroke volume of mitral regurgitation, despite a normal level of
contractility, is representative of this behavior[26,27].

In the setting of myocardial failure, afterload mismatch can be
identified when the level of systolic wall stress is too high for the
depressed level of myocardial contractility. When acute pressor
stress is given, the ventricle responds as if the preload is fixed
(i.e. there is no detectable change in fiber length or end-diastolic
volume), and the stroke volume and velocity of shortening fall[28].
During this latter circumstance, the end-diastolic pressure may
increase significantly even though no detectable change is noted in
end-diastolic fiber length (ventricle operating at the limit of
preload reserve on the steep portion of its diastolic pressure-volume
curve).

Mismatch mechanics may also be detected by a salutary response
of the failing ventricle to afterload reduction. For example, administration of a mixed venous and arterial dilator will lead to an
augmentation of stroke volume (and cardiac output) while at the same
time end-diastolic volume and pressure decline[29]. The increase in
the stroke volume can only be explained under these circumstances by
lowering of wall stress during ejection. By contrast, if the inotropic state was appropriate to the level of afterload, there would
be no significant improvement in the stroke volume as afterload was
reduced; in fact, the stroke volume would fall.

ATFERLOAD MISMATCH IN VALVULAR HEART DISEASE

At our institution we have become aware of afterload mismatch in
patients with chronic pressure or volume overload of th left ventricle. For example, in patients with calcific aortic stenosis, we
have utilized simultaneous left ventricular angiography and high-
fidelity micromanometry to obtain the raw data necessary to calculate
wall forces throughout systole, Figure 1. We have found significant
increases in peak as well as mean stress during ejection in these
subjects, as compared to a normal group, despite the presence of
concentric hypertrophy, Figure 2, panel A, [30]. Moreover, as shown
in Figure 2, panel B, the profile of wall stress development and

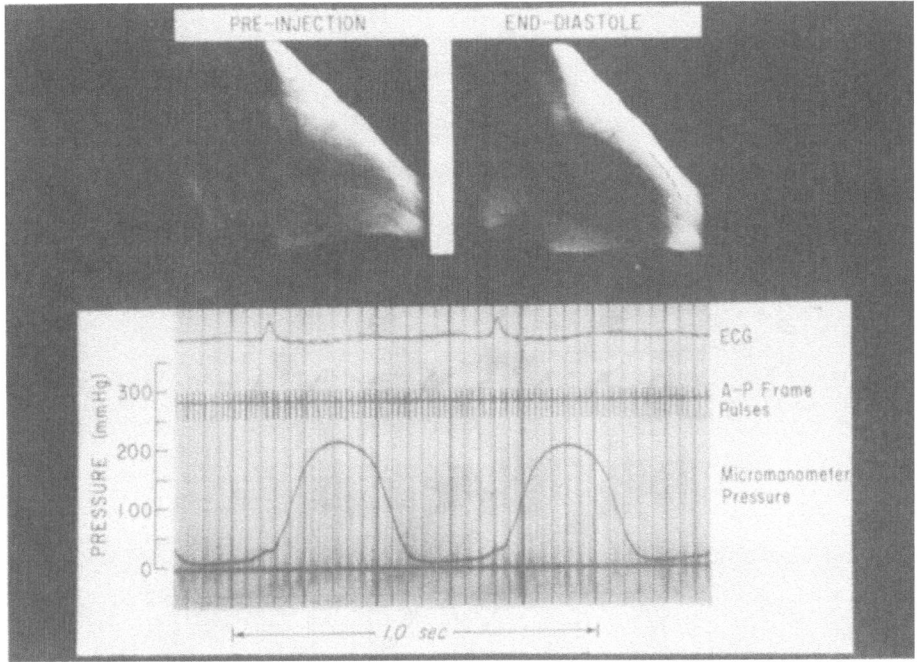

Fig. 1. Simultaneous registration of high-fidelity left ventricular
         pressure and a cine-angiogram of the left ventricle in a
         patient with calcific aortic stenosis. The volume and wall
         thickness (dotted line) are calculated and measured, respec-
         tively, for each cine frame and correlated with the instan-
         taneous pressure. See text for discussion of calculation of
         wall forces throughout systole.

dissipation is quite different in the patient with aortic stenosis
and a normal peak isovolumic dP/dt (closed circles, presumably normal
contractility) as compared to the normal subject (closed triangles);
the rate of development of stress is comparable in these two subjects, but both peak stress and end-ejection stress are higher in the
presence of significant outflow tract obstruction. Note also that
the patient with aortic stenosis and a depressed isovolumic dP/dt
(open circles, presumably depressed myocardial contractility) shows a
retarded rate of development of stress in early systole, a somewhat
lower peak stress, but a significantly higher wall force at the end
of ejection. Analysis of mid-wall shortening in these same patients
reveals depression of the extent and speed of wall thickening and
shortening and an inverse relation between the shortening indices and
calculated mean stress, Figure 2, panel C.

     It might be expected that aortic valve replacement with coinci-
dent relief of outflow obstruction and the afterload excess would

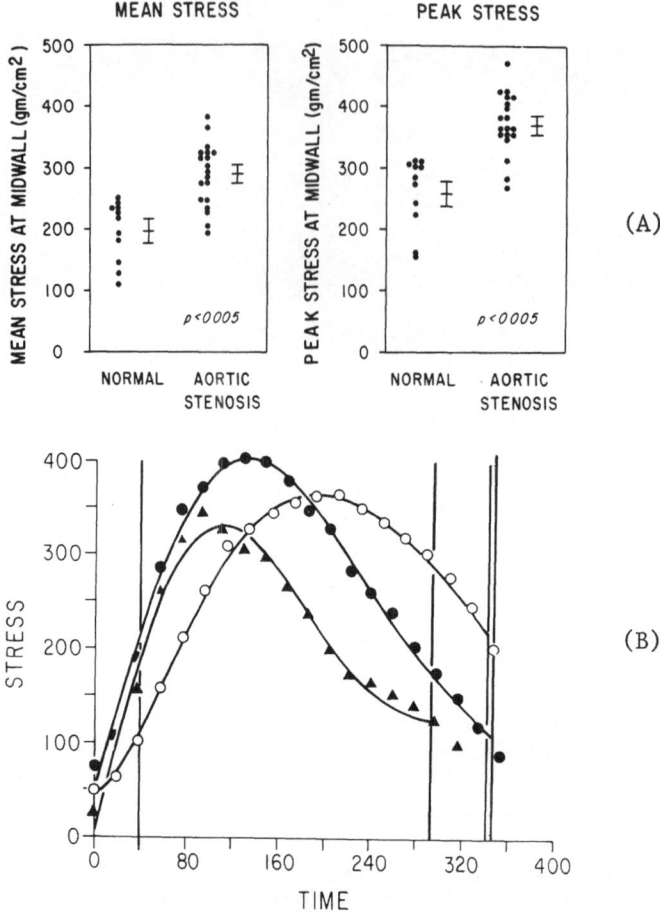

Fig. 2.   Panel A: Measurements of mean left ventricular circumfer-
          ential midwall stress (left) and peak stress (right) in a
          group of normal subjects as compared to a group with
          calcific aortic stenosis. See text for discussion.
          Panel B: Comparison of profiles of circumferential midwall
          stress in a normal subject (closed triangles), patient with
          aortic stenosis and normal myocardial contractility (closed
          circles), and patient with aortic stenosis with depressed
          myocardial contractility (open circles). See text for
          discussion.

lead to improved ventricular wall shortening even in the patient with
depressed myocardial contractility. In fact, this has been shown
both experimentally and by analyses of the ejection fraction before
and after surgery. After aortic valve replacement, such patients
exhibit a significant improvement in both the ejection fraction and
the mean velocity of circumferential fiber shortening [31].

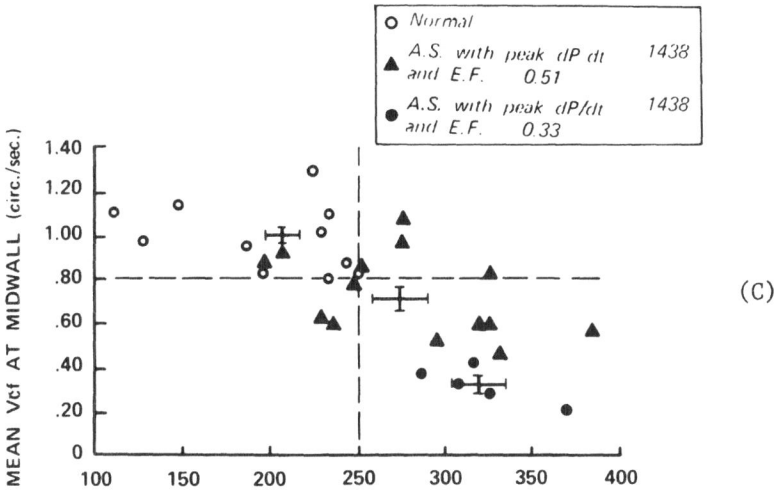

Fig. 2. Panel C: Mean velocity of circumferential fiber shortening (mean Vcf) at mid-wall plotted as a function of mean stress at mid-wall in normal subjects, patients with aortic stenosis who have an ejection fraction greater than 0.51 and peak dP/dt greater than 1438 mm Hg/sec., and patients with aortic stenosis who have an ejection fraction less than 0.33 and peak dP/dt less than 1438 mm Hg/sec. The mean and standard error are shown for each group of patients. The analysis suggests that both myocardial failure and afterload excess are operative in depression of mid-wall circumferential velocity of shortening. Reproduced with permission of author from reference 30.

The converse is noted in patients with chronic mitral regurgitation with left ventricular volume overload and marked left ventricular dilatation. In this circumstance, the left ventricle unloads rapidly into the left atrium even though myocardial contractility might be depressed. However, once the mitral valve is replaced and there is correction of the low impedance leak into the left atrium the left ventricle is forced to eject solely against the systemic circuit. If the left ventricular chamber is sufficiently dilated preoperatively (greater than 7.0 cm. in diameter by M-mode echocardiography), the reduction of the regurgitant volume is often not sufficient to substantially alter ventricular volume, the forces resisting shortening in the ventricle augment, and wall shortening declines[32,33]. Thus, in this valve lesion, the afterload is kept inappropriately low or mismatched to the intrinsic contractile state, and corrective valve surgery uncovers depressed contractility once physiologic afterload conditions are restores (see associated article on "Left Ventricular Function in Mitral Regurgitation" in this volume).

A final example of afterload mismatch in valvular heart disease
has been noted in an investigation of right ventricular function in
patients with mitral valve disease. In a number of subjects, preoperative radionuclide angiographic studies revealed dilatation of
the right ventricular chamber and depression of the ejection fraction, Figure 3. Since this chamber, as compared to the left ventricle, is relatively thin-walled, it might be expected that systolic
pulmonary arterial hypertension would cause an afterload excess for a
given level of intrinsic contractile state of the myocardium. We
have found that post mitral valve surgery, and coincident with the
fall in pulmonary artery pressure, there is a significant and consistent improvement in the extent of shortening of the right ventricle, even in those patients with relatively normal ejection
fraction values preoperatively, Figure 3. Again, this prompt
improvement in right ventricular shortening supports the notion that
an afterload mismatch (in this case an excess) causes the fall in
ejection fraction despite dilatation of the chamber.

SUMMARY

Over the past decade a number of experimental and clinical
studies have served to clarify the importance of afterload in the
pathogenesis of cardiac dysfunction. While no single parameter or
construct has yet been uncovered which uniquely describes myocardial
contractile state, it is possible to assess this important physiologic property of heart muscle by accounting for the afterload con-

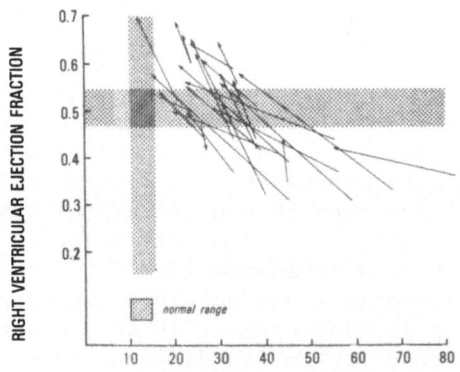

MEAN PULMONARY ARTERY PRESSURE [mm Hg]

Fig. 3. Plot of right ventricular ejection fraction versus the
         measured mean pulmonary artery pressure (mm Hg) in group of
         patients undergoing mitral valve surgery. The beginning of
         the vector represents preoperative values and the tip of the
         vector (arrow) shows postoperative values. See text for
         discussion.

ditions against which it operates. The relation between the ejection
fraction or mean velocity of circumferential fiber shortening and the
mean wall stress during ejection, over a broad range of afterload,
provides a comprehensive construct for assessment of myocardial
contractility. The end-systolic pressure-volume and stress-volume
relations, determined over a broad range of afterload, are also now
showing promise for evaluating myocardial function. Finally, study
of the afterload conditions of the left and right ventricles in a
number of valvular lesions has given us improved understanding of the
mechanics of myocardial shortening both before and after corrective
valve surgery.

REFERENCES

1. H. Dodge, H. Sandler, D. Ballow, and J. Lord, The use of biplane
    angiography for measurement of left ventricular volume in
    man, Am.Heart J., 60:762 (1960).
2. H. Feigenbaum, R. Popp, R. Wolfe, B. Trot, J. Pombo, C. Haine,
    and H. Doge, Ultrasound measurements of the left ventricle: A
    correlative study with angiography, Arch.Intern.Med., 179:461
    (1972).
3. R. Slutsky, J. Karliner, D. Ricci, R. Kaiser, M. Pfisterer, D.
    Gordon, K. Peterson, and W. Ashburn, Left ventricular volumes
    calculated by gated radionuclide angiography: A new method,
    Circulation, 60:556 (1979).
4. H. Sandler and H. T. Dodge, Left ventricular tension and stress
    in man, Circ.Res., 13:91 (1963).
5. Mirsky, I., Left ventricular stresses in the intact human heart,
    Biophys.J., 9:189 (1969).
6. J. W. Burns, J. W. Covell, R. Myers, and J. Ross, Jr., Compari-
    son of directly measured left ventricular wall stress and
    stress calculated from geometric reference figures, Circ.Res.
    28:611 (1971).
7. E. S. Imperial, M. N. Levy, and H. Zieske, Jr., Outflow resist-
    ance as an independent determinant of cardiac performance,
    Circ.Res., 9:1148 (1961).
8. W. R. Milnor, Arterial impedance as ventricular afterload,
    Circ.Res., 36:565 (1975).
9. W. W. Nichols, C. R. Conti, W. E. Walker, and W. R. Milnor,
    Input impedance of the systemic circulation in man,
    Circ.Res., 40:451 (1977).
10. H. Pouleur, J. W. Covell, and J. Ross, Jr., Effects of alter-
    ations in aortic input impedance on the force-velocity-length
    relationship in the intact canine heart, Circ.Res., 45:126
    (1979).
11. J. Ross, Jr., and K. L. Peterson, On the assessment of cardiac
    inotropic state, Circulation, 47:435 (1973).
12. K. L. Peterson, D. Skloven, P. Ludbrook, J. Uther, and J. Ross,
    Jr., Comparison of isovolumic and ejection phase indices of
    myocardial performance, Circulation, 49:1088 (1974).

13. J. Ross, Jr., Afterload mismatch and preload reserve: A con-
     ceptual framework for the analysis of ventricular function,
     Prog.Cardiovasc.Dis., 18:225 (1976).
14. J. H. Gault, J. Ross, Jr., and E. Braunwald, Contractile state
     of the left ventricle in man: Instantaneous tension-velocity-
     length relations in patients with and without disease of the
     left ventricular myocardium, Circ.Res., 22:451 (1968).
15. K. L. Peterson, J. B. Uther, R. Shabetai, and E. Braunwald,
     Assessment of left ventricular performance in man: Instantan-
     eous tension-velocity-length relations obtained with the aid
     of an electromagnetic velocity catheter in the ascending
     aorta, Circulation, 47:924 (1973).
16. K. T. Weber and J. S. Janicki, Instantaneous force-velocity-
     length relations in isolated dog heart, Am.J.Physiol., 232:
     H241 (1977).
17. M. A. Quinones, W. H. Gaasch, and J. K. Alexander, Influence of
     acute changes in preload, afterload, contractile state and
     heart rate on ejection and isovolumic indices of myocardial
     contractility in man, Circulation, 53:293 (1976).
18. S. Sasayama, D. Franklin, and J. Ross, Jr., Hyperfunction with
     normal inotropic state of the hypertrophied left ventricle,
     Am.J.Physiol., 232:H418 (1977).
19. H. Suga, K. Sagawa, and A. A. Shoukas, Load independence of the
     instantaneous pressure-volume ratio of the canine left
     ventricle and effects of epinephrine and heart rate on the
     ratio, Circ.Res., 32:314 (1973).
20. K. Sagawa, The end-systolic pressure-volume relation of the
     ventricle: definition, modifications and clinical use,
     Circulation, 63:1223 (1981).
21. F. Mahler, J. W. Covell, and J. Ross, Jr., Systolic pressure-
     diameter relations in the normal conscious dog, Cardiovasc.
     Res., 9:447 (1975).
22. W. Grossman, E. Braunwald, T. Mann, L. P. McLaurin, and L. H.
     Green, Contractile state of the left ventricle in man as
     evaluated from end-systolic pressure-volume relations,
     Circulation, 56:845 (1977).
23. T. Tajimi, T. F. Widmann, M. Matsuzaki, and K. L. Peterson,
     Differing effects of Angiotensin II and phenylephrine on end-
     systolic pressure-volume relationship in conscious dogs,
     J.Am.Coll.Cardiol., 3:523 (1984).
24. T. Tajimi, T. F. Widmann, M. Matsuzaki, and K. L. Peterson,
     The sensitivity of Emax in comparison to other parameters of
     ventricular contractility, Clin.Res., 32, No.1:13A (1984).
25. C. Yoran, J. W. Covell, and J. Ross, Jr., Structural basis for
     the ascending limb of left ventricular function, Circ.Res.,
     32:297 (1973).
26. C. W. Urschel, J. W. Covell, E. H. Sonnenblick, J. Ross, Jr.,
     and E. Braunwald, Myocardial mechanics in aortic and mitral
     valvular regurgitation. The concept of instantaneous
     impedance as a determinant of the performance of the intact
     heart, J.Clin.Invest., 47:867 (1968).

27. D. L. Eckberg, J. H. Gault, R. L. Bouchard, J. S. Karliner, and
     J. Ross, Jr., Mechanics of left ventricular contraction in
     chronic severe mitral regurgitation, Circulation, 47:1252
     (1973).

28. D. Ricci, Afterload mismatch and preload reserve in chronic
     aortic regurgitation, Circulation, 66:827 (1982).

29. J. N. Cohn and J. A. Franciosa, Vasodilator therapy of cardiac
     failure, N.Engl.J.Med., 297:27,254 (1977).

30. K. L. Peterson, Instantaneous force-velocity-length relations
     of the left ventricle: Methods, limitations, and applications
     in man, in: "Heart Failure," A. Fishman, ed., Hemisphere
     Publishing Corp., ch.9 (1978).

31. N. Smith, J. H. McAnulty, and S. H. Rahimtoola, Severe aortic
     stenosis with impaired left ventricular function and clinical
     heart failure: Results of valve replacement, Circulation,
     58:255 (1978).

32. G. Schuler, K. L. Peterson, A. Johnson, G. Francis, G. Dennish,
     J. Utley, P. O. Daily, W. Ashburn, and J. Ross, Jr., Temporal
     response of left ventricular performance to mitral valve
     surgery, Circulation, 59:1218 (1979).

33. C. Y. H. Wong and H. M. Spotnitz, Systolic and diastolic
     properties of the human left ventricle during valve replace-
     ment for chronic mitral regurgitation, Am.J.Cardiol., 47:40
     (1981).

STATIC AND DYNAMIC DETERMINANTS OF LEFT

VENTRICULAR CHAMBER STIFFNESS AND FILLING

W. H. Gaasch, Y. Arial and A. S. Blaustein

Dept. of Medicine, Tufts University School of
Medicine and the Veterans Adm. Medical Center
Boston, Massachusetts, USA

Ventricular filling occurs as a consequence of a left atrial to
left ventricular pressure gradient which has generally been though
to be primarily due to passive mechanisms influencing both chambers.
Left atral pressure is determined by the compliance of the left
atrium and pulmonary venous system, the central blood volume, and
to some extent by the contractile strength of the atrial myocardium.
Left ventricular diastolic pressure is determined by all of the
static and dynamic factors which influence chamber stiffness. In
this article we will outline the factors which influence the diastolic properties of the left ventricular chamber[1-6]. The discussion will be divided into a review of static factors (i.e. chamber
volume, wall mass, and stiffness of the wall) and dynamic factors
(i.e. pericardial and right ventricular effects, a hydraulic effect
of the coronary vasculature, and the process of myocardial relaxation). This distinction is based on the notion that change in the
static factors evolve very slowly, while dynamic factors may change
from moment to moment. It should be emphasized, however, that these
factors are interdependent and the effect of any single mechanisms is
difficult to isolate and evaluate.

As early as the second century, Galen hypothesized that active
dilatation of the right ventricle contributed to he transfer of blood
from the vena cava to the heart and by implication, suggested that
diastolic suction was important to ventricular filling[7]. Over a
thousand years later, Harvey concluded that suction could not be
responsible for ventricular filling and he wrote that most filling
occurred during atrial systole[8]. In 1877 Francois-Franck concluded
that most of ventricular filling occurred in early diastole[9], and
shortly thereafter Henderson separated diastole into the three phases
of early rapid filling, diastasis, and atrial contraction[10]. By

17

1921 Wiggers and Katz found wide variations in the contribution of atrial systole[11], and it is now generally recognized that substantial alterations in the pattern of early or late filling can be seen in diseased hearts.

The marked differences in heart size seen in patients with cardiac disease have, in the past, been related to differences in cardiac "tone" and an understanding of this topic has likewise shown considerable evolution. In 1927 Meek defined tone as "a sustained partial contraction," independent of systolic contraction, by which the muscle fibers resist distension during diastole more than they would because of their mere physical properties[12]. Wiggers preferred the term "inherent elasticity" to "mere physical properties"[13], but both authors were referring to the passive properties of viable myocardium. Meek further emphasized that the "contraction remainder" could extend into diastole and it is now accepted that delayed or slow relaxation can influence measurements of the passive properties of the ventricle as well as the filling rate of the chamber.

Basic to a balanced view of diastole and the assessment of resistance to ventricular filling is the curvilinear left ventricular diastolic pressure-volume curve. During chronic steady state conditions, the major determinants of the diastolic pressure-volume relation are the volume of the chamber, the left ventricular wall volume (or myocardial mass), and the physical properties (i.e. stiffness) of the myocardium. During abrupt hemodynamic and other acute interventions, the influence of several dynamic factors may be superimposed on the passive (static) diastolic pressure-volume curve. These include the effects of the pericardium and the right ventricle, the hydraulic or erectile effect of the coronary vasculature, the process of relaxation, and others. Thus, alterations in many factors, alone or in concert, may contribute to changes in the diastolic properties of the left ventricle.

CHAMBER AND MYOCARDIAL STIFFNESS

Throughout this article we will distinguish left ventricular chamber stiffness from myocardial stiffness[1]. The pathophysiology of pulmonary venous hypertension and congestion is best understood by examining the factors responsible for alterations in chamber stiffness (derived from left ventricle diastolic pressure-volume data). Insight into myocardial functional and structural defects can be provided by assessing myocardial stiffness (derived from the stress-strain characteristics of the left ventricular wall). Intrinsic myocardial stiffness and a number of extramyocardial factors influence left ventricular chamber stiffness, some on a chronic, often irreversible (static) basis and others on an acute or beat-to-beat (dynamic) basis.

Left Ventricular Chamber Stiffness

Chamber stiffness is defined by analyzing the curvilinear diastolic pressure-volume relations. The slope of a tangent (dP/dV) to this curvilinear relation defines chamber stiffness at a given filling pressure. An increase in dP/dV secondary to an increase in filling pressure (preload dependent change in stiffness) is shown diagrammatically in Figure 1. A leftward shift to a different pressure-volume curve can cause a similar increase in dP/dV; in the example shown, the tangent is steeper at the same diastolic pressure. Thus, chamber stiffness may change by virtue of a change in filling pressure _or_ through a shift to a different pressure-volume curve. To optimally compare chamber stiffness in hearts of different sizes, dP/dV must be normalized for volume (see below).

Since diastolic pressure-volume data can be fit by an exponential function (assuming a monoexponential curve fit), a chamber stiffness constant or modulus of chamber stiffness (k_c) may be derived from the slope of the linear relation between dP/dV and pressure. Thus, characterization of the diastolic properties of the left ventricular chamber includes calculation of the chamber stiffness constant (k_c) and chamber stiffness (dP/dV) at a specified pressure (i.e. end diastolic pressure). In practice, this is accomplished by simultaneously measuring LV pressure and volume at frequent intervals throughout diastole; the pressure-volume coordinates are fit by an exponential function and values for k_c and dP/dV are

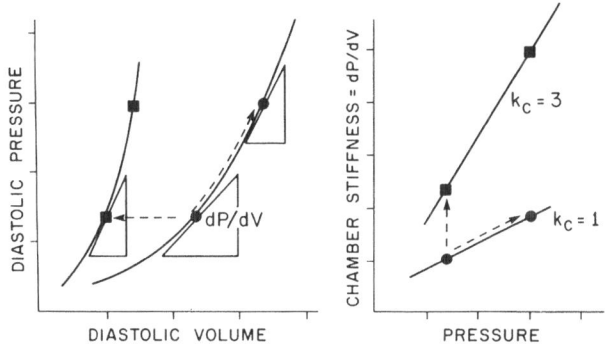

Fig. 1. Left ventricular diastolic pressure-volume relations and chamber stiffness-pressure relations. In the panel on the left, operating chamber stiffness (dP/dV) will increase with an increment in volume; this has been called a preload-dependent change in stiffness. Alternatively, a leftward shift of the pressure-volume relation may cause an increase in dP/dV. In the panel on the right, if the diastolic pressure-volume relation is exponential, the relation between dP/dV and pressure is linear; the slope of this line is the chamber stiffness constant (k_c).

found. Early diastolic coordinates should not be used because these
data are likely to be influenced by the intensity of myocardial
relaxation, elastic recoil, and probably other dynamic factors.

Myocardial Stiffness

The intrinsic stiffness of heart muscle is described by cal-
culating myocardial stiffness constants; these constants are derived
from an analysis of myocardial stress-strain data.

Stress is defined as force per unit cross-sectional area of a
material and the units employed are dynes or grams per square centi-
meter. Midwall fibers at the equator are primarily oriented circum-
ferentially and thus, calculated values values for circumferential
wall stress are approximately equal to fiber stress. The average
stress formulae developed by Mirsky[3] are widely used for clinical
and experimental purposes.

Strain is defined as the deformation of a material that is
produced by the application of a force (stress). Although strain is
usually expressed as a percentage change from the unstressed dimension, Mirsky has concluded that the natural strain definition is best
suited for analysis of myocardial stress-strain relations because it
does not require measurement of length at zero stress.

Plots of diastolic stress-strain data are curvilinear and, much
like the diastolic pressure-volume data, they can be fit by an exponential function. Thus, the slope of a tangent to this relation
describes myocardial elastic stiffness at a given level of wall
stress (Figure 2). Since the stress-strain relation is nearly exponential, a myocardial stiffness constant (k_m) can be derived as
the slope of the linear relation between muscle elastic stiffness and
left ventricular wall stress. Myocardial stiffness in the intact
heart is generally characterized by assessing the stiffness constant
(k_m) as well as stress and elastic stiffness at end diastole.

These definitions of chamber and myocardial stiffness assume
elastic diastolic pressure-volume and stress-strain relations that
are nearly exponential. It is recognized that this simplified approach ignores viscous and inertial drag, stress relaxation, creep,
and elastic recoil. Stress relaxation refers to the time-dependent
decline in stress when a material is held at constant length (after
being subjected to a rapid increase in length). Creep refers to the
gradual increase in length that occurs when a material is held at
constant stress (after a rapid increase in stress). These properties
of heart muscle probably have little significance in the interpretation of diastolic pressure-volume data in the intact heart. Visco-
elasticity refers to that property of material which results in
higher levels of stress when the lengthening rate (strain rate) is

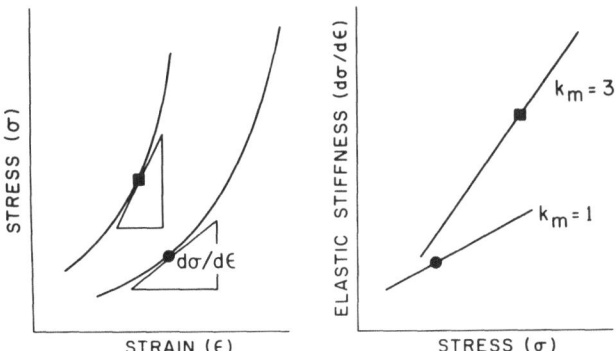

Fig. 2. Myocardial stress-strain relations (panel on the left) and
elastic stiffness-stress relations (panel on the right).
Similar to the pressure-volume analysis, myocardial elastic
stiffness will increase as strain increases. Alternatively,
a leftward shift of the stress-strain relation results in a
high elastic stiffness. The relation between elastic stiffness and stress is linear and the slope of this relation is
the myocardial stiffness constant (k_m).

rapid. In addition, viscosity depends on the intrinsic properties of
the material (the myocardium) and its length[14]. Early diastolic
elastic recoil can cause a diastolic suction effect which is created
by systolic contraction (much like a tennis ball recoils to its
original shape after being squeezed or indented). The magnitude of
the recoil is inversely related to end systolic volume and thus the
greater the systolic emptying and the smaller the end systolic
volume, the greater will be the elastic recoil and suction in early
diastole[15].

The assumption of a purely passive elastic and exponential
pressure-volume or stress-strain relation also ignores several other
potential determinants of these relationships. Some of these are
extrinsic to the LV (i.e. pericardial and right ventricular pressure)
and some are intrinsic to the LV itself (i.e. the "erectile" effect
and the process of myocardial relaxation). These factors will be
considered below, in the section on dynamic determinants of filling;
however, it is appropriate at this time to emphasize that the effects
of these dynamic factors can be superimposed on the static pressure-
volume curve, and can result in substantial displacement from a
purely exponential pressure-volume curve.

Another problem involves the difficulties in defining and inter-
preting so-called "parallel shifts" in the diastolic pressure-volume
relation. Many, if not most, of the dramatic shifts in the diastolic
pressure-volume relation that occur during acute interventions consist primarily of vertical displacements (parallel shifts) of the

press-volume relation; in this situation, the slope of the log
pressure-volume relation may change only minimally, but the ventricle
operates at higher diastolic pressure and the pressure intercept (at
zero volume) is increased. Conceptually, the chamber is less com-
pliant (higher pressure at the same volume), but calculated values
for chamber stiffness are unchanged. This apparent paradox has not
been completely resolved, but for clarity and consistency, the
general term "decreased ventricular compliance" can be used to indicate higher pressure at the same volume. As will be seen, these
parallel shifts are due primarily to external (pericardial, pleural,
etc.) pressures.

STATIC DETERMINANTS OF LV FILLING

     The "static" factors (LV volume, mass, and myocardial stiffness)
are related, but substantial changes in any one may or may not be
associated with changes in the others. For example, chronic left
ventricular pressure overload can result in large increments in
myocardial mass with little change in chamber volume; myocardial
stiffness may be normal or increased. On the other hand, chronic
volume overload can cause large increments in left ventricular mass
and volume; again, myocardial stiffness may be normal or increased.
Thus, while they are related, these static determinants of ventricular filling are not tightly coupled.

     The relationship between left ventricular volume, mass, and
myocardial stiffness is especially important when one attempts to
evaluate chamber stiffness in different patients. Chamber stiffness
from different hearts can be compared directly only if the volumes of
the two hearts are comparable. This is so because the distensibility
of a chamber depends on its size. For example, the heart of a normal
adult will be more distensible than the heart of a normal child in
part because of differences in diastolic volume. Thus, if one wishes
to compare the chamber stiffness of two hearts that differ in volume,
mass, and intrinsic myocardial stiffness, the analysis should include
volume normalization and the comparison should be made at equal
levels of diastolic pressure[16].

     A quantitative relation between chamber stiffness, chamber
volume and wall mass, and myocardial stiffness has been suggested by
Mirsky[3]. From the theory of elasticity, operative chamber stiffness (dP/dV) is directly proportional to $E/V(1 + V/M)$, where E is
myocardial elastic stiffness, V is chamber volume, and M is wall
mass. Elastic stiffness (E) is approximately equal to the product of
the myocardial stiffness constant (k_m) and operation wall stress.
These relations indicate that normalized chamber stiffness (VdP/dV)
is directly proportional to $E/(1 + V/M)$. Thus, a large ventricle
with normal or decreased chamber stiffness may have increased myocardial stiffness if the V/M is sufficiently increased. On the other

hand, chamber stiffness may be increased in the presence of normal
myocardial stiffness if the V/M is low. This analysis, shown diagrammatically in Figure 3, is especially important in understanding,
chamber and myocardial properties in patients with myocardial and/or
valvular heart disease.

   Changes in the static determinants of filling are at least in
part responsible for the chronic alterations in the pattern of filling seen in patients with left ventricular <u>hypertrophy</u>. The increased myocardial mass and the low volume/mass ratio (and in some
cases, abnormal intrinsic myocardial stiffness) cause a leftward
displacement of the diastolic pressure-volume curve and increased
chamber stiffness. It is likely that increased chamber stiffness,
by virtue of increased passive resistance, results in a reduced rate
of early diastolic filling <u>and</u> changes in the contribution of atrial
contraction; however, alterations in several dynamic factors (i.e.
viscosity, relaxation, etc.) may also contribute. Likewise, changes
in the static determinants of filling cause the filling abnormalities
seen in chronic <u>coronary heart disease</u>; scar and fibrosis secondary
to infarction are responsible for these changes.

DYNAMIC DETERMINANTS OF LV FILLING

Pericardial Effects

   In the absence of the pericardium, the diastolic pressure-volume
curve of the left ventricle, much like the resting length-tension

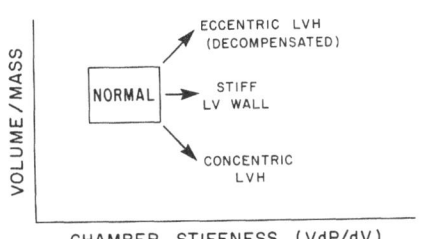

Fig. 3. Left ventricular end diastolic volume/mass radio is plotted
         against normalized chamber stiffness (VdP/dV). In decompen-
         sated eccentric left ventricular hypertrophy (LVH), both
         volume/mass and VdP/dV are increased; this indicates in-
         creased myocardial stiffness. In some restrictive cardio-
         myopathies, chamber stiffness may be increased in the
         presence of a normal volume/mass ratio; this indicates
         increased intrinsic myocardial stiffness. A low volume/mass
         ratio in concentric LVH indicates that the increased chamber
         stiff- ness is at least in part due to the abnormal volume/
         mass and not entirely due to increased myocardial stiffness.
         See text for discussion of normalized chamber stiffness.

relation of isolated papillary muscle, does not shift significantly during acute pharmacologic (i.e. digitalis) or hemodynamic
(i.e. volume loading) interventions. However, when the pericardium
is intact, several factors alone or in combination may contribute
to significant changes in left ventricular diastolic compliance[5,
17,18].

    An example of a substantial pericardial effect is shown in
Figure 4. In this acute dog experiment, an acute increase in preloa
(volume infusion) was followed by unloading (nitroprusside); these
interventions were performed before and after pericardectomy. With
an open pericardium, the diastolic pressure-length coordinates are
displaced from the baseline curvilinear pressure-length relation
during the loading and unloading interventions. The pressure-length
coordinates merely move up or down the same curve and there is no
change in the chamber stiffness constant (k_c); this has been called
preload dependent change in operating stiffness[1]. When the pericardium is intact, a similar volume infusion results in a substantia
upward displacement of the pressure-length relation (a "parallel"
upward shift). Again, there is little change in k_c, but the diastolic pressures are much higher and the chamber is less complaint.
Nitroprusside results in a parallel downward shift. These "parallel
shifts would not be present if left ventricular transmural pressure
(rather than absolute pressure) had been used.

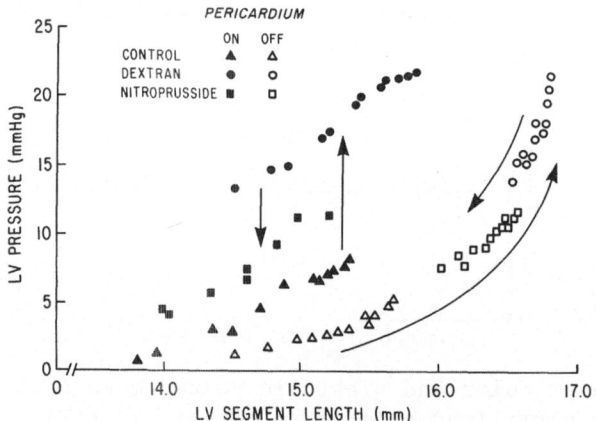

Fig. 4. Left ventricular (LV) diastolic pressure-segment length
         relations before and after removal of the pericardium.
         In the absence of the pericardium (open symbols), dextran
         infusion followed by nitroprusside infusion produced
         pressure-length coordinates that appear to be on the
         same curvi- linear pressure-length curve. When the peri-
         cardium is intact, similar interventions cause verticle
         ("parallel") displacements of the pressure-length re-
         lationship.

In this regard, Mirsky has developed theoretical arguments that transmural pressure-volume relations are not altered by acute pharmacologic interventions[19]. This analysis indicates that the upward "parallel" shift of the diastolic pressure-volume curve, which is seen during the administration of methoxamine, is not present if LV transmural pressure is substituted for absolute pressure. Furthermore, he argues that a substantial (acute) increase in left ventricular mass, an increase in myocardial stiffness, or a change in geometry cannot reproduce the changes observed with methoxamine. It appears, therefore, that the acute "parallel" shift in the diastolic pressure-volume curve seen during acute pharmacologic and other hemodynamic interventions requires an intact pericardium. Indeed, it is likely that the pericardium plays a role in modulating most acute changes in diastolic stiffness, including those seen during angina pertoris.

The pericardium also influences the effect of the right ventricle on left ventricular diastolic pressure-volume relations. Right ventricular distension has a distinct but small effect on the left ventricle when the pericardium is absent; however, changes in right ventricular volume can cause substantial shifts in the left ventricular diastolic pressure-volume relations when the pericardium is intact[20]. Because the pericardium modulates right and left ventricular (and probably atrial) diastolic interaction, the presence and stiffness of the pericardium must be considered in any analysis of the diastolic properties of the left ventricle.

Erectile Effects

In 1960, Salisbury demonstrated a direct relation between coronary perfusion pressure and left ventricular diastolic pressure, and he suggested that this result was due to alterations in pressure and volume within the coronary tree; he considered the phenomenon to be an "erectile effect"[21]. It is now well established that myocardial turgor due to coronary filling, whether secondary to changes in coronary perfusion pressure or to direct coronary vasodilatation, can influence left ventricular diastolic wall thickness and stiffness[22, 23]. It has also been shown that the effect is much more prominent in damaged hearts with high filling pressures than in normal hearts (Figure 5). The extent to which this erectile or hydraulic effect influences cardiovascular function remains to be determined, but the effect should be considered in the interpretation of changes in diastolic pressure in both experimental and clinical situations.

Myocardial Relaxation

Relaxation refers to the process by which the myocardium return to its initial length and tension; it is energy dependent and influ-

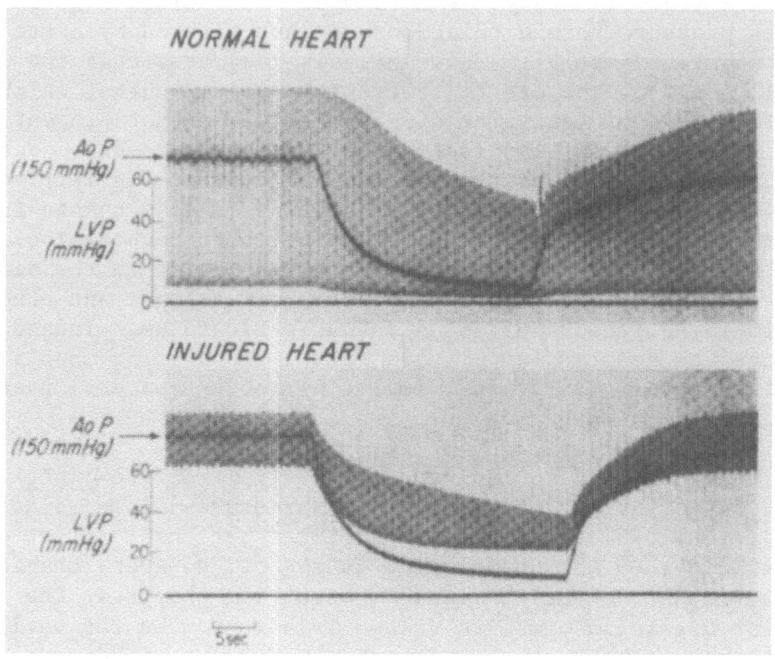

Fig. 5. Effect of a decrease in coronary perfusion pressure on left
         ventricular diastolic pressure before and after ischemic
         injury. Left ventricular pressure (LVP) was recorded from
         an isolated rabbit heart with an intraventricular balloon
         held at constant volume. In the upper panel (normal heart),
         coronary perfusion pressure (AoP) was adjusted to 150 mm Hg
         and the coronary perfusion pump was transiently turned off;
         left ventricular diastolic pressure decreased by 5 mm Hg.
         The lower panel shows the result of a similar experiment in
         an injured heart (90 minutes of severe ischemia); the de-
         crease in coronary perfusion resulted in a 42 mm Hg decrease
         in left ventricular diastolic pressure.

enced by the interaction of deactivation (the decay of active force
generating capacity) and loading conditions (forces which affect
myocardial fiber length). These forces may be thought of as those
which are applied early in the cardiac cycle and those which are
abruptly late in systole. The application of an early systolic load
(when adequate activating calcium is available to the contractile
proteins) results in the formation of more crossbridges, and a prolonged contraction-relaxation. In contrast, late systolic loads
result in an early and more rapid relaxation; this latter phenomenon
has been called "load-dependent relaxation"[6]. Late in the cardiac
cycle, myoplasmic calcium is below the threshold necessary for activating new crossbridges; existing bridges cannot support the new load,

the muscle yields, and relaxation is premature and more rapid. These speculations are consonant with myocardial intracellular calcium transients[24].

    We[25,26] and others[27,28] have studied the effects of acute (early) hemodynamic loads on isovolumic relaxation rate in the intact dog heart; over a wide range of systolic loads, relaxation rate was inversely related to LV systolic pressure and length. Our data are summarized in Figure 6. Thus, an (early) increase in left ventricular afterload results in a slower isovolumic relaxation rate. In distinct contrast, the application of a late systolic load (LV volume increment) results in premature and more rapid relaxation. Using a computer-controlled servo pump connected to the left ventricular apex, we have been able to rapidly infuse a small volume into the ventricle at specific times throughout the cardiac cycle. As shown in Figure 7, a small volume increment in early systole caused a slight delay in the onset of relaxation, while an identical volume increment in late systole caused an early and more rapid relaxation. This ventricular pressure response to a late systolic load (quick stretch) is the manifestation of "load-dependent relaxation" in the intact heart. Others have made similar observations[29, 20].

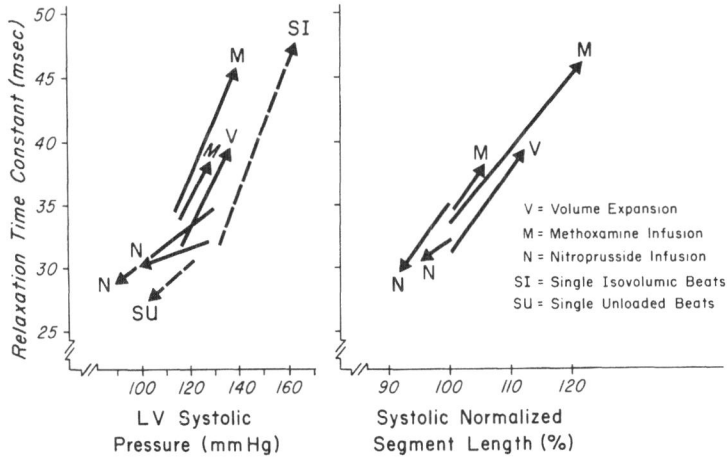

Fig. 6. Effects of acute alterations in left ventricular (LV) loading conditions on the LV isovolumic relaxation time constant. Changes in the relaxation time constant are plotted against LV systolic pressure (left panel) and systolic segment length (right panel). At higher LV systolic pressure and longer segment length, relaxation is prolonged (increased time constant); at lower systolic pressure and shorter systolic length, relaxation is more rapid (decreased time constant).

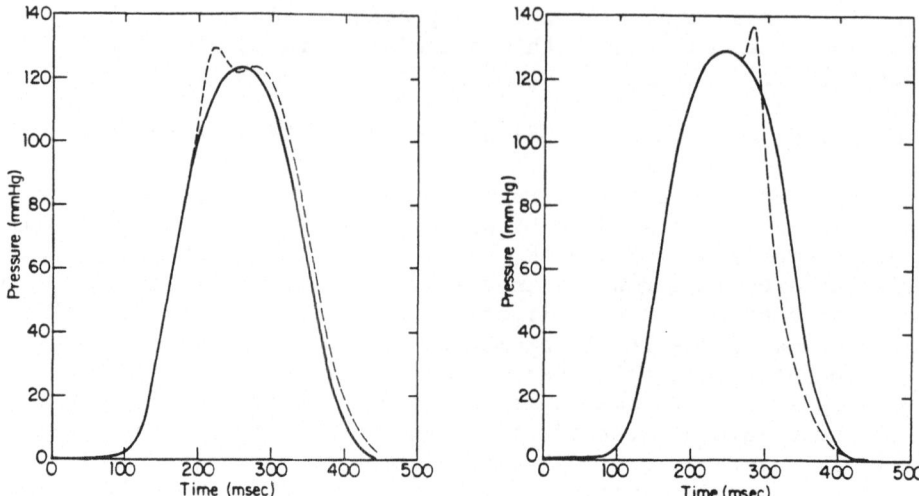

Fig. 7.   The effect of an abrupt volume increment on the time course
          of left ventricular pressure. The solid lines are control
          isovolumic beats and the broken lines are intervention
          (quick stretch) beats. In the panel on the left, the volume
          increment is given early in the contraction; in the panel on
          the right the volume increment is given late in the contrac-
          tion. In both experiments, 6 ml of warm blood was infused
          within 15 msec. The early volume infusion caused an abrupt
          increase in pressure, followed by a steep decline, then a
          plateau before pressure begins to drop toward the diastolic
          level. The onset of pressure decline was 12 msec later in
          the early intervention experiment. By comparison, the same
          volume increment given in late systole produces a quite
          different result. Here the pressure decline is earlier and
          much more rapid; peak negative dP/dt increased from 1742
          mm Hg/sec in the control beat to 3005 mm Hg/sec in the
          intervention beat.

    The effects of these early and late systolic loads and the
history of contraction[31], depend on which effect predominates; the
dominant loading condition then interacts with the energy-dependent
process of deactivation to control relaxation. Under some circumstances, temporal and spatial non-uniformities can influence relaxation rate[26], and these interrelationships are modulated by
neurohumeral, metabolic, and pharmacologic influences. Thus, left
ventricular relaxation may vary on a beat-to-beat, minute-to-minute,
and hour-to-hour basis.

    It is now established that prolonged or delayed myocardial
relaxation can, under certain circumstances, cause higher diastolic
pressures (at a given volume) than would be expected from a purely
passive diastolic pressure-volume relation[32, 33]. Not only can

prolonged relaxation influence early diastolic pressures, but relaxation may be incomplete at end diastole if the process is sufficiently slow (as with hypothermia) or if diastole is sufficiently short (as with rapid heart rates). Thus, experiments or clinical studies which are designed to assess passive stiffness of the chamber or muscle, must consider the variable and dynamic nature of altered myocardial relaxation.

While abnormal relaxation of the left ventricle undoubtedly causes altered patterns of diastolic filling, the factors governing isotonic relaxation rate and filling are not well understood. Early diastolic filling rate is determined by the left atrial pressure and the multiple factors which affect left ventricular early diastolic pressure. The factors which influence these interactions are a fertile area for future research.

VENTRICULAR FILLING IN DISEASED HEARTS

The static and dynamic factors responsible for abnormal left ventricular filling in diseased hearts are summarized in Figure 8. Disordered filling in chronic heart disease (hypertrophy and post infarction coronary heart disease) is primarily caused by alterations in the static determinants of left ventricular filling, whereas filling abnormalities in acute interventions of disease (i.e. angina pectoris) are related primarily to dynamic changes in the diastolic properties of the left ventricle.

	STATIC	DYNAMIC
LV HYPERTROPHY	↓ VOLUME/MASS RATIO ↑ MYOCARIAL STIFFNESS	ABNORMAL RELAXATION
CORONARY DISEASE CHRONIC CHD	↑ MYOCARDIAL STIFFNESS	ABNORMAL RELAXATION
ACUTE ISCHEMIA	——	ABNORMAL RELAXATION PERICARDIUM, RIGHT VENT CORONARY VASC VOLUME

Fig. 8. Factors responsible for abnormal left ventricular chamber stiffness and filling in diseased hearts. Abnormal filling in left ventricular hypertrophy and chronic coronary heart disease (CHD) is primarily related to alterations in the static determinants. Filling abnormalities during angina pectoris are primarily related to dynamic changes in the diastolic properties of the left ventricle. The extent to which alterations in myocardial relaxation influence ventricular filling in patients with heart disease is not well defined.

Left Ventricular Hypertrophy

There is general agreement that altered diastolic properties of
the left ventricle represent an important functional abnormality in
cardiac hypertrophy. Indeed, recent studies have described decreased
early diastolic filling rates in hypertrophic ventricles[34,35,36];
similarly, segment lengthening rates and wall thinning rates may be
reduced in cardiac hypertrophy. The mechanisms responsible for these
changes, however, are not well understood, but it is likely that both
static and dynamic factors contribute.

Increased passive stiffness of the hypertrophic heart is due in
part to the increased myocardial mass and the low volume/mass ratio;
abnormal intrinsic myocardial stiffness may also contribute to increased chamber stiffness[37,38]. Thus, a reduced early diastolic
filling rate could be a consequence of increased static resistance
to filling. Whether abnormal mycardial relaxation contributes to
slow filling rates in these hearts is uncertain; it has been shown,
however, that myocardial hypertrophy is associated with a depressed
calcium binding ability of the sarcoplasmic reticulum[39]. In
theory, this could influence relaxation in hypertrophic hearts.
Incoordinate or asynchronous relaxation of the ventricle (or even
spacial inhomogenities) could also contribute to many of the diastolic abnormalities seen in hypertrophic hearts[26,40].

The concept of load-dependent relaxation might also explain some
of the filling abnormalities seen in hypertrophic ventricles. Relaxation is very sensitive to load (wall stress) at the time of mitral
valve opening; this mechanism is said to be responsible at least in
part for the explosive rapid filling that occurs after mitral valve
opening in the normal heart[6]. In hypertrophic hearts, wall stress
at the time of mitral valve opening is exceedingly low and thus, the
load which contributes to fiber lengthening (load-dependent relaxation) can be very low despite normal or increased left atrial pressure. Thus, a purely mechanical and normal phenomenon could explain
reduced filling rates in left ventricular hypertrophy.

Coronary Heart Disease

Following myocardial infarction the damaged segment heals in
part by a process of fibrosis and consequently the diastolic
pressure-volume curve gradually shifts to the left indicating an
increase in chamber stiffness[1-4]. Left ventricular filling patterns are commonly abnormal in these ventricles and it is likely
that both static and dynamic factors contribute to this finding.
Increased chamber stiffness secondary to scar and fibrosis will cause
increased passive resistance to ventricular filling and a reduced
rate of early diastolic filling. The presence of hypertrophy (in the
uninfarcted segment), asynchronous relaxation (due to regional con-

duction abnormalities), and possibly even subclinical or mild ischemia might also contribute to these findings in damaged ventricles.
It is more difficult to explain the finding of a low filling rate
in patients with no regional wall motion abnormalities and normal
ejection fraction[41,42]. It should be noted, however, that abnormal
left ventricular filling has not been universally observed in such
patients[43]. The reason for these conflicting results in unknown,
but it is possible that filling is maintained by elevated left atrial
pressure in some, but not all hearts.

During acute myocardial ischemia (angina pectoris) substantial
changes in the diastolic properties of the left ventricle are
seen[44-47]. These changes are most likely related to alterations in
the dynamic determinants of filling and several mechanisms have been
proposed to be responsible; these include incomplete or impaired
relaxation[48] and tension prolongation during recovery from ischemia[49]. However, a pericardial and/or right ventricular or other
effects have not been excluded as contributing factors[5]. It is
likely that several dynamic factors acting in concert are responsible
for these acute ischemic changes.

CONCLUSIONS

Alterations in the diastolic properties of the left ventricle
have only recently been recognized as early markers of cardiac
disease; both the static and dynamic determinants of chamber stiffness and filling appear to be useful in this regard. Methods for the
quantitative assessment of these properties, especially the process
of relaxation, are evolving rapidly and it is likely that a careful
assessment of the static and dynamic properties of the left ventricular chamber will lead to an earlier diagnosis of cardiac disease; a
better understanding of the mechanisms of disease processes and
cardiovascular drug actions will follow.

Acknowledgements

The authors wish to acknowledge the assistance of Doctor Ralph
Shabetai who provided the data shown in Figure 4, and Ms Susan Pierce
who assisted in the preparation of this manuscript.

Supported by American Heart Association research grant No.81-850
and by Veteran Administration Medical Research funds.

REFERENCES

1. W. H. Gaasch, H. J. Levine, M. A. Quinones, and J. K. Alexander,
    Left ventricular compliance: mechanisms and clinical implic-
    ations, Am.J.Cardiol., 38:645-653 (1976).

2. W. Grossman and L. P. McLaurin, Diastolic properties of the left
    ventricle, Ann.Intern.Med., 84:316-326 (1976).
3. I. Mirsky, Assessment of passive elastic stiffness of cardiac
    muscle: mathematical concepts, physiologic and clinical con-
    siderations, direction of future research, Prog.Cardiovasc.
    Dis., 18:277-308 (1976).
4. S. A. Glantz and W. W. Parmley, Factors which affect the
    diastolic pressure-volume curve, Circ.Res., 42:171-180
    (1978).
5. J. Ross, Jr., Acute displacement of the diastolic pressure-
    volume curve of the left ventricle: role of the pericardium
    and the right ventricle, Circulation, 59:32-37 (1979).
6. D. L. Brutsaert, P. R. Housmans, and M. A. Goethals, Dual
    control of relaxation: its role in the ventricular function
    in the mammalian heart, Circ.Res., 47:637-652 (1980).
7. D. Fleming, Galen on the motions of the blood in the heart and
    lungs, Isis., 46:14-21 (1955).
8. W. Harvey, Exercitatio anatomica de motu cordis et sanguinis in
    animalibus (translated from the latin by C. D. Leake),
    Thomas, Springfield, Ill. (1928).
9. C. E. Francois-Frank, Sur les effects de la systole des
    orellettes, Arch.Physiol., 22(2):395-410 (1890).
10. Y. Henderson, The volume curve of the ventricles of the mam-
    malian heart, and the significance of this curve in respect
    to the mechanics of the heart beat and the filling of the
    ventricles, Am.J.Physiol., 16:325-367 (1906).
11. C. J. Wiggers and L. N. Katz, The contour of the ventricular
    volume curves under different conditions, Am.J.Physiol.,
    58:439-475 (1922).
12. W. J. Meek, Cardiac Tonus, Phys.Rev., 7:259-287 (1927).
13. C. J. Wiggers, "Physiology in Health and Disease," 5th Edition,
    Lea, Philadelphia, 740 (1949).
14. H. Pouleur, J. S. Karliner, M. M. LeWinter, and J. W. Covel,
    Diastolic viscous properties of the intact canine left ven-
    tricle, Circ.Res., 45:410-419 (1979).
15. E. H. Sonnenblick, The structural basis and importance of res-
    toring forces and elastic recoil for the filling of the
    heart, Eur.Heart J., 1(Suppl.A):107-110 (1980).
16. W. H. Gaasch, W. E. Battle, A. A. Oboler, J. S. Banas, and H. J.
    Levine, Left ventricular stress and compliance in man: with
    special reference to normalized ventricular function curves,
    Circulation, 45:746-762 (1972).
17. K. Shirato, R. Shabetai, V. Bhargava, D. Franklin, and J. Ross,
    Jr., Alteration of the left ventricular diastolic pressure-
    segment length relation produced by the pericardium: effects
    of cardiac distension and afterload reduction in conscious
    dogs, Circulation, 57:1191-1198 (1978).
18. T. Linderer, K. Chatterjee, W. W. Parmely, R. E. Sievers, S. A.
    Glantz, and J. V. Tyber, Influence of atrial systole on the
    Frank-Starling relation and the end-diastolic pressure-volume

relation of the left ventricle, Circulation, 67:1045-1053 (1983).

19. I. Mirsky and J. S. Rankin, The effects of geometry, elasticity and external pressures on the diastolic pressure-volume and stiffness-stress relations. How important is the pericardium? Circ.Res., 44:601-611 (1979).

20. J. Spadaro, O. H. L. Bing, W. H. Gaasch, A. Franklin, J. Clement, D. Rhodes, and R. M. Weintraub, Pericardial modulation of right and left ventricular diastolic interaction, Circ.Res., 48:233-238 (1981).

21. P. F. Salisbury, C. E. Cross, and P. A. Rieben, Influence of coronary artery pressure upon myocardial elasticity, Circ. Res., 8:794-800 (1960).

22. W. H. Gaasch, O. H. L. Bing, A. Franklin, D. Rhodes, S. A. Bernard, and R. M. Weintraub, The influence of acute alterations in coronary blood flow on left ventricular diastolic compliance and wall thickness, Eur.J.Cardiol., 7(Suppl.1): 147-161 (1978).

23. W. M. Voegel, C. S. Apstein, L. L. Briggs, W. H. Gaasch, and J. Ahn, Acute alterations in left ventricular diastolic chamber stiffness. Role of the erectile effect of coronary arterial pressure and flow in normal and damaged hearts, Circ.Res., 51:465-478 (1982).

24. D. G. Allen and S. Kurihara, Calcium transients in mammalian ventricular muscle, Eur.Heart J., 1(Suppl.A):5-15 (1980).

25. W. H. Gaasch, A. S. Blaustein, C. W. Adnrias, and B. Avitall, Myocardial relaxation II. Hemodynamic determinants of the rate of left ventricular isovolumic pressure decline, Am.J.Physiol., 239:H1-H6 (1980).

26. A. S. Blaustein and W. H. Gaasch, Myocardial relaxation VI. Effects of beta adrenergic tone and asynchrony on LVP and relaxation rate, Am.J.Physiol., 244:H417-H422 (1983).

27. J. S. Karliner, M. M. LeWinter, F. Mahler, R. Engler, and R. A. O'Rourke, Pharmacologic and hemodynamic influences on the rate of isovolumic left ventricular relaxation in the normal conscious dog, J.Clin.Invest., 60:511-521 (1977).

28. G. L. Raff and S. A. Glantz, Volume loading slows left ventricular isovolumic relaxation rate: evidence of load-dependent relaxation in the intact dog heart, Circ.Res., 48:813-824 (1981).

29. M. I. M. Nobel, The contribution of blood momentum to left ventricular ejection in the dog, Circ.Res., 23:663-670 (1968).

30. N. A. Goethals, I. E. Kersschat, V. A. Claes, C. F. Hermans, A. H. Jageneau, and D. L. Brutsaert, Influence of abrupt pressure increments on left ventricular relaxation (abstract), Am.J.Cardiol., 45:392 (1980).

31. M. Hori, M. Inoue, M. Fukunami, Y. Ishida, S. Nakajima, M. Kitakaze, M. Kitabatake, and H. Abe, Influence of ejection timing on left ventricular relaxation in isolated canine heart (abstract), Circulation, 66:304 (1982).

32. M. L. Weisfeldt, J. W. Frederiksen, F. C. P. Yin, and J. L. Weiss, Evidence of incomplete left ventricular relaxation in the dog, J.Clin.Invest., 62:1296-1302 (1978).

33. A. S. Blaustein and W. H. Gaasch, Myocardial relaxation III. Reoxygenation mechanics in the intact dog heart, Circ.Res., 49:633-639 (1981).

34. J. E. Sanderson, D. G. Gibson, D. J. Brown, and J. F. Goodwin, Left ventricular filling in hypertrophic cardiomyopathy: an angiographic study, Br.Heart J., 39:661-670 (1977).

35. P. Hanrath, D. G. Mathey, R. Siegert, and W. Bleifeld, Left ventricular relaxation and filling in different forms of left ventricular hypertrophy: an echocardiographic study, Am.J. Cardiol., 45:15-23 (1980).

36. R. O. Bonow, D. R. Rosing, and S. L. Bacharach, Effects of verapamil on left ventricular systolic function and diastolic filling in patient with hypertrophic cardiomyopathy, Circulation, 64:787-796 (1981).

37. K. L. Peterson, J. Tsuji, A. Johnson, J. DiDonna, and M. M. LeWinter, Diastolic left ventricular pressure-volume and stress-strain relations in patients with valvular aortic stenosis and left ventricular hypertropy, Circulation, 58:77-89 (1978).

38. W. H. Gaasch, O. H. L. Bing, and I. Mirsky, Chamber compliance and myocardial stiffness in left ventricular hypertrophy, Eur.Heart J., 3(Suppl.A):139-145 (1982).

39. L. A. Sordahl, W. B. McCollum, W. G. Wood, and A. Schwartz, Mitochondria and sarcoplasmic reticulum function in cardiac hypertrophy and failure, Am.J.Physiol., 224:479-487 (1973).

40. J. E. Sanderson, T. A. Traill, M. G. Sutton, D. J. Brown, D. G. Gibson, and J. F. Goodwin, Left ventricular relaxation and filling in hypertrophic cardiomyopathy: an echocardiographic study, Br.Heart J., 40:596-601 (1978).

41. L. A. Reduto, W. J. Wickemeyer, J. B. Young, L. A. DelVentura, J. W. Reid, D. H. Glaeser, M. A. Quinones, and R. R. Miller, Left ventricular diastolic performance at rest and during exercise in patients with coronary artery disease: assessment with first pass radionuclide angiography, Circulation, 63: 1228-1237 (1981).

42. R. O. Bonow, S. L. Bacharach, M. V. Green, K. M. Kent, D. R. Rosing, L. C. Lipson, M. B. Leon, and S. E. Epstein, Impaired left ventricular diastolic filling in patients with coronary artery disease: assessment with radionuclide angiography, Circulation, 64:315-323 (1981).

43. J. D. Carroll, O. M. Hess, H. O. Hirzel, and H. P. Krayenbuehl, Dynamics of left ventricular filling at rest and during exercise, Circulation, 68:59-67 (1983).

44. E. M. Dwyer, Left ventricular pressure-volume alterations and regional disorders of contraction during myocardial ischemia induced by atrial pacing, Circulation, 42:1111-1122 (1970).

45. W. H. Barry, J. Z. Brooker, E. L. Alderman, and D. C. Harrison, Changes in diastolic stiffness and tone of the left ventricle during angina pectoris, Circulation, 49:255-263 (1974).

46. T. Mann, S. Goldberg, G. H. Mudge, and W. Grossman, Factors contributing to altered left ventricular diastolic properties during angina pectoris, Circulation, 59:14-20 (1979).

47. P. D. Bourdillon, B. H. Lorell, I. Mirsky, W. J. Paulus, J. Wynne, and W. Grossman, Increased regional myocardial stiffness of the left ventricle during pacing induced angina in man, Circulation, 67:316-323 (1983).

48. W. Grossman and J. T. Mann, Evidence for impaired left ventricular relaxation during acute ischemia in man, Eur.J.Cardiol., 7(Suppl.1):239-249 (1978).

49. O. H. L. Bing, J. F. Keefe, M. J. Wold, L. J. Finkelstein, and H. J. Levine, Tension prolongation during recovery from hypoxia, J.Clin.Invest., 50:660-668 (1971).

CRITICAL EVALUATION OF STRESS-VOLUME IN SYSTOLE

L. Bonandi

Cattedra di Cardiologia
Università degli Studi di Brescia
Italy

The extent of the myocardial fiber shortening is influenced by the initial length (preload), by the forces that oppose its shortening (afterload) and by its contractile state which - as it has demonstrated - can be described in terms of instantaneous force-velocity-length relations of the myocardial fiber.

Mitchell and Wildenthal (1972) found in the animal that the variations of the cardiac volume at end-systole could be used for assessing the contractile state of the left ventricle. In 1957 Holt, in dogs analyzing the factor which the ejection phase of the left ventricle found a linear relationship between the length and the force of the myocardium at end-systole. Monroe and French (1961) investigated the pressure-volume relationship of the left ventricle with an experimental model in which the systolic load may be changed, and found that changes of the systole load are followed by changes of the systolic pressure-volume curve, and the relationship of the points of the end-systolic pressure-volume ratio was linear, and that inotropic drugs - like digitals and epinephrine, induced an increase and a shift to the left of the slope determined by end-systolic pressure-volume ratio. More recently Suga and Sagawa, Mahler et al. (1973), Weber et al. (1976), confirmed the linearity of this relationship in the intact canine left ventricle and its sensitivity to changes of the inotropic state of the left ventricle. Following these and other studies the end-systolic force-length relationship of the left ventricle has been proposed as an ideal index for assessing myocardial contractility because of three major advantages compared to other conventional ejection phase indexes. 1) This relationship is preload independent, and 2) The afterload is incorporated into the analysis so that observed changes assess contractility rather than usual mixed changes of contractility and loading conditions, as it is the case with the conventional ejection phase indexes (VCF, EF, etc.).

Many investigators have examined the end-systolic force-length relationship of the left ventricle to assess myocardial contractility in various heart disease: the left ventricular end-systole pressure or wall stress could be taken as a measure of the force, and the end-systolic volume or diameter as a measure of the length.

A potential problem in using this relationship in man is represented by the autonomic reflexes, which could induce significant changes of the inotropic state of the left ventricle during the data collection; a way to circumvent this problem could be the use of the wall stress instead of the pressure as a measure of the force. Very small pressure changes (of the order of few millimeters of mercury) with concomitant variations of wall thickness and internal diameter of the left ventricle, are followed by relevant changes of the size of wall stress (Reichek and coll., 1982). Thus, in this way not only the effect of the autonomic reflexes could be minimized but the aortic impedance also could be less relevant for the assessment of this index. Nevertheless this method also has the following important limitations:

1) the contractility of the left ventricle should be homogeneous – its use is then limited in patients with coronary artery disease and regional wall motion disturbances.
2) in patients with significant mitral regurgitation, in which isometric conditions may not be obtained, identification of the end-systole is ambiguous (Reichek and coll., 1982).
3) furthermore, to obtain a contractility index from the force-length relationship such as the "Suga and Sagawa" index, several data points are needed. This requires a series of ventriculograms at the different loading conditions. This, however, could be circumvent by using M-mode echocardiography together with high-fidelity pressure recordings, even though objections could arise concerning the use of meridional instead of the circumferential wall stress (Marsh and coll., 1979).

Recently, Pouleur and coworkers (1982) proposed a new index of myocardial contractility, using the late systolic stress-volume relationship of a single ventriculogram as a substitute for the end-systole pressure-volume relationship of several ventriculograms.

The slope of the late stress-volume relation was found directly related to the length-tension relationship of the left ventricle.

Therefore, we evaluated the late systolic stress-diameter relationship as a new index of myocardial contractility in patients with aortic valve disease before and after successful valve replacement, in order to determine the sensitivity of this index for assessment of left ventricular contractility. This study has been performed in collaboration with Prof. Krayenbuehl at the University Hospital in Zurich (Bonandi and coll., in press).

In this study we included 25 patients with aortic valve disease before and after valve replacement (mean interval 18 months) and 10 control patients with no cardiovascular disease. All patients underwent a left and right catheterization, they were all in sinus rhythm and had no abnormalities of left ventricular wall motion. Selective coronary arteriography was carried out in all patients with aortic valve disease.

M-mode echocardiogram was obtained simultaneously with the pressure recording, with an Ekoline 20 A interfaced to an DR 16 recorder, at the paper speed of 100 mm/sec. (Figure 1).

Left ventricular high-fidelity pressure was record with a transseptal transducer Millar micromanometer; conventional aortic pressure was recorded simultaneously with a pigtail catheter.

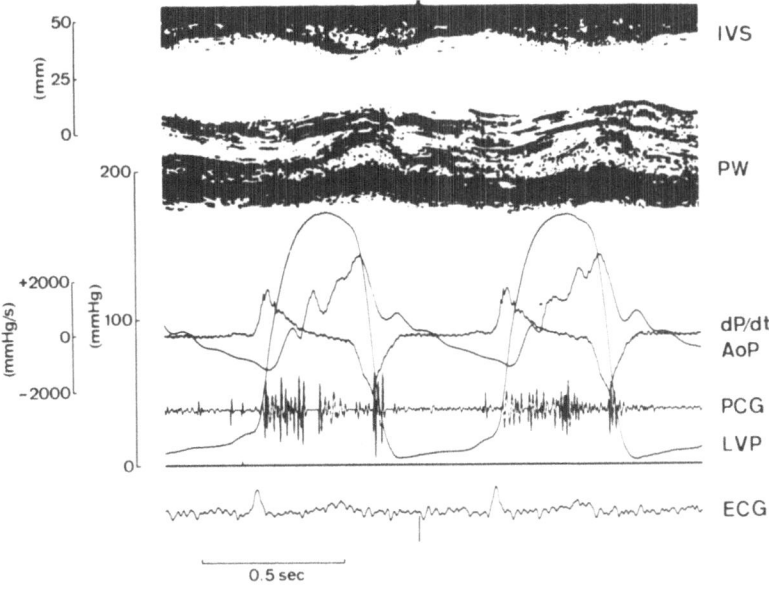

L.M.♀,1941   27.1.78

Fig. 1. Original tracings in a patient with severe aortic stenosis. At the top of the tracing there is the M-mode echocardiogram with the interventricular septum (IVS) and the posterior wall (PW). Left ventricular pressure (LVP) and conventional aortic pressure (AoP) as well as the first derivative of LVP (dP/dt) were recorded at a paper speed of 100 mm/sec. ECG: Electrocardiogram; PCG: Phonocardiogram.

M-mode echocardiography was carried out using a 2.25 MHz transducer with the patient in supine or slight lateral position. The
echo-pressure tracings were digitized manually by an electronic
digitizer with a resolution of 0.25 mm. The following parameters
were calculated and printed out at time intervals of 3 to 5 msec:
left ventricular diameter (D), posterior wall thickness (h), left
ventricular pressure(P), dP/dt, left ventricular meridional stress
(σ) using the equation of Brodie for calculation of meridional wall
stress:

$$\sigma = \frac{P \cdot D}{4h \cdot (1 + \frac{h}{D})}$$

and dσ/dt.

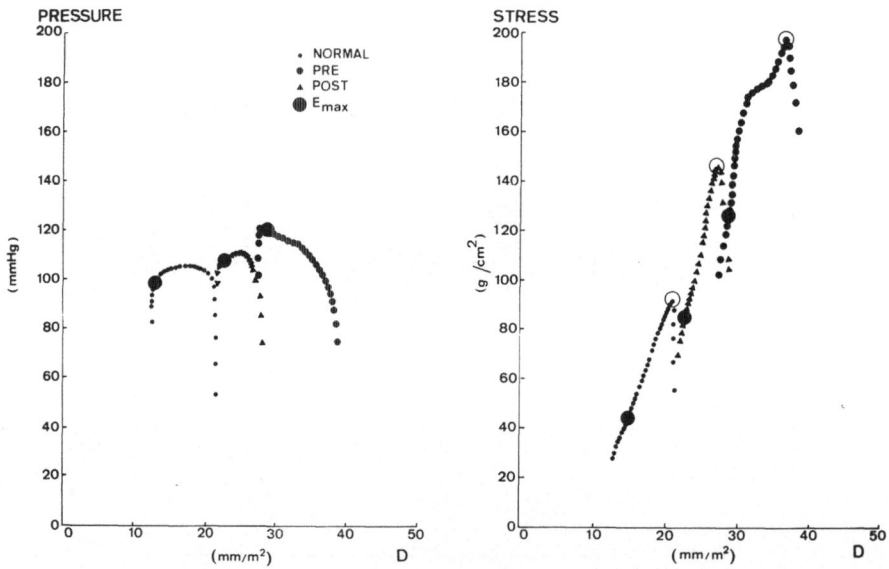

Fig. 2.   Pressure-diameter (left hand panel) and stress-diameter
          (right hand panel) relationship in a control patient
          (asterisk) and in a patient with severe aortic insufficiency
          before (circles) and after (triangles) successful valve
          replacement. The maximal pressure-diameter ratio (E_{max}) is
          marked in all three patients by a big closed circle and peak
          stress is marked by a big open circle. In all patients E_{max}
          was determined from the pressure-diameter relationship (left
          hand panel) and then the slope of the stress-diameter
          relationship was calculated from E_{max} to peak stress (right
          hand panel) using a linear regression analysis. D: left
          ventricular internal diameter; Pre: preoperative; Post:
          postoperative.

Left ventricular volume (area-length method) and ejection fraction were calculated by biplane cineangiography.

In each patient we plotted left ventricular pressure versus
diameter (normalized for b.s.a.) during the systolic ejection phase.
Then, the pressure-diameter ratio was calculated for each coordinate
and the largest ratio, E_{max}, and the ratio at end-systole (that was
defined from the aortic component of the second heart sound from
the phonocardiogram (Figure 1)) E_{es}, were used as indexes of left
ventricular contractility (Figure 2); the slope of the stress-
diameter relationship was calculated from E_{max} back to the highest
stress value using a linear regression equation as proposed by
Pouleur and coworkers (1982). The correlation coefficents were
generally good, the lowest observed value being 0.928, and the intercept at 0 stress (D_o), the so called "dead diameter".

The results of the standard hemodynamic evaluation are summarized in Table 1. E_{max} and E_{es} did not differ significantly in the
three groups, suggesting normal contractility in patients with
aortic valve disease before and after valve replacement (Figure 3).

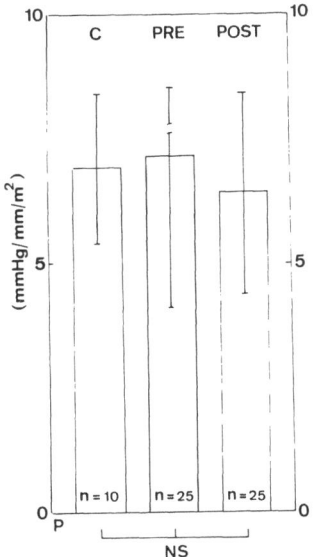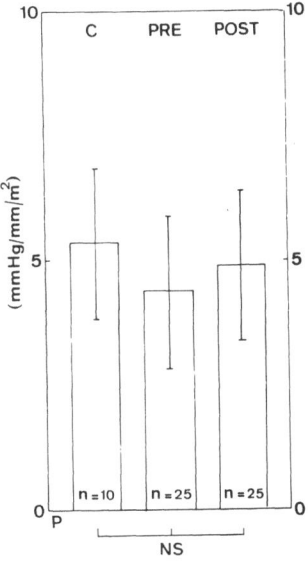

Fig. 3.   Maximal (E_{max}; left hand panel) and end-systolic (E_{es}; right
          hand panel) pressure-diameter ratio in 10 control patients
          (C) and 25 patients with aortic valve disease before (PRE)
          and after (POST) aortic valve replacement. There was no
          significant difference in E_{max} and E_{es} between the three
          groups. P: probability; NS: not significant.

Table 1.

	HR	LVEDP	LVSP	Sp	Sed	EDVI	EF
CONTROLS :	70 ± 11	8.6 ± 3	115 ± 11	139 ± 61	11 ± 6	84 ± 21	69 ± 6
		**	**	**	**	**	
AVD pre :	72 ± 12	17.4 ± 8	175 ± 42	198 ± 50	25 ± 12	169 ± 73	58 ± 13
		**	**	*	**	**	
AVD post :	77 ± 12	11.7 ± 6	138 ± 19	164 ± 52	15 ± 10	118 ± 58	60 ± 13

HR   : Heart rate (beats/min)
LVEDP : Left ventricular end-diastolic pressure (mmHg)
LVSP : Left ventricular end-systolic pressure (mmHg)
S_p : Left ventricular meridional peak stress (g/cm^2)
S_{ed} : Left ventricular meridional end-diastolic stress (g/cm^2)
EDVI : End-diastolic volume index (ml/m^2)
EF   : Left ventricular ejection fraction (%)

*   : $P < 0.05$
** : $P < 0.01$

The slope of stress-diameter relationship and the intercept (D_o) also showed no significant differences between the three groups (Figure 4).

To further test the sensitivity of the late systolic stress-diameter relation we divided the patients with aortic valve disease in two subgroups: subgroup 1 consisted of 16 patients with normal preoperative ejection fraction (EF$\geq$57), subgroup 2 of 9 patients with decreased ejection fraction preoperatively (EF<57).

When EF was plotted against peak systolic wall stress, it could be observed that in subgroup 2 EF is lower at the same stress level, when compared to subgroup 1 (Figure 5). Thus, the depressed left ventricular ejection performance of subgroup 2 could not be solely due to an increased afterload, but due to a reduced myocardial contractility.

We evaluated in the two subgroups also the following standard hemodynamic parameters (Table 2).

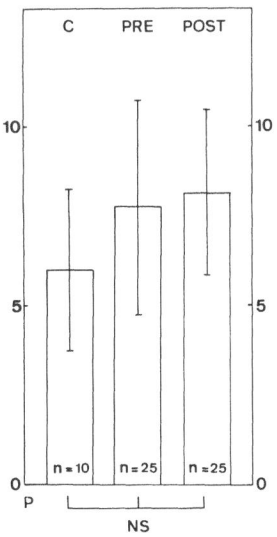
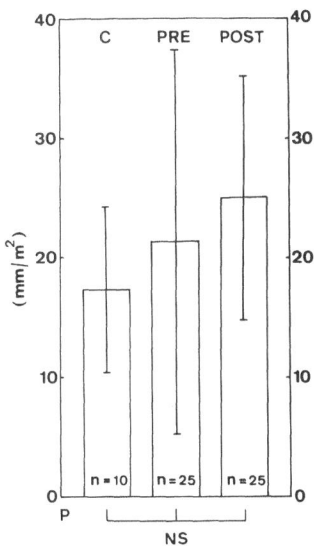

Fig. 4. Slope (β; left hand panel) and intercept at zero wall stress
(D_o; right hand panel) of the late systolic stress-diameter
relationship in 10 control patients (C) and 25 patients with
aortic valve disease before (PRE) as well as after (POST)
successful valve replacement. Note that there is no significant difference in β or D_o between the three groups. P:
probability; NS: not significant.

Table 2.

	HR	LVEDP	LVSP	S_p	S_{ed}	EDVI	ESVI	EF
GROUP 1 pre:	76±16	18.4±9	186±43	208±52	23±12	137±46	45±23	67±6
post:	73±13	11.0±5	135±18	153±51	15±8	105±24	39±15	65±6
GROUP 2 pre:	69±8	15.5±8	155±34	184±46	29±13	225±83	116±62	45±10
post:	78±11	11.5±6	140±23	177±53	17±14	150±86	80±72	51±10

HR   : Heart rate (beats/min)
LVEDP : Left ventricular end-diastolic pressure (mmHg)
LVSP : Left ventricular end-systolic pressure (mmHg)
S_p : Left ventricular meridional peak stress (g/cm^2)
S_{ed} : Left ventricular meridional end-diastolic stress (g/cm^2)
EDVI : End-diastolic volume index (ml/m^2)
ESVI : End-systolic volume index (ml/m^2)
EF   : Left ventricular ejection fraction (%)

 * : $P\ 0.05$
 ** : $P\ 0.01$

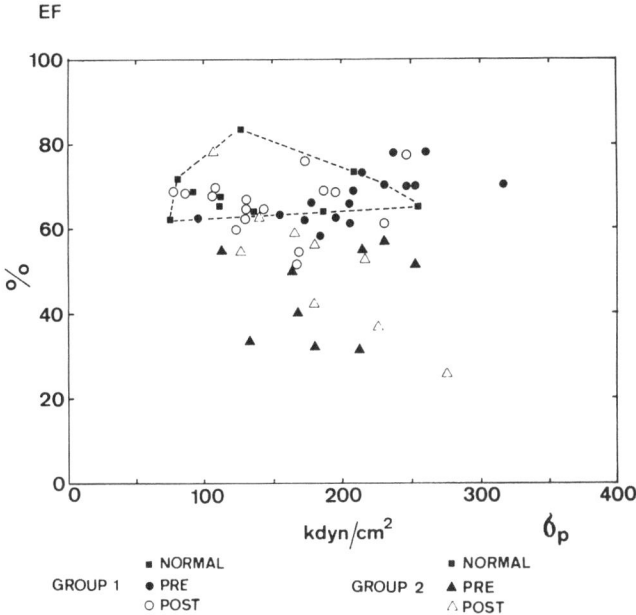

Fig. 5. Relationship between angiographic ejection fraction (EF) and
         left ventricular meridional peak stress (σ_p) in 10 control
         patients (squares) and 16 patients with normal preoperative
         ejection fraction (group 1; circles) and 9 patients with
         depressed preoperative ejection fraction (group 2; tri-
         angles) before (closed symbols) as well as after (open
         symbols) successful valve replacement. The normal range is
         indicated by the dashed line. Note that only one patient of
         group 2 is within the normal limits after successful valve
         replacement but no one before aortic valve surgery. How-
         ever, in group 1 16 patients (10 postoperative and 6 pre-
         operative patients) are within the normal limits, whereas
         only 2 postoperative patients had a really depressed left
         ventricular ejection fraction (<57%) in group 1. Thus, at
         a similar afterload left ventricular ejection performance
         is clearly lower in group 2 patients than in patients of
         group 1.

        In the last figure (Figure 6), E_{max} and E_{es} as well as the
"slope" of the stress-diameter relationship are summarized; the bars
with the hatched area represent the preoperative and the open bars
the postoperative evaluation, the range of control patients is indicated by the dashed-line.

        E_{max} and E_{es} are significantly decreased preoperatively in
subgroup 2 when compared to subgroup 1, but are not significantly
different in the two subgroups after successful valve replacement.

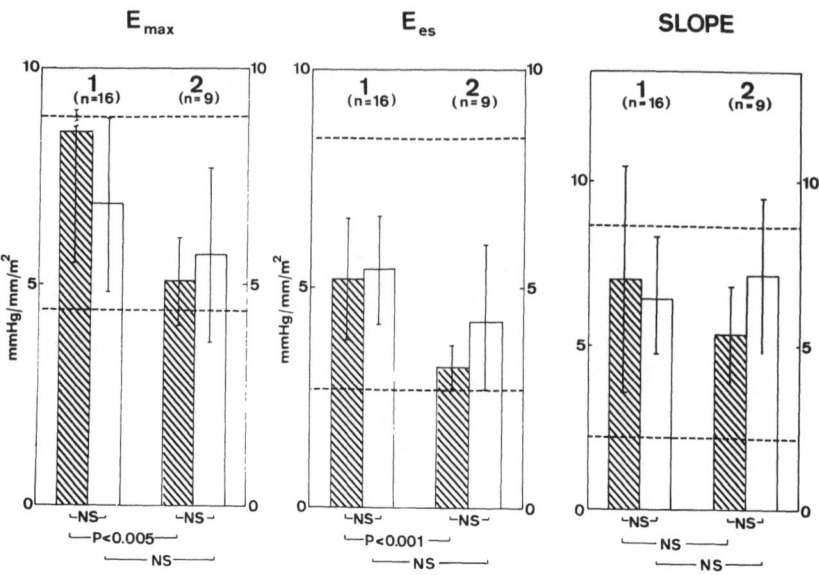

Fig. 6. E_{max}, E_{es} and the slope of the late systolic stress-diameter relationship in patients with normal preoperative ejection fraction (group 1) and in patients with depressed pre-operative ejection fraction (group 2) before (hatched area) and after successful valve replacement. The upper and lower limit of normality is indicated by the horizontal dashed lines. Note that the maximal (E_{max}) and the end-systolic (E_{es}) pressure-diameter ratio is significantly decreased preoperatively in group 2 as compared to group 1. However, both E_{max} and E_{es} are no longer significantly different following successful valve replacement. The slope of the late systolic stress-diameter relationship showed no significant different between the two groups pre- and post-operatively. P: probability; NS: not significant.

The slope of the late systolic stress-diameter relationship did not show any difference pre- and post-operatively in both subgroups.

Thus, the "slope" of the late systolic stess-diameter relation" seems to be very insensitive to change in left ventricular ejection performance in patients with aortic valve disease; this insensitivity can be attributed to the intrinsic relation between stress and diameter, because the left ventricular internal diameter is plotted both on the X and Y axis of the diagram (Figure 2) (Bonandi and coll., in press).

REFERENCES

Bonandi, L., Hess, O. M., Turina, M., and Krayenbuehl, H. P., Late
    systolic stress-diameter relation in patients with aortic
    valve disease before and after valve replacement, Basic Res.
    Cardiol., in press.
Holt, J. P., 1957, Regulation of the degree of emptying of the left
    ventricle by the force of ventricular contraction, Circ.Res.,
    5:281.
Marsh, J. D., Green, L. H., Wynne, J., Cohn, P. F., and Grossman, W.,
    1979, Left ventricular end-systolic pressure-dimension and
    stress- length relation in normal human subjects, Am.J.Card.,
    44:1311.
Michell, J. H., and Wildenthal, K., 1972, Analysis of left
    ventricular function, Proc.Roy.Soc.Med., 65:542.
Monroe, R. G., and French, G. N., 1961, Left ventricular pressure-
    volume relationship and myocardial oxygen consumption in the
    isolated heart, Circ.Res., 9:362.
Pouleur, H., Rousseau, M. F., Eyll, C., Mechelen, H., Brasseur, L.,
    and Charlier, A., 1982, Assessment of left ventricular con-
    tractility from late systolic stress-volume relation,
    Circulation, 65:1204.
Reichek, N., Wilson, J., Sutton, M., Plappert, T. A., Goldberg, G.,
    and Hirshfeld, J. W., 1982, Noninvasive determination of the
    left ventricular and end-systolic stress: validation of the
    initial application, Circulation, 65:99.
Suga, S., Sagawa, K., and Shoukas, A. A., 1973, Load independence of
    the instantaneous pressure-volume ratio of the canine left
    ventricle and effects of epinephrine and heart rate on the
    ratio, Circ.Res., 32:314.
Weber, K. T., Janicki, J. S., and Hefner, L. L., 1976, Left ven-
    tricular force-length relations of isovolumic and ejecting
    beats, Am.J.Physiol., 231:337.

CHAPTER I:

DISCUSSION

DENOLIN

I think that in spite of what was said by Harvey, we understand a little more and Dr Peterson has opened at least a small door of the sky, for the understanding of the ventricular function.

It appears clearly that the known classical concept of regulation of ventricular function by preload, afterload, contractility, remains acceptable. All these parameters are closely associated, closely related. You can not probably isolate one without considering the others and this is probably a very important conclusion.

The next problem is that preload is a very important concept and not only systolic function should be considered but also diastolic function of the heart which is probably responsible for regulation of preload. I would appreciate a comment on that.

Another question is the problem of the right ventricle: a few words were said on that and my question would be: are the general principles of regulation the same on the right and the left side of the heart?

Now, going on to the clinical data presented on mitral regurgitation, where we saw a decrease of ejection fraction, what does this mean from the clinical point of view and what is the status of the patient in such conditions? This is very important because this will probably give the opportunity of discussing the significance of ejection fraction, which is considered today as one of the best indicators of L.V. function.

PETERSON

The assessment of preload is perhaps almost as complex as afterload. The reason being that we know from ultrastructural studies that the sarcomeres stretch from their normal operating length of

around 1.9 microns to a maximum of approximately 2.2 microns. We
know that the augmentation in the sarcomere length is inadequate to
explain some of the increases in end diastolic volume that we see in
clinical heart disease. I think it is reasonably well accepted,
although not proved today, that there is fiber slippage which occurs
in the dilated ventricle, and therefore when we measure end-diastolic
volume as an index of fiber length it may be inappropriate for what
the true preload of the muscle is.

You asked about right ventricular function and whether the same
principles of physiology and pathophysiology apply to the right
ventricle as those that we have observed in the left ventricle. I do
not think we can say that with absolute certainty.

We know that there is a considerable amount of interaction
between the left ventricle and the right ventricle, one can be influencing the function of the other, and sorting out what those
influences are, can be quite difficult.

Nevertheless, I think, in general, the right ventricle being a
thin wall chamber is the one that we predict and it will be more
sensitive to afterload, than is the left ventricle. One other point
we can make about the right ventricle is that it is a very difficult
chamber to model from a geometric standpoint, and therefore it is
difficult to study.

The ejection fraction is a relatively poor reflection of the
contractile state in mitral regurgitation, and you must look at other
indeces of ventricular function besides ejection fraction to make
your assessment of what the response on the ventricle will be post-
operatively. And up to this point we have felt that some simple
measurements, such as the size of the ventricle which ultimately
affects its post-operative afterload, are as predictive as any in
making that decision about what the effect the valve replacement is
going to be.

If you are faced with a patient who has already a markedly
diminished ejection fraction, a very dilated ventricle, are you doing
that patient a favor by subjecting him to a mitral valve replacement?
Our answer to that is yes, we think that you are doing the patient a
favor provided there is still a very large regurgitation fraction,
even though the myocardium is already failing, if there is still a
very large regurgitant volume, then you have some prediction of what
the volume change of the ventricle will be post-operatively.

GAASCH

I just wanted to expand a little bit on what Doctor Peterson said on the question of preload and afterload.

When we use the term "preload", we do not use it to imply a
volume or pressure; we tend to use the word "afterload" to imply wall
stress during systole, and "preload" meaning wall stress during
diastole, so in that fashion preload is determined by afterload, it
is just a wall stress in this sphere, or the ellipsoid at a different
point in time.

I saw in the abstracts that you include under "afterload mismatch" the idea that afterload could be inappropriately low. I
wondered if you could expand on that for me.

PETERSON

I think that the best example of inappropriately low afterload
is in the setting of acute mitral regurgitation. It is difficult to
measure but, immediately after rupturing a cordal structure in a dog,
you can actually show a marked increase in the shortening. You know
that afterload remains normal or actually wall stresses slightly
diminish. So the best definition I find for afterload mismatch is
that there is an inappropriate degree of shortening of the ventricle
for what the given afterload is. And the inappropriate degree of
shortening can be either enhanced or can be diminished from what it
should be.

RAZZOLINI

Do you think that a certain degree of fibrosis with the mechanisms of limitation of increased cavity volume/mass volume ratio could
be a protective mechanism in a sense?

GAASCH

Usually we would think in most forms of clinical disease that
fibrosis once developed would be to a great extent irreversible and
would preclude normalization of ventricular function. But in some
situations I suppose that I have to agree with you, it is possible
that it would be protective.

VISIOLI

From the lectures up to now, we have a clear vision of left
ventricular function which needs all parameters: ventricular volume,

sfigmomanometric pressure and ventricular wall thickness. In my
opinion we cannot speak generally of "left ventricular function" but
of "left ventricular functions" discussing valvular disease and,
above all, chronic ischemic cardiopathy. Moreover when we speak of
diastole and systole, we must also speak of regional diastole and
systole.

GAASCH

I agree with everything you said. I think I could amplify,
however, your comment about asynchrony. I think it is something
which we tend to overlook in experimental and clinical cases of our
diseases. It is extremely important not only to consider that it is
possible that by echography or some other technique we are looking
only at one segment and missing many others. In addition we must
remember that every segment is being acted upon by all the others.

PETERSON

I think that Dr. Gaasch's demonstration of the effect of late
systolic loading on relaxation was quite interesting. Could you say
that by using balloon counterpulsation in late systole it actually
augments early diastolic relaxation? And so my question is, have you
been able to show in any way that diastolic relaxation is enhanced by
causing late systolic loading?

GAASCH

Sequencing an intra-aortic balloon to expand proximally at the
time the aortic valve closes, or perhaps a little before that, in
mammalian heart muscle results in an extremely rapid relaxation,
whether it is measured as a time versus tension decline or as filling.

LEACHMAN

Claude Bernard in the 1930s investigated ventricular function
and it was of great interest that the major argument in the French
School of physiology at that time was the problem with regard to
whether all of the forces leading the blood were pumping the blood
out or the blood was in fact sucked in.

GAASCH

In the normal heart I think suction has very little importance.
However, under some circumstances, we know that volume can rapidly
enter the ventricle as in a post extra systolic beat, or any other

intervention that determines a very small end-systolic volume. In
this case the next diastole does manifest suction. Moreover I think
that in more chronic or sub-acute situations, perhaps mitral
regurgitation, there is an example of suction.

VISIOLI

I only wanted to add something to Suga Sagawa's relation. This
is impracticable not only from a conceptual but especially from a
clinical point of view because measurements at different afterloads
are needed and this is difficult to perform. Nevertheless afterload
changes may cause neurovegetative reflexes.

Based on Pouleur data we wanted to verify the accuracy of an
evaluation on a single systole in a single afterload condition during
a systole. This is the spirit of research and the spirit of those
who take advantage of "Emax" - in my opinion in an unjustifiable way.

SCOGNAMIGLIO

I'd like to add something about Emax. I agree on the theoretical limits of this contractility index but it seems to me that we
should discuss the usefulness of this index from the clinical point
of view. It is incorrect to ask a contractility index to predict
postsurgical pump function because contractility doesn't change after
surgery and, on the contrary, pump function depends on the type of
pre-surgical left ventricular hypertrophy: if contractility is depressed, but hypertrophy is adequate, volume changes after surgery
will cause an increment of pump function; on the other hand this will
not happen if hypertrophy is not adequate. Emax gives a very useful
piece of information about post-surgical survival; among 80 patients
under our observation those with a critically depressed value of Emax
(under 1) showed a worse 5 years post surgical survival with respect
to non surgical patients with the same degree of contractility
derangement. I'd like to ask Dr Gaasch if the isometric relaxation
changes he showed following a late systolic load were similiar to
those happening during short duration ischemia. Do you think that
this is only an occasional morfologic analogy or that there may be a
relationship between late load, coronary perfusion and isometric
relaxation?

GAASCH

Allow me to just mention my impression. I think that if you
look at a left ventricle angiogram within 5 to 10 seconds after
inflating a balloon during coronary angioplasty, you will see, under
those circumstances, segmental early relaxation phenomena, during
very mild or very early ischemia. Now we have to remember that what

we are seeing is a segment that is relaxing early, but that segment
is being acted on by a large mass of myocardium which is normal. So
perhaps, mild ischemia or very early coronary occlusion weaken the
muscle and the remaining normal myocardium stretches the ischemic
segment which relaxes early. If you, in fact, wait for a period of
1, 2, or 3 minutes and then look at regional motion, now the ischemic
segment is showing paradoxical motion and the normal segment is
showing segmental early relaxation.

        I would like to take a specific slide that Dr. Peterson showed
about an apparent increase in stiffness of the ventricle during
angina. You recall the upward shift of the pressure volume relationship and I would like to postulate that it is difficult to accept it
as a change in muscle stiffness due to ischemia because the majority
of the muscle of that heart remains normal, so we would have to
conclude that the change in the stiffness of the muscle is a sum or
an average of the normal muscle and the abnormal muscle, where the
normal muscle dominates. And I think that the observation must also
concentrate on the effects of the pericardium. Now, the conclusion
that I draw is that if we want to go further with this type of analysis, we are going to have to analyse regional compliance or regional
function and I believe that to do that we are going to have to visul-
ize the whole ventricle rather than a segment as we might do by echo.

PETERSON

        I was trying to be cautious about that compliance shift that we
saw this morning on the anginous slide. I have been trying to keep
an open mind about the factors responsible for that shift but I must
say I tend to agree with you, that it can not only be a segment of
the myocardium that has been ischemic and is relaxing abnormally.
Some people have modelled a ventricle and assigned 30% of the myocardium an infinite stiffness and you do not see any global change.
And for that reason or for others there must be more than just an
ischemic segment causing that big shift. Although I would like to
think it might be pericardium, there is one study of Grossmann and
his associates on dogs during pacing induced angina in the absence
of the pericardium and they obtained some shift in some of their
animals, it was not a uniform finding, however. So I think that is
an unanswered question.

CHAPTER II

SPECIFIC STUDIES OF VENTRICULAR FUNCTION USING

INVASIVE AND NON INVASIVE TECHNIQUES

CHAPTER II:

INTRODUCTORY REMARKS

H. Denolin

Hôspital Universitaire Saint-Pierre
Bruxelles, Belgium

A correct evaluation of the ventricular function is important, in cardiac diseases, giving important insights into both prognosis and treatment. However the assessment of left ventricular function in clinical practice is complex, and this explains the number of indexes utilized to evaluate pump and muscle function of the heart in various types of cardiac diseases.

In many cases, it is probably possible to evaluate ventricular function by simple, unexpensive and innocuous procedures at bedside, and some authors consider that this simple approach remains the more useful (see Burch in Rappaport's "Current controversies in cardio-vascular disease" 1980). But apparently in some clinical situations a more sophisticated approach of the ventricular function seems to be a necessity.

The most widely used methods for assessing left ventricular performance are cine-ventriculography, nuclear angiography, echo-cardiography and systolic time intervals. In the following paper, these methods will be discussed in detail.

The discussion should answer to the following questions:

- what are the important variables to be measured in specific clinical conditions.

- what is the best method to evaluate these variables.

- what are the normal values.

- what is the clinical implication of these measurements for diagnosis, prognosis and treatment in the different diseases.

Probably no single measurement or no single index for the evaluation of systolic and diastolic function - and many were proposed - are ideal and they have limitation of either methodology, practicability or concept.

The relative value of the available methods need more research, for specific application, as well as the determination of normal values; in many papers, only mean values are presented. The usefulness and practical application of the left ventricular performance in clinical conditions remain to be specified: the conclusion of many papers is that the methods proposed by the authors is of great importance for diagnosis or prognosis, but generally no proofs are given to such a statement.

Finally, the relative value of the proposed methods need a comment, as well as the use of combined methods and their complementarity, a problem already discussed in some excellent papers (see Boudoulas et al., Int.J.Cardiol., 2:493 (1983), and Kotler, id. p.503).

Whatever the methods used for the determination of the ventricular function, exercise will improve the value of the information. Measurements of left ventricular function at rest have apparently only a limited correlation with exercise capacity.

But here again, the following problems should be considered:

- type of exercise: dynamic or static

- level of exercise

- variables to be measured

- methods

- clinical applications.

The exercise protocol seems to be important in the evaluation of left ventricular response to exercise (see Foster et al., Am.J. Cardiol., 51:859 (1983).

We hope that the papers to be presented will clarify the important problem of the best method - or methods - to be used for evaluation of patients with suspected regional global ventricular dysfunction and that they will demonstate the usefulness of such measurements in different clinical conditions.

EXERCISE IN IMPAIRED VENTRICULAR FUNCTION

Jan J. Kellermann

The Chaim-Sheba Medical Center
Tel-Hashomer, Israel

INTRODUCTION

The concept of comprehensive coronary care is being accepted broadly as a most efficient tool of management of a chronic disease. Cardiac rehabilitation aimed to improve the physiological, psychological and social condition of the patient using a multifactorial therapeutic approach. Physical training is a major component of this approach and therefore it seems of importance whether or not such a measure can be implemented in coronary patients with impaired ventricular function. The aim of the following review is to examine the possible implementation and clinical justification for a physical training program in patients suffering from various degrees of ventricular dysfunction.

Psychological Benefits Of Physical Training

Evidence was given that physical training improves the emotional stability of an individual. It has been found that there is no correlation between training intensity i.e. physiological improvement and the psychological effect. In coronary patients fear, anxiety and frustration decreases as consequence of training. Furthermore it enhances the return to work of the patient, his self-esteem. His general motivation to life improves[1]

Physiological Benefits Of A Physical Training Program (1,2,3,4,) (see Table 1)

It is generally accepted that physical training improves circulatory conditions and diminishes cardiac work. As a response

59

Table 1. Effects of Physical Training

ESTABLISHED	NOT ESTABLISHED
* P.W.C. ↑	Collateral Growth
* χ H.R. ↓	Acceleration of Collateral Circulation
Stroke Volume ↑	Rise in Ventricular Ejection Fraction
* Systolic B.P. ↓	
* Diastolic B.P. ↓	NO EFFECT ON
* χ Rate Pressure Product ↓	
* χ Oxygen Pulse mℓ/min ↑	Deteriorated Ventricular Function
* χ Total Cholesterol mg% ↓	Regression of Atherosclerotic Process
* * A.T.H.R. ↑	
* * Target Rate - Pressure Prod. ↑	
H.D.L. ↑ L.D.L. ↓	

EFFECT ON MORTALITY

Beneficial in Non-Ramdomized Trails (Longterm)

Beneficial in Randomized Trails without Statistical Significance
(Short term)

---

*   p< 0.00ℓ

χ   for a given work load

* *   Ocasionally

to effective physical training heart rate and rate-pressure product
for given work levels are decreased. The proper training effect includes:

(1) Decrease of H.R., Systolic blood pressure, Muscular blood flow,
     Lactic acid concentration and Myocardial oxygen demand (for
     given tasks).
(2) Training increases the arteriovenous oxygen difference at max-
     imal work levels, stroke volume, maximal oxygen uptake, maximal
     work performance and the concentration of oxydative enzymes.
(3) Training improves the maximal oxygen potential, indicated by a
     larger mitochondrial mass. In healthy individuals vigorous
     training may increase maximal oxygen uptake by 15% or more.

     According to Paul the "adaptation to long-lasting exercise
should not be regarded as effecting important changes in only one or
few tissues, enzymes, or substrates, but as producing changes in the
body as a whole depending on type, serverity and length of training,

all adaptive changes work synergistically to achieve high levels of
energy expenditure for a longer time." The implementation of physical training in patients with coronary heart disease must be based
on clinical and conceptual considerations. Contraindications should
be strictly observed and an effective follow up should always be
available. The aim of physical training in the coronary patient is
the improvement of work performance and the achievement of a training
effect. It has been found that training effect is not limited to the
central circulation only, but that there are also peripheral circulatory alternations as a result of training. Furthermore it should be
mentioned that there is accumulating evidence that physical training
does not change left ventricular end-diastolic pressure, end-
diastolic volume or ejection fraction. Segmental contractility was
found uneffected. No deterioration after training has been observed
also in patients with end-diastolic pressures above 20 mm Hg and
ejection fraction below 45%. These latter observations has been made
after short term training only. Interestingly, it was found that the
coronary patient has a much better tolerance for physical exertion
than believed. Certainly, in assessing the physiological response of
training one must differentiate between the various type of training.
The effect of physical training is dependent on sufficient intensity,
duration and frequency, in order to produce measurable effects and
enhanced overall performance. We have found that the cardiocirculatory response in arm training is especially beneficial in patients
with angina pectoris[5].

In our experience physical training in coronary patients should
be supervised and implemented only according to possible benefits.

Hemodynamic Response To Physical Training

In the late nineteen sixties, Frick and Katila[6] found that in
a trained group of patients there was a trend to larger stroke volume
and better left ventricular functions when compared to controls.
They found smaller A-V oxygen differences during exercise which, they
presumed, reflects enhanced forward flow due to better L.V. function.
It was that the mechanism evoking larger stroke volumes and improved
L.V. functions is myocardial hypertrophy and/or a more synchronous
contractile pattern. Hagberg et al.[7] found in a recently published
study an increase in left ventricular stroke volume and stroke work
in patients with CHD. At the same time, percentage of VO_2 maximal
and mean B.P. was the same before and after training, while left
ventricular stroke work increased by 18% ($p < 0.01$). The authors
concluded that in patients with CAD, prolonged intensive exercise
causes an increase in stroke volume and this as a result of cardiac
rather than of peripherial adaptations.

Varnauskas[8] examined the coronary circulatory during heavy
exercise in controls and coronary patients. He found that the

coronary flow response to moderate exercise was identical in patients
and controls. It was found, that during heavy exercise the increase
of coronary flow was lower in relation to HRxSBP product in the
patients group when compared to the control subjects. Also vascular
resistance did not decrease to the same level as in the controls.
Others have found that patients with CAD can exhibit signs of left
ventricular failure during exercise including a decrease in stroke
volume at higher work loads, reduced myocardial contractility and
increased LVEDP. There is little doubt that peripherial circulation
plays a decisive role in the effectivness of a physiologic training
effect. The reduction in cardiac output/oxygen consumption relation,
observed especially under submaximal conditions in coronary patients
may be considered as a peripherial training effect. It has been
shown that exercise therapy can result in a precipitation of cardiac
failure. Vatner and coworkers[9] found that the occlusion of the
left main coronary artery in dogs caused marginal global myocardial
ischemia, which was tolerated well at rest. However when animals
began to exercise rise in myocardial metabolic demands could not be
met by an appropriate elevation of coronary flow. The unbalance
between myocardial oxygen supply and demand resulted in a precipitation of acute cardiac failure.

It is believed that physical training is a stimulus for a coronary collateral development and that exercise may induce or acceler-
ate coronary collateral circulation[10]. It is questionable whether
collateral circulation can be stimulated by severity of hypoxia or by
physical training per/se. Needless to underline that there certainly
may be a causal influence of both stimuli. In animals physical
training was shown to increase coronary collateral growth. However
such an influence has never been clearly demonstrated in men.

The Effect Of Exercise Training On Impaired Ventricular Function
(see Table 2)

It is only in the past few years that more information was
available concerning the effect of exercise on ventricular dimensions.

The following overview based on summaries of a number of studies
published in recent years will enable us to reach some conclusions as
to what can be accepted scientifically at the present state of our
knowledge in this field. Williams et al.[11] analyzed the treadmill
performance before and after a 2 to 57 months of physical conditioning in 121 coronary patients.

The subjects were stratified by resting LVEF into 3 Groups:

Group 1) 14 patients E.F.8 - 26%; Group 2) 23 patients E.F.3C
- 49%; Group 3) 84 patients E.F.51 - 86%.

Table 2. Review of Literature

AUTHOR	N	DURATION OF TRAINING	TRAINING EFFECT	E.F. PERCENTAGE
Williams et al. (11)	121	2 - 57 (Months)	Positive	8-26 (14) 30-49 (23) 51-86 (84)
Denniss et al. (12)	32	Short term	Positive	18 (mean)
Lee et al. (13)	18	12- 42 (Months)	Positive	40 or less unchanged also other dim.
Jensen et al. (15)	16	6 (Months)	Positive	Unchanged 6 pts. increased
Verani et al. (14)	16	12 (Weeks)	15 out of 16 Positive	52.4 (mean) increase $p < 0.02$
Letac et al. (16)	15	2 (Months)	Positive	Unchanged
Conn et al. (17)	10	4 - 37 (Months)	Positive	Less than 27
Sklar et al. (20)	21	12 (Weeks)	Positive	MBFD + E.F. Unchanged

The following results were obtained: There was a significant increase of aerobic capacity in each group. The mean oxygen pulse increased by 27% in group 1, 14% in group 2, 16% in group 3. The authors concluded that physical training in selected patients with coronary heart disease can enhance exercise performance regardless of the extent of myocardial damage.

Denniss et al.[12] examined 32 patients with a mean age of 53± years. The mean resting left ventricular ejection fraction (LVEF) was 18% as assessed by radionuclide ventriculography. They found that patients with a low EF after myocaridal infarction (MI) can be as active as symptoms permit. Furthermore they observed a significant training effect with activity levels corresponding to NYHA class I. The authors came to the final conclusion that EF alone may not be a good indicator of mortality in the first month after MI.

Lee et al.[13] examined 18 patients with CHD and EF of 40% or less. Maximal symptom limited exercise testing and cardiac catheterization were performed initially and after 12 to 42 months (mean 8.5 months) of exercise training. During follow up the mean functional aerobic impairment improved p<0.01 and the resting and submaximal heart rated were significantly lower p<0.01 and <0.05. There was no change in the left ventricular end-diastolic pressure, cardiac index, stroke index, left ventricular end-diastolic volume or EF. The authors concluded that exercise training can be beneficial even in patients with impaired ventricular function.

Verani et al.[14] determined in their study the effect of exercise training on left ventricular performance in a group of 16 patients with CHD. After a 12 weeks training period, 15 of the 16 patients had improved exercise tolerance. Total treadmill exercise duration increased and so did the estimated maximal consumption per kg. body weight. A further observation was the increase of LVEF from 52±4 to 57±4 (p<0.02) after training. Jensen et al.[15] studied a group of 16 patients with coronary heart disease by radionuclide ventriculography before and after a 6 months exercise training (mean time). After training there was no change in mean EF at rest or during maximal exercise, but a higher maximal mean systolic BP., heart rate and work load were achieved. At equivalent submaximal work loads after training similar levels of heart rate and systolic BP were reached, but a statistically greater mean EF was obtained.

Letac et al.[16] examined 15 subjects after MI or suffering from angina pectoris before and after a physical training program. Maximal exercise tests showed that physical capacity increased by 17% (p<0.02) and heart rate decreased by 13% and BP by 7% (p<0.01). Left ventricular end-diastolic pressure, ventricular volume, EF in the total group and segmental contractility also in those whose contractility was considerably impaired, or in those who had large dyskinetic areas or widespread akinesia were unchanged. The authors concluded that physical training had no influence on the myocardium.

Conn et al.[17] examined the ability of patients with severely impaired left ventricular function to participate in a cardiac rehabilitation program. 10 Patients after MI with resting LVEF of less than 27% participated in a supervised exercise program, with a follow up period of 4 to 37 months (mean 12,7 months). These authors also used the oxygen pulse before and after conditioning for the assessment of the training effect. They found a significant increase in the oxygen pulse at the level of p<0.01. There was no exercise related morbidity or mortality, although 2 patients died during the study period. The authors concluded that selected patients with severly impaired left ventricular function can participate safely in a condition program and also achieve a training effect.

Litchfield and coworkers[18] stated that about one third of patients who have severe left ventricular dysfunction can achieve normal levels of exercise.

Nagle and coworkers[19] found that physical activity appear to enhance left ventricular function at lower exercise levels by its effects on peripheral circulation and was evident in patients with abnormal LV function. At higher exercise level LV function was maintained despite increased metabolic demands.

Special attention should be paid to a study conducted by Sklar and coworkers[20]. The authors determined whether the changes in ventricular function seen after exercise training are related to changes in myocardial blood flow distribution. They compared maximal stress radionuclide angiography and treadmill exercise with quantitative thallium imaging. 21 subjects with coronary artery disease underwent a 12 weeks standardized exercise program. An improvement of exercise capacity was achieved, resting EF was unchanged and there was no overall change in either stress EF or stress myocardial blood flow distribution. There was no correlation between changes in stress MBFD and EF. The findings suggest that wherever stress EF are improved after a training program, they are not due to an improvement in MBFD and that improvement in functional capacity is not caused by either. In a recently published paper by Carroll and coworkers[21] the authors examined the systolic function during exercise in patients with CAD. It was found that systolic function in CAD is determined by acute and chronic alterations in regional functions. During exercise there is an interplay between regional disfunction from ischemia or infarction and regional hyperfunction of non-ischemic myocardium, which determines global performance. Their conclusions seems to us especially important in the light of the results of another study by Braunwald and coworkers[22] namely, that the severity and duration of post ischemic ventricular dysfunction depend on the length and intensity of the ischemia, as well as on the condition of the myocardium at the onset of the ischemic episode. The authors conclude that it is likely that when the myocardium is repeatedly "stunned" it may cause chronic post-ischemic left dysfunction. All of us involved in exercise therapy should consider the importance of the interactions between ischemic and normal or hyperperfused areas and the possibility that exercise therapy may enhance in selected patients myocardial perfusion. All these speculations are of course dependent on the accuracy of informations gained by various methods and techniques of assessment. Of late there is more hesitation as to the importance of single parameter in the assessment of ventricular function.

Osbakken et al.[23] found that here is a wide spectrum of left ventricular EF responses to supine exercise and that EF alone was an insensitive and non-specific marker of CAD. Moreover, one should take into consideration that all those measurements of regional functions may contribute significantly to the assessment of ventricular function. The measurement by gated blood pool imaging remains unsatisfactory. Therefore it is mandatory to use other parameters of ventricular dimensions, such as end-systolic and end-diastolic volume, in order achieve valuable information.

SUMMARY

    Data available to date point to a wide range of conclusions
concerning the effect of physical training in patients with impaired
ventricular function. While some authors speculate that ventricular
function is being improved after training others found unchanged
left ventricular dimensions and concluded that training has no
direct influence on the myocardium, either beneficial or detrimental.
It seems indeed that a controlled well dosed physical training
applied in coronary patients with mild to moderate impaired ventricular function has no deteriorating effect on the myocardium. On
the other hand, there is serious doubt as to whether exercise has a
beneficial influence on myocardial perfusion i.e. increased coronary
blood flow.

    In our experience[1,24,25] an individually adapted supervised
physical training program can be recommended also in well selected
coronary patients with mildly impaired left ventricular function. A
beneficial training effect has been demonstrated also by our group in
this category of patients. Finally it should be pointed out that
methods to measure distinctive changes in left ventricular dimensions
and especially the assessment of regional function remains somewhat
unsatisfactory[26]. To date there is little doubt that the forth-
coming introduction of new techniques in myocardial imaging such as
the nuclear magnetic resonance will further enlighten and broaden our
scope of knowledge in this important field.

REFERENCES

1. J. J. Kellermann, Rehabilitation of patients with coronary heart
    disease, Prog.Cardiovasc.Dis., XVII (4):303-328 (1975).
2. J. J. Kellermann, E. Ben-Ari, M. Chayet, C. Lapidot, Y. Drory,
    and E. Fisman, Cardiocirculatory response to different types
    of training in patients with angina pectoris, Cardiology, 62:
    218-231 (1977).
3. E. Ben-Ari, J. J. Kellerman, C. Lapidot, Y. Drory, E. Fisman,
    and M. Hayet, Effect of prolonged intensive training on
    cardiovascular response in patients with angina pectoris,
    Brit.Heart J., 40:1143-1148 (1978).
4. J. J. Kellermann and H. Denolin, eds., "Critical Evaluation
    of Cardiac Rehabilitation," Bibliotheca Cardiologica No.36,
    S. Karger, Basel - London - New York (1977).
5. E. Ben-Ari and J. J. Kellermann, Comparison of cardiocirculatory
    responses to intensive arm and leg training in patients with
    angina pectoris, Heart and Lung, 12 (4):337-341 (1983).
6. M. Frick, M. Katila, and A. Sjogren, Cardiac function and
    physical training after myocardial infarction, in: "Patients
    in CHD and Physical Fitness," O. A. Larsen and R. O.
    Malmborg, eds., Copenhagen, Munksgaard, pp.43-47 (1971).

7. J. M. Hagberg, A. Ehsani, and J. O. Holloszy, Effect of 12 months of intense exercise training on stroke volume in patients with coronary artery disease, Circulation, 67 (6): 1194-1199 (1983).

8. E. Varnauskas, H. Bergman, P. Houk et al., Hemodynamic effects of physical training in coronary patients, Lancet, 2:8 (1966).

9. S. F. Vanter and M. Pagani, Cardiovascular adjustments to exercise: hemodynamics and mechanism, in: "Exercise and Heart Disease," E. H. Sonnenblick and M. Lesch, eds., Grune and Stratton, New York, pp.127-144 (197),

10. J. Scheuer, Effects of physical training on myocardial vascularity and perfusion, Circulation, 66 (3):491-495 (1982).

11. R. S. Williams, E. H. Conn, and A. G. Wallace, Enhances exercise performance following physical training in coronary patients stratified by left ventricular ejection fraction, Circulation, Abstracts, 64, Supp.IV:IV-186 (1983).

12. A. R. Denniss, D. A. Ross, and P. A. Russell, Early exercise testing physical training and mortality in patients with severe left ventricular dysfunction, J.Amer.Coll.Cardiol., Supp.1, p.718 (1983).

13. A. P. Lee, R. Ice, R. Blessey, and M. E. Sanmarco, Long term effects of physical training on coronary patients with impaired ventricular function, Circulation, 60 (7):1519-1526 (1979).

14. M. S. Verani, G. H. Hartung et al., Effects of exercise training on left ventricular performance and myocardial perfusion in patients with coronary artery disease, Am.J.Cardiol., 47:797-803 (1981).

15. D. Jensen, J. E. Atwood, V. Froelicher et al., Improvement in ventricular function during exercise studied with radionuclide ventriculography after cardiac rehabilitation, Am.J.Cardiol., 46:770-777 (1980).

16. B. Letac, A. Cribier, and J. F. Desplanches, A study of left ventricular function in coronary patients before and after physical training, Circulation, 56 (3):375-378 (1977).

17. E. H. Conn, R. S. Williams, and A. G. Wallace, Exercise responses before and after physical conditioning in patients with severely depressed left ventricular function, Am.J.Cardiol., 49:298-300 (1982).

18. R. L. Litchfield, R. E. Kerber, J. W. Benge et al., Normal exercise capacity in patients with severe left ventricular dysfunction compensatory mechanisms, Circulation, 66 (1):129-134 (1981).

19. R. E. Nagle and R. G. Murray, The effect of physical training on left ventricular function after acute myocardial infarction, Circulation, Abstracts, 68 (4):III-377 (1983).

20. J. Sklar, A. Niccoli, M. Leithner, B. Groves, and H. Brammell, Changes in ventricular function after cardiac rehabilitation

are not related to changes in myocardial perfusion,
Circulation, Abstracts, 66 (4):II-187 (1982).

21. J. D. Carroll, O. M. Hess, N. P. Studer et al., Systolic
    function during exercise in patients with coronary artery
    disease, J.Amer.Coll.Cardiol., 2 (2):206-216 (1983).

22. E. Braunwald and R. A. Kloner, The stunned myocardium:
    prolonged, postischemic ventricular function, Circulation,
    66 (6):1146-1149 (1982).

23. M. D. Osbakken, C. A. Boucher, and R. D. Okada, Spectrum of
    global left ventricular responses to supine exercise,
    Am.J.Cardiol., 51 (1):28-35 (1983).

24. J. J. Kellermann, The secondary preventive effect of comprehen-
    sive coronary care (C.C.C), in: "Selected Topics in
    Preventive Cardiology," A. Raineri and J. J. Kellermann,
    eds., Ettore Mojorana International Science Series, Plenum
    Press, New York - London, pp.165-172 (1983).

25. J. J. Kellermann, Cardiac rehabilitation as a secondary
    preventive measure - endpoints. The logig of desirability
    and availability, in: "Comprehensive Cardiac Rehabilitation,"
    J. J. Kellermann, ed., Adv.Cardiol., 31:134-137, S. Karger,
    Basel - London - New York (1982).

26. K. L. Gould, Quntitative imaging in nuclear cardiology,
    Circulation., 66 (6):1141 (1982).

EXERCISE TESTING IN IMPAIRED VENTRICULAR FUNCTION

M. Niederberger, D. Glogar and H. Zilcher

Kardiologische Univ. Klinik Wien
Vienna, Austria

INTRODUCTION

The interrelation of central (cardiac output, Q, stroke volume, SV, heart rate, HR) and peripheral hemodynamic parameters (blood pressure, BP, total peripheral vascular resistance, TPVR) have to be thoroughly understood in judging left ventricular function from exercise testing in cardiac patients.

It is the intention of this paper to enhance such understanding in two clinical situations:

1) in coronary heart disease (CHD), when ventricular function is little impaired or even unimpaired at rest but acute left ventricular (LV) dysfunction may occur during the exercise test due to exercise induced myocardial ischemia
2) in dilatative cardiomyopathy (CMP), in which myocardial function is depressed under resting conditions and cardiac failure is induced during physical exercise by increased preload and afterload.

Methods Of Exercise Testing:

Differences should be noted in the hemodynamic responses to various forms of dynamic exercise.

During treadmill exercise with stepwise increases of aerobic requirements, BP is lower at any relative ($\%\dot{V}_{O_2max}$) or absolute level of work than during upright bicycle exercise with stepwise increases of workload. The same peak HR are attained with both methods, but on

the treadmill, higher values of $\dot{Q}$, arterial-mixed venous oxygen
difference (DavO$_2$) and oxygen uptake ($\dot{V}_{O_2}$) can be achieved. This is
the case in normal subjects and also in patients with CHD (1,2,3).
No differences are encountered in preload between treadmill exercise
and bicycle exercise in the upright position, as estimated from
pulmonary pressure measurements. However, bicycle exercise in the
recumbent position is associated with higher pulmonary arterial and
higher left ventricular end-diastolic pressures and lower HR are
attained at symptom limited (SL) maximal exercise[4]. Accordingly,
bicycle exercise in the recumbent position is associated with an
approximately 50% lower risk of life threatening arrhythmais but with
a fourfold risk of acute left ventricular failure and pulmonary
oedema than bicycle exercise in the upright position[5,6].

Independant of the form of dynamic exercise (treadmill upright
or supine bicycle exercise), the risk of exercise testing depends
mainly on the observation of contraindications and criteria for
termination of exercise and on the availability of electrical de-
fibrillation, but not per se on the use of a "submaximal" or a
symptom limited maximal protocol[5].

Haemodynamic Responses To Upright Bicycle Exercise

1) In coronary heart disease

During upright bicycle exercise with a low initial workload
(25 watts) and stepwise increases of workload by 25-50 watts every
2-3 minutes, $\dot{V}_{O_2}$ increases near linearly with time until maximal
aerobic power or a symptomatic limitation is reached.

In normals, there is a linear and directly proportional relation
between $\dot{V}_{O_2}$ and $\dot{Q}$. SV increases initially but levels off at more
strenuous exercise. DavO$_2$ increases steadily while TPVR drops at low
work loads but levels off at strenuous exertion.

In patients with CHD, we find restrictions of aerobic capacity
during dynamic exercise due to restricted SV and HR. Patients who
are physically more restricted due to functional left ventricular
impairment can be identified by their reduced exercise capacity and
by their lower maximal values of HR and BP[7,8] (Figure 1).

A fall of systolic BP (SBP) at near maximal exercise is thought
to signal acute left ventricular failure due to ischemia of a large
wall portion in severe coronary disease[9]. A complementary role of
peripheral vascular regulation is illustrated by the results of a
study, in which a group of coronary patients, whose SBP rose from
near maximal to SL maximal exercise was compared with another group
of otherwise similar patients, whose SBP dropped at this point by at
least 10 mm Hg. Both groups did not differ in terms of age and sex

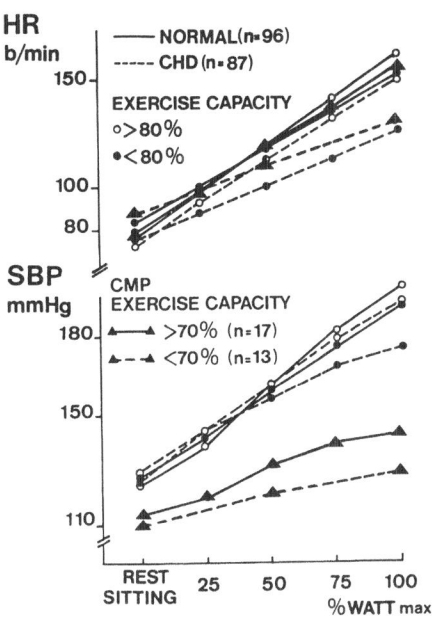

Fig. 1.   Measurements of heart rate (HR) and of systolic blood pressure (SBP) (cuff method) at rest and during SL maximal
upright bicycle exercise of normals, of patients with CHD
and of patients with CMP.   In each subject, the measurement
points were chosen to represent relative intensities of work
of 25%, 50%, 75% and 100% of the highest attained work load.
Relative exercise capacity (%) was calculated in each subject as a percentage of sex, age and body surface adjusted
normal values.   Subgrouping was performed according to an
individual exercise capacity below or above 80% (normals,
CHD) or below or above 70% (CMP).

distribution or functional aerobic capacity.   The group with a
further SBP increase tended to a concomitant increase of SV from near
maximal to SL maximal exercise by an average of 3% whereas the group
with a fall of SBP demonstrated a slight but statistically significant fall of SV by an average of 5%.

This slight difference in the evolution of SV - minus 5% versus
plus 3% from near maximal to SL maximal exercise - , accompanied by
an equal rise of HR in both groups, could alone not explain the
marked difference in the SBP response.   Therefore, a major contrib-
utory factor for the fall in SBP was to be sought in the changes of
TPVR.   Patients who developed a fall of SBP not only showed a lower
initial value of TPVR at rest but also a steeper slope during exercise and a lower end value at SL maximal exercise than the other
patients, which points to an impairment to vaso-constrict adequately

for maintaining or raising BP. This mechanism seems to play a role
in addition to impairment of LV function, i.e. loss of SV due to
development of akinesia and the interrelation of both mechanisms may
provide insights into the adaptation to and the severity of coronary
disease.

Also the fact, that mean pulmonary arterial pressure (PAP) at SL
maximal exercise did not differ between both groups (52 vs. 54 mm Hg,
mean values) shows that development of akinesia and LV failure was
less important in these patients than has hitherto been thought.
However, the risk of a drop of BP for initiation of ventricular
fibrillation is not challenged by these findings: a drop of BP may be
accompanied by a reduction of coronary flow and the danger of occur-
rance of ventricular fibrillation[10].

2)   In dilatative cardiomyopathy

In our study groups, patients with dilatative cardiomyopathy
(CMP) had significantly lower values of SBP and TPVR during exercise
than CHD patients with a similar exercise capacity, whereas no significant differences existed between CMP and CHD patients with respect to HR, $\dot{Q}$, SV or PAP at rest, at near maximal and at SL maximal
exercise.

In CMP, the reduction of TPVR does not seem to play a role in
limiting exercise capacity and a further fall at higher relative work
loads was - as an average - not accompanied by a drop in SBP. Hypo-
thetically, the finding of relatively low TPVR during exercise indicates a circulatory adaptation to the systolic LV dysfunction in CMP.
Obviously, in these CMP patients, a reduction of afterload with low
TPVR and low BP enhanced Q but even then, SV could not be maintained
at strenuous exercise despite the pathologic increase in preload.

CMP patients with a severely reduced exercise capacity, when
compared to patients with less impairment, have a higher HR at rest
but terminate exercise - in most cases with dyspnoea - at a lower HR
while PAP does not differ but stroke index is lower. Impairment of
LV compliance (diastolic dysfunction) in CMP is unmasked when diastole shortens with increasing HR. The more severe the disease, the
more impaired is LV compliance and the lower maximal HR at which
dyspnoea or signs of peripheral hypoperfusion, i.e. of inadequate $\dot{Q}$
for a given workload, limit exercise capacity. This is shown by an
average maximal cardiac index of 4.8. 1/min in the patients with
exercise capacity of less than 70% of (sex, age and body surface
adjusted) normal values, versus 5.61/min in the patients with exercise capacity of more than 70%, with accompanying differences in peak
HR (128 vs. 153 b/min, mean values) but not in PAP (46 vs. 44 mm Hg,
mean values).

Results of standard bicycle exercise testing[11] of 17 less impaired (exercise capacity less than 70%) and 13 more impaired (exercise capacity more than 70% of sex, age and body surface adjusted normal values) CMP patients are related to those of CHD patients are related to those of CHD patients in Figure 1. In the less impaired CMP patients, the relative exercise capacity averaged 85±13%, in the more impaired patients 51±15%. With 156±15b/min, HR at SL maximal exercise was significantly higher in the less impaired group than in the more impaired group, in whom it averaged 131±19b/min, whereas SBP was 144±22 vs. 129±28 mm Hg.

Comparing the patients with CMP to those with CHD it is obvious that the HR response does not differ significantly. In both groups, HR that can be attained during SL maximal exercise are reduced in more severe disease.

At rest and at all levels of exercise, SBP is significantly lower even in the less impaired CMP patients than in CHD patients and normals.

CONCLUSIONS

Exercise testing, preferably by a SL maximal protocol on the treadmill or the bicycle ergometer, enables differentiation between acute LV dysfunction due to ischemia in CHD and due to chronic LV impairment in CMP. Once the diagnosis of CHD has been established, such exercise testing will help in estimating the functional and prognostic severity of disease, in identifying mechanisms that limit exercise capacity and in evaluating effects of treatment.

SUMMARY

The interrelations of central and peripheral hemodynamic parameters have to be thoroughly understood in judging left ventricular function from exercise testing in cardiac patients.

In coronary heart disease, exercise capacity is reduced due to restricted stroke volume and heart rate. The rate of increase of arterial-mixed venous oxygen difference and a higher peripheral resistance in more impaired patients at all work loads indicate, that other mechanisms, such as redistribution of blood and increased peripheral resistance compensate in part for left ventricular impairment. A fall of blood pressure with myocardial ischemia at near maximal exercise is due to development of akinesia and left ventricular failure and to inadequate regulation of total peripheral vascular resistance.

In dilatative cardiomyopathy, blood pressure is reduced at rest and at all work loads. Reduction of total peripheral vascular resistance enhances stroke volume in the presence of systolic left ventricular dysfunction. Impairment of left ventricular compliance is unmasked when diastole shortens with increasing heart rate. More impaired patients are symptomatically limited at lower heart rates than are less impaired patients.

Therefore, symptom limited maximal exercise testing enables in some cases to differentiate by easily obtained measurements of heart rate and blood pressure between acute left ventricular dysfunction due to ischemia and chronic left ventricular impairment due to myocardial disease.

Once the diagnosis of coronary heart disease or of dilatative cardiomyopathy has been established, exercise testing will help in estimating the severity of disease with respect to function and to prognosis, in identifying mechanisms that limit exercise capacity and in evaluating effects of treatment.

REFERENCES

1. L. Hermansen and B. Saltin, Oxygen uptake during maximal tread-mill and bicycle exercise, J.Appl.Physiol., 26:31 (1969).
2. L. Hermansen, B. Ekblom, and B. Saltin, Cardiac output during submaximal and maximal treadmill and bicycle exercise, J.Appl.Physiol., 29:82 (1970).
3. M. Niederberger, R. A. Bruce, F. Kusumi, and S. Whitkanack, Disparities in ventilatory and circulatory responses to bicycle and treadmill exercise, Br.Heart J., 36:377 (1974).
4. F. Kubicek and G. Gaul, Comparison of supine and sitting body position during a triangular exercise test, Europ.J.appl. Physiol., 36:275 (1977).
5. M. Kaltenbach, D. Scherer, and S. Dowinsky, Complications of exercise testing. A survey in three German-speaking countries, Eur.Heart J., 3:199 (1982).
6. M. Niederberger, R. Ehrenböck, H. Böhm, R. Bürklen, F. Dienstl, G. Gaul, H. Herbinger, E. Kiss, F. Kubicek, and P. Kühn, Komplikationen der Standardergometrie. Ergebnisse der Österreichischen Multicenterstudie 1979-1982, Österr. Ärztezeitung (1983).
7. R. A. Bruce, F. Kusumi, M. Niederberger, and J. L. Petersen, Cardiovascular mechanisms of functional aerobic impairment in patients with coronary heart disease, Circulation, 49:696 (1974).
8. P. Haber and M. Niederberger, Einschätzung der kardialen Reserve und der Kreislaufregulation mittels einfacher ergometrischer Meßwerte, Herz/Kreislauf, 9:453 (1977).

9. P. D. Thomson, M. H. Kelemen, Hypotension accompanying the onset
    of exertional angina. A sign of severe compromise of left
    ventricular blood supply, Circulation, 52:28 (1975).
10. J. B. Irving and R. A. Bruce, Exertional hypotension and post-
    exertional ventricular fibrillation in stress testing,
    Am.J.Cardiol., 39:849 (1977).
11. Arbeitsgemeinschaft für Ergometrie der Österreichischen Kardio-
    logischen Gesellschaft (Koordinator: M. Niederberger),
    Empfehlungen für eine standardisierte Ergometrie, Osterr.
    Arztezeitung, 33:333 (1978).

ASSESSMENT OF CARDIAC PERFORMANCE BY DIGITAL ANGIOGRAPHY

J. Forrester

Cedars-Sinai Medical Center
University of California
USA

Computer enhanced digital angiography (CEDA) allows the cardiologist to quantitate ejection fraction, coronary anatomy, and regional myocardial perfusion. This chapter describes the method by which digital images of the heart are obtained and processed, and the application of this new technology to evaluation of the cardiovascular system.

MAKING IMAGE AND COMPUTER COMPATIBLE: ANALOG-TO-DIGITAL CONVERSION

Digital computers process numbers. Cardiovascular images, however, exhibit continuous gradations of brightness in two dimensions. Therefore, the first step in all image-processing methods is to convert the black-and-white analog image to digits.

A photograph or analog image is composed of a series of discrete dots or picture elements, called pixels. When the distance between pixels is very small, the eye perceives the pixels as a continuous merging of shades. The images of the planet Saturn transmitted to Earth by Voyager space probe, for instance, are digital in origin: each photograph is a square digital image composed of 800 x 800 pixels. The standard TV screen used in medicine produces an analog image on a 512 x 512 matrix composed of approximately 250,000 pixels.

To convert the continuously varying brightness of the analog image into discrete numerical values, the possible range of values is divided into a number of intervals called quantization levels. All values of pixel brightness that fall within a given interval are represented by the numerical value of that quantization level. The number of quantization levels defines the gray scale, or the number

77

of shades that can be represented. In CEDA there are 256 discrete
quantization levels in the gray scale. This range allows detection
of small contrast differences between adjacent structures.

The pixel is converted to a binary number and either stored on
videotape or video disc, or entered into the computer by an analog-
to-digital converter that can digitize more than 8 million pixels per
second. All 250,000 pixels from a TV screen are represented in the
computer's memory within milliseconds. In the computer, the digitized image is processed to obtain a new set of digital values which
improve the picture. The processed image is then transferred back to
the TV screen for viewing. For moving images, the standard TV input
rate is 30 images per second. After computer processing, the image
is recreated on the TV screen as 512 horizontal scan lines by using
the numerical value of each pixel to continuously vary the brightness
of the corresponding scan line.

BASIC IMAGE-PROCESSING TECHNIQUES

The computer enhances an image by point processes, neighborhood
processes, and geometric processes. Point processes modify a pixel
value without reference to adjacent points (pixels). For instance,
in an underexposed photograph a simple point operation called linear
amplification adds a constant to each pixel value; this is identical
to increasing the exposure time and compensates for diminished
brightness. Alternatively, by selectively increasing or decreasing
pixel values it is possible to increase contrast ultimate producing
a completely black-on-white image.

Neighborhood processes are substantially more complex, since
each new pixel value is determined by its original value and by the
values of neighboring pixels. The latter are varied according to the
desired result. Neighborhood operations can be used to smooth edges,
such as the border of a left ventricular image. On specific type of
neighborhood operation, Fourier transformation, is particularly
useful in restoring blurred images, since it can identify a significant difference in image brightness corresponding to the edge of a
structure.

In geometric operations, distortions are intentionally introduced to compensate for known image distorions. The Voyager images,
for instance, are a composite of several photographs obtained at different distances and angles from Saturn. Geometric processing con-
verts these images into composite photographs. In cardiology, geometric processing can eliminate the so-called pincushion distortion
introduced by the image amplifier of an x-ray system.

Although the algorithms for image processing can become quite
complex, particularly in nuclear scintigraphy, computer enhanced

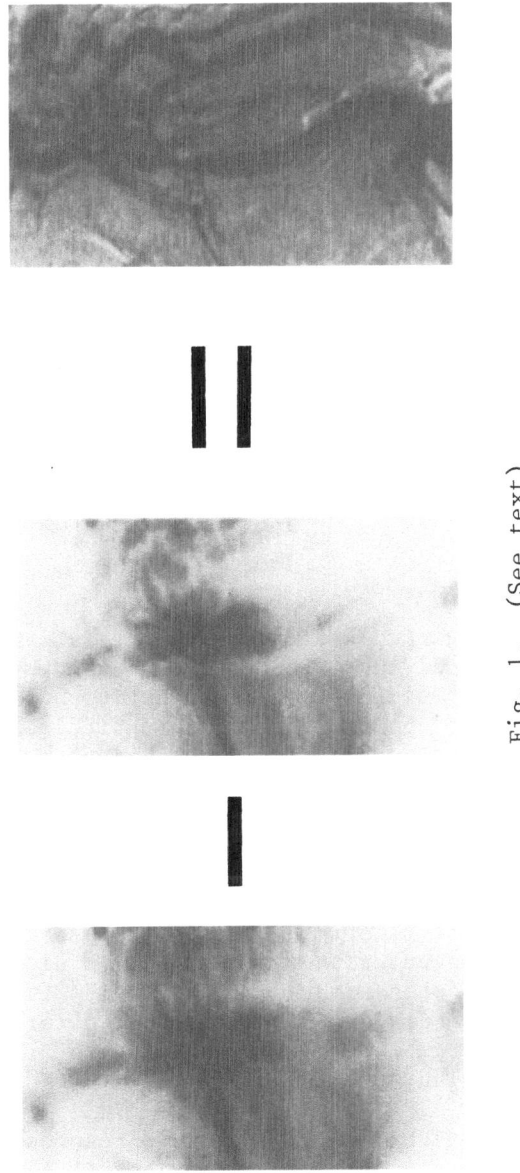

Fig. 1. (See text)

digital angiography relies primarily on a combination of straight-
forward point-processing techniques. The three most important point
processes are logarithmic amplification, digital subtraction, and
contrast enhancement.

LOGARITHMIC AMPLIFICATION

     The attenuation coefficient for all material through which an
x-ray passes is exponential. This means that the absorption of
x-rays passing through iodinated contrast material is not linearly
related to iodine concentration. In practical terms, it is necessary
to correct this nonlinearity before processing. This is done by
logarithmic amplification; each pixel value is changed to its logarithm and then rescaled to the 0-255 dynamic range of the computer
image memory. For instance, a pixel value of 128 would be changed
to 223 (255 x log128/log255) and a value of 32 would become 160 (255
x log32/log255). Multiplication of the logarithmic value by 255/
log255 restores the range of pixel values to 256, but now the relationship of image brightness to iodine concentration is linear.
Logarithmic amplification is critical for measurement of cardiac
volumes and blood flow using the technique of videodensitometry,
discussed later.

DIGITAL SUBTRACTION

     In routine angiography, contrast material passing through a
cardiac chamber or a vessel is partly obscured by surrounding and
overlying bone and muscle tissue. Digital subtraction eliminates
the structures that obscure the angiographic image, by substracting
pre-injection digital values from the digital values of the dye-
filled vessel without the confounding structures surrounding it
(Figure 1).

     In clinical application, pre-injection image is entered in
digital form into the computer. This digitized image is called the
mask. When contrast material subsequently passes through the chamber
of vessel being imaged, each video frames is digitized and subtracted
from the mask, and then returned to a TV display before the next
frame arrives approximately 30 millisecond later. The processed
image is therefore viewed on-line and also recorded for subsequent
additional processing.

     Most masks are the sum of 30 images recorded in one second.
These masks have a signal/noise ratio that is about four times
greater than that of a normal digitized image. After the first frame
is entered in memory, 1/16 of each pixel value from the next incoming
video frame is added to 15/16 of the pixel value in memory. Although
the mask obviously represents a blurred summation of the heart's
movement throughout the cardiac cycle, mask subtraction still produces an image with an order of magnitude increase in contrast.

Because it is a temporal process, however, mask subtraction is susceptible to motion artifact. Even slight movement will mismatch pixels from image to image, effectively destroying the picture. Motion artifact can be eliminated by time-interval differencing, which progressively alters each mask by fixed-interval recording of incoming frames. For instance, if the interval is 30 milliseconds, the mask for the second frame is the first frame. The second frame then becomes the subtraction mask of the third frame. If the time interval between updating the mask is brief, then the artifactual effect of patient movement or respiratory motion is eliminated.

CONTRAST ENHANCEMENT

Because the concentration of contrast medium in the heart or blood vessels frequently is significantly less in computer enhanced digital angiography images than in images obtained by direct injection, the computer enhanced digital angiography image is dominated by low pixel values, as would be seen in underexposed, dark film. The level of brightness can be increased by a uniform increase in all pixel values; this linear amplification will improve a sharp but dark image. Contrast is sharpened by increasing only those pixel values within a certain numerical range; such logarithmic amplification affects larger pixel values more than small ones, thereby increasing the contrast between the black background the the gray-white, iodine-containing cardiovascular image. The technique of contrast enhancement, therefore is a combination of linear and logarithmic amplification. Extended to its limit, contrast enhancement produces a pure white image on a black background, particularly useful for visualizing the contraction and relaxation pattern of the heart during the cardiac cycle.

CLINICAL APPLICATION OF DIGITAL ANGIOGRAHPIC IMAGE AND PROCESSING

There are three types of cardiac images presently being obtained by computer enhanced digital angiography: the cardiac chambers, bypass grafts, and myocardial perfusion. Images of the coronary arteries by aortic root injection in man will probably be obtained within the next year (Figure 2).

VISUALIZATION OF THE CARDIAC CHAMBERS AND QUANTIFICATION OF SEGMENTAL WALL MOTION

Images comparable in quality to those obtained by direct-injection left ventricular (LV) angiography can be obtained by venous injection computer enhanced digital angiography. The quality of the cardiac image varies with the injection site, and images improve with higher volumes of contrast material, e.g. 40 ml. We have shown that

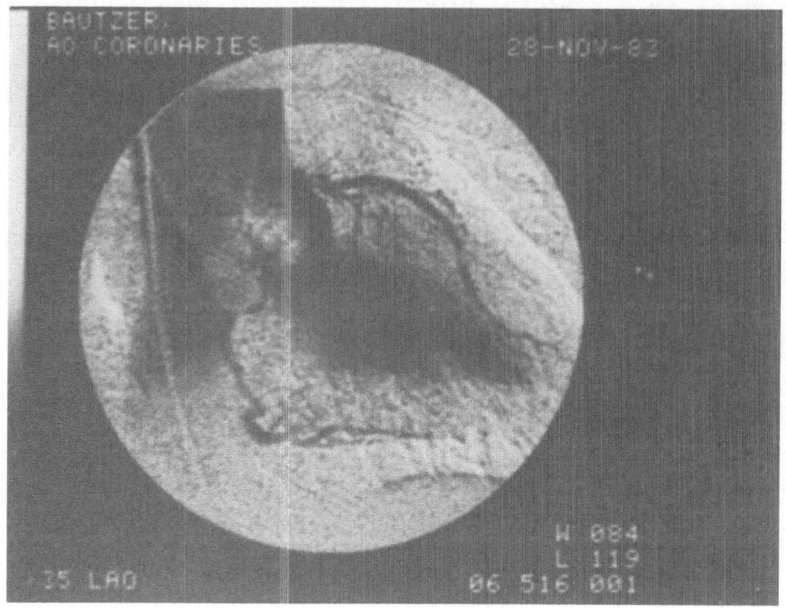

Fig. 2. (See text)

the calculated LV end-diastolic volume, LV end-systolic volume, and LV ejection fraction for practical purposes compare well with identical measurements from direct-injection[1].

When direct-injection LV angiography is used with digital processing, the contrast dose can be reduced from 40 ml to 10 ml so that serial studies can be performed during cardiac stress. Serial injections before and immediately postpacing have been used to localize disease and assess it severity. Tobis et al., have reported that cardiac pacing has a 93% sensitivity and 83% specificity for detecting coronary artery disease[2].

The advantage of a digitized LV image extends to the display and reporting of results. In Figure 3, the LV end-systolic frame is superimposed on the end-diastolic frame after maximal contrast enhancement. The end-diastolic frame has been reversed from white to black, allowing the cardiologist to review the symmetry of cardiac contraction in a manner not feasible with routine cineangiography. A high-quality reproduction of this image can be immediately generated for the patient's chart, thus providing visual information previously only available by a time-consuming visit to the catheterization laboratory.

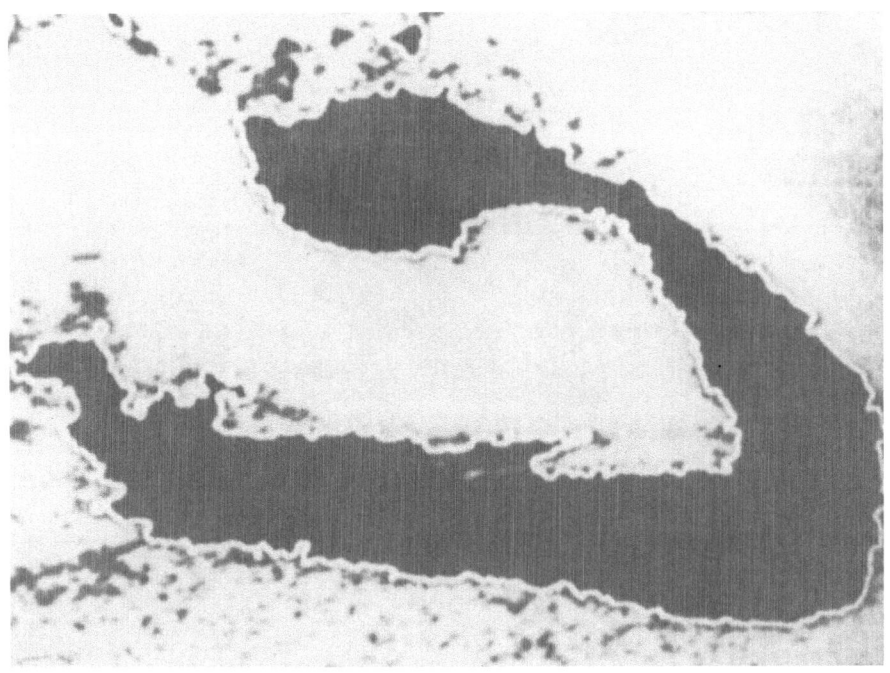

Fig. 3.   (See text)

SEGMENTAL WALL MOTION

In a study of segment wall motion in LV angiograms, we found that reproducibility of readings was only 41%, and that there was no difference in measured segmental wall motion between those segments classified as dyskinetic and akinetic[3]. Classification of the motion of a cardiac segment, even when performed by highly experienced angiographers, therefore, is not reproducible. However, the digitized cardiac silhouette at end-diastole and end-systole is ideally suited for quantitative analysis of point or region movement during the cardiac cycle. Figure 4 shows a computer-generated series of radii to the cardiac center of area at end-diastole and end-systole. Movement along any radius can be calculated and compared to the established normal range. Because the LV image can be presented in white and black, linear edge-detection algorithms cam automate the entire analysis of segmental wall motion (Figure 5), thus significantly increasing its accuracy and reproducibility.

CORONARY ARTERY BYPASS GRAFTS

Bypass grafts can be visualized by nonselective injection into the venous circulation or onto the aortic root (Figure 6). For intra-aortic studies, 20 to 40 ml of the contrast material

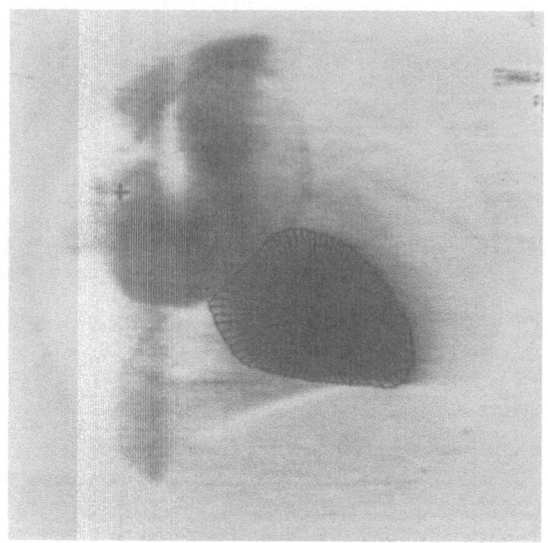

Fig. 4.   (See text)

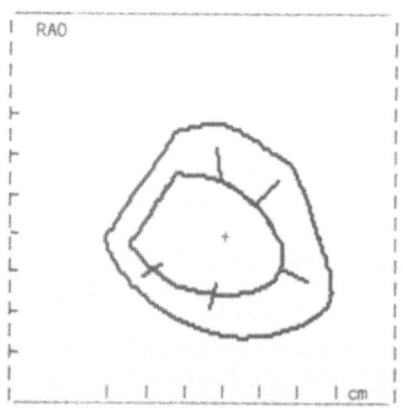

```
                --------------------
                   SUMMARY REPORT
                --------------------
```

GLOBAL		REGIONAL	
EDV = 77 ml		I Antero-Basal	Normal
ESV = 18 ml		R I Antero-Lateral	Normal
SV = 59 ml		A I Apical	Normal
EF = 0.77		O I Inferior	Normal
		I Infero-Basal	Borderline

Fig. 5.   (See text)

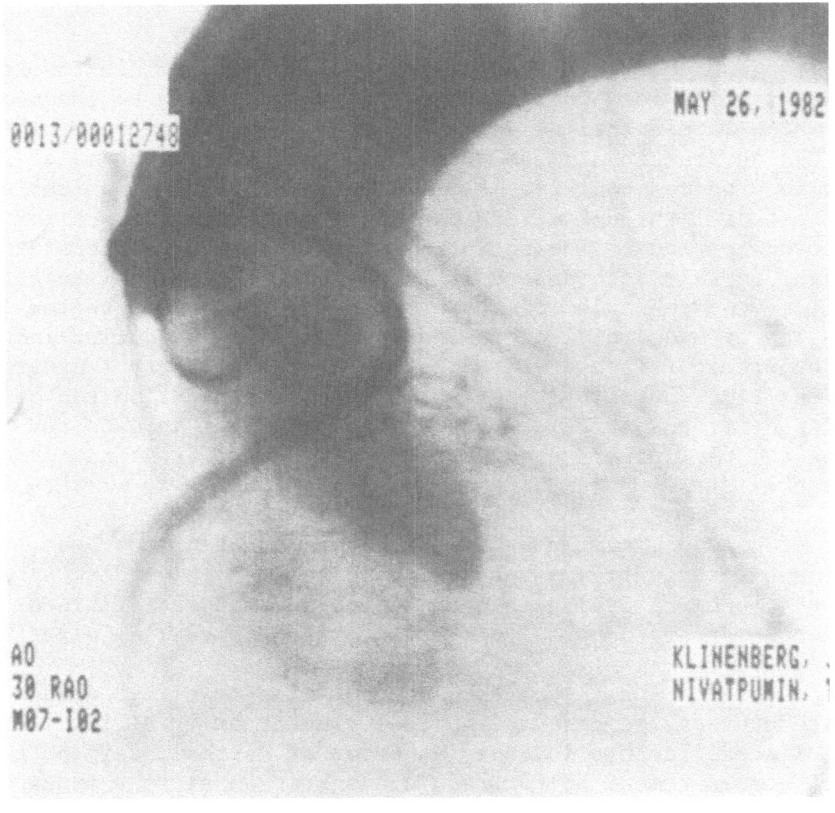

Fig. 6. (See text)

Renografin-76 is injected at a rate of 10 to 20 ml/sec; for inferior
vena cava studies, 40 ml is injected at 25 ml/sec. We found that
when intra-aortic computer enhanced digital angiography was used to
assess the patency of 17 grafts in 11 patients using fluoroscpic
x-ray doses, 11 grafts were interpreted as patent and six as closed.
Selective graft angiography confirmed every interpretation. One
graft, initially thought to be closed, was shown to be patent by
selective injection only after review of the CEDA image[4]. Coronary
bypass grafts can therefore be identified as patent or closed by
nonselective aortic root injection. The images obtained second
generation imaging devices using higher x-ray exposures are substantially superior to the early graft images, suggesting that study
of both left ventricular function and bypass graft patency by intravenous injection could become a routine predischarge procedure following bypass surgery.

MYOCARDIAL PERFUSION IMAGING

The images acquired at coronary angiography contain information
about the distribution and rate of coronary blood flow to the myocardium which in conventional cine films is so marginally perceptible
that is has no practical value. Digital image processing makes it
possible to produce quantifiable images of myocardial distribution of
contrast material throughout its passage from the coronary arteries
to the coronary sinus. The potential significance of successful
development of this methodology is substantial for several reasons.
First, it may be possible to obtain serial images of the regional
distribution of myocardial perfusion both at rest and during induced
maximal hyperemic response at routine selective coronary angiography.
Second, the time course of contrast material distribution can be
quantified by videodensitometry, possibly providing a means for
measuring the regional rate of perfusion as part of routine coronary
angiography.

Our studies of beam hardening performed using narrow beam
geometry and a range from a very thin and very thick cardiac patient
suggest that beam hardening is not an important source of video-
densitometric error compared to x-ray scatter and veiling glare[6].

X-ray scatter, on the other hand, has a major effect upon densitometric accuracy, as does its optical counterpart, veiling glare.
Neglect of these factors leads to an error of approximately 50 per
cent in the measurement of contrast concentration at peak myocardial
opacification, which translates to an error in derived washout rate
of up to 25 per cent. These effects can be corrected by subtracting
their contributions to image intensity in each pixel before logarith
mic transformation of the exponentially attenuated primary x-ray
beam. The most elegant approach to this correction seems to be
application of sophisticated image processing methods to determine
what fraction of the intensity at each pixel is due to the scatter
mechanisms and what fraction is due to the useful primary beam.
However, practical implementation of this approach appears to be som
time off. An approximate method which we have found to reduce densitometric error to about 10 per cent is to directly measure the
scatter and glare component by totally blocking the primary beam wit
a thick lead disk. If the disk is placed in the center of the myocardial silhouette and its intensity subtracted from the surrounding
pixels, error from this source in the measurement of rate of washout
of contrast material can be limited to under 5 per cent[5].

ACQUISITION OF ROI T-D CURVES

Figure 7 shows typical time-intensity curves following injectio
of a 4 ml bolus of Renographin 76 into the left main coronary artery

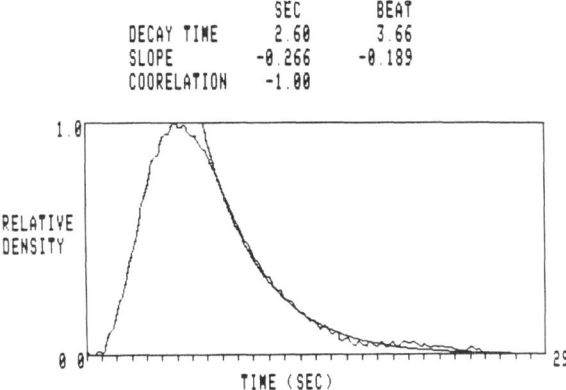

Fig. 7.   (See text)

The curves represent the contrast intensity over the anterior wall of
the myocardium.   There are several features of these curves which
were considered in the design of the analysis algorithm.   From the
continuous acquisition curves it is clear that there is a substantial
cyclic variation in myocardial videodensity due to cardiac motion.
Elimination of this cyclic variation by temporal averaging makes the
minimum time resolution approximately equal to the period of the
cardiac cycle.   For this reason only measurements insensitive to the
averaging technique, such as time to peak intensity and washout
half-time were analyzed.

   The washout portion of the curve was assumed to consist of an
exponentially decreasing term plus a constant.   The constant term was
included in the analysis to account for the observed failure of the
curve to return to the pre-injection intensity value.   The program
allows the operator to choose the value of this constant to fit the
curve as it reaches plateau at the end of the washout.   This value is
then subtracted from each data point in the curve.   The logarithm of
these subtracted data are then fit by the least squares straight
line, and the washout decay constant is obtained as the slope of this
line.

   In summary, image processing of the heart involves three basic
processes: digitization, subtraction and contrast enhancement.
These processes allow visualization of the cardiac chambers, bypass
grafts and myocardial perfusion.   Because the images are in digital
form, a broad range of automatic calculations are possible.   The
improvement in image quality and automatic processing suggest that
digital images processing will become part of most catheterization
laboratories in the next five years.

REFERENCES

1. R. Vas, G. A. Diamond, J. S. Whiting, J. S. Forrester, and H. J
    C. Swan, Computer enhancement of direct and venous injected
    left ventricular contrast angiography, Am.Heart J., 102:719
    (1981).
2. J. Tobis, O. Nalcioglu, W. D. Johnston et al., Digital
    angiography in assessment of ventricular function and wall
    motion during pacing in patients with coronary artery
    disease, Am.J.Cardiol., 51:668-675 (1983).
3. R. Vas, G. A. Diamond, J. S. Forrester, J. S. Whiting, M. J.
    Pfaff, J. A. Levisman, F. S. Nakano, and H. J. C. Swan,
    Computer-enhanced digital angiography: correlation of
    clinical assessment of left ventricular ejection fraction an
    regional wall motion, Am.Heart J., 104:732-9 (1982).
4. J. D. Drury et al., Computer enhanced digital angiography
    visualizes coronary bypass grafts without need for selective
    injection, Circulation, 66(11):917 (1982).
5. J. S. Whiting, T. Nivatpumin, M. Pfaff, R. Vas, K. Drury, G.
    Diamond, H. J. C. Swan, and J. S. Forrester, Assessing the
    coronary circulation by digital angiography: Bypass graft
    and myocardial perfusion imaging, in: "Digital Imaging in
    Cardiovascular Radiology," P. H. Heintzen and R. Brenneck,
    eds., Thieme, Stuttgart - New York, pp.205-211 (1983).

EVALUATION OF LEFT VENTRICULAR FUNCTION IN CHRONIC

REGURGITATION BEFORE AND AFTER VALVE REPLACEMENT

M. G. Modena, A. Benassi and G. Mattioli

Cattedra di Malattie Cardiovascolari
Université di Modena, Modena, Italy

INTRODUCTION

Controversy continues regarding the appropriate timing of operative intervention in patients with chronic aortic regurgitation (CAR) [1,2,3,4]. Once symptoms develop, surgical replacement of the aortic valve is recommended because continued medical therapy is associated with progressive clinical deterioration. However, if surgery is delayed until the occurrence of symptoms and marked cardiomegaly, some patients have left ventricular dysfunction and do not improve symptomatically postoperatively.

Different indexes have been proposed to identify the optimal time for surgical treatment before a definitive myocardium damage[5, 6,7]. We studied left ventricular contractility, pump function, and wall stress in CAR using echocardiography to determine the clinical utility of some indexes, to examine the relation with post-operative persistent left ventricular dilatation, and clinical status.

METHODS

The study population consisted of 52 patients 34 men and 8 woman with isolated CAR. The mean age was 48±14.4, 20 were in class III of the New York Heart Association, 18 in class II and 14 in class I. 20 of these patients (18 men + 2 women mean age 44.1±14.3 years) underwent aortic valve replacement.

Aortic insufficiency was defined as dense opacification of the left ventricle after power injection of contrast medium into the ascending aorta. All patients had normal coronary arteries, and there were no associated valve lesions.

Patients with an aortic valve gradient more than 10 mm Hg were excluded. The post-operative study was carried out 30 or more days after surgery, 30 normal subjects were used as a control group. Echocardiograms were performed in the left lateral decubitus position. All left ventricular measurements were obtained with the ultrasonic beam directed just below the level of the mitral valve with the transducer placed at the "standard interspace" as proposed by Popp et al.[8].

All studies were of good diagnostic quality. The following indexes were calculated from the previous measurements:

1) End diastolic diameter (EDD).
2) End systolic diameter (ESD).
3) Shortening fraction (FS) derived as (EDD-ESD)/EDD.
4) Mean velocity of circumferential fiber shortening (Vcf) derived as (EDD-ESD)/EDD x ejection time).
5) Ejection Fraction as (End diastolic volume - End systolic volume/End diastolic volume).
6) End-diastolic radius/thickness ratio derived as EDD/2/Thd, where Thd represents the average of the end diastolic poterior wall and septal thicknesses.
7) End systolic Pressure/Diameter ratio (P/D) [9].
8) End systolic stress (ESS) derived as ESS = (PxESD)/4 Ths (1 + Ths/ESD) where P is the systolic pressure and Ths is the average of the end systolic posterior wall and septal thicknesses[10].
9) FS/ESS ratio[11].

The last three indexes were derived by combining systolic blood pressure (obtained by cuff) with echocardiographic measurement.

RESULTS

In Table 1 the values of some indexes in the different classes NYHA are summarized.

In relation to the ventricular dimensions, we noted as the diameters, especially ESD, resulted significantly modified already in patients in Class I but as there was significant overlapping of values (Figure 1) mainly in patients in Classes I and II.

Nevertheless it was clear that the ESD result was useful in discriminating patients in Class III and when greater than 5 cm it was a good index of the status of myocardial damage. The EDD appeared to be of less value, because of the wide range of variability (Figure 2).

The classical indexes of ejection phase (FS, VCF, EF) did not give a result of great interest in our experience since, as shown in

Table 1. Pre-op values (mean ± SD) of the different parameters con-
          sidered in control patients and in CAR patients, EDD = End
          diastolic diameter, ESD = End-systolic diameter, FS% =
          fractional shortening, Vcf = Velocity of circumferential
          fiber shortening, EF% = Ejection fraction, R/thD = dia-
          stolic radius/thickness, P/D = Pressure diameter, ESS =
          End-systolic stress, FS/ESS = Fractional shortening/end-
          systolic stress. 'p<0.05; ''p<0.01; '''p<0.001.

	Control	Class I	Class II	Class III
EDD cm	4.87± 0.56	5.40± 0.45'	6.10± 1.00''	7.40± 1.40'''
ESD cm	3.02± 0.42	3.52± 0.46'	4.30± 0.70'''	5.80± 1.10'''
FS %	37.59± 4.40	35.40± 5.40	30.10± 6.20''	20.80± 6.50'''
Vcf cir/sec	1.33± 0.32	1.22± 0.17	0.98± 0.23''	0.85± 0.25'''
EF %	68.50± 6.70	63.90± 7.60	56.10± 9.10''	40.70±10.70'''
R/ThD	2.80± 0.50	2.70± 0.50	2.60± 0.51	3.29± 0.53''
P/D g/cm³	57.70± 6.70	52.10±12.90	49.30±11.60	31.20± 6.80'''
ESS g/cm²	67.70±12.40	76.30±17.50	109.60±41.35'	142.00±29.70'''
FS/ESS	0.58± 0.17	0.49± 0.17	0.31± 0.16''	0.15± 0.07'''

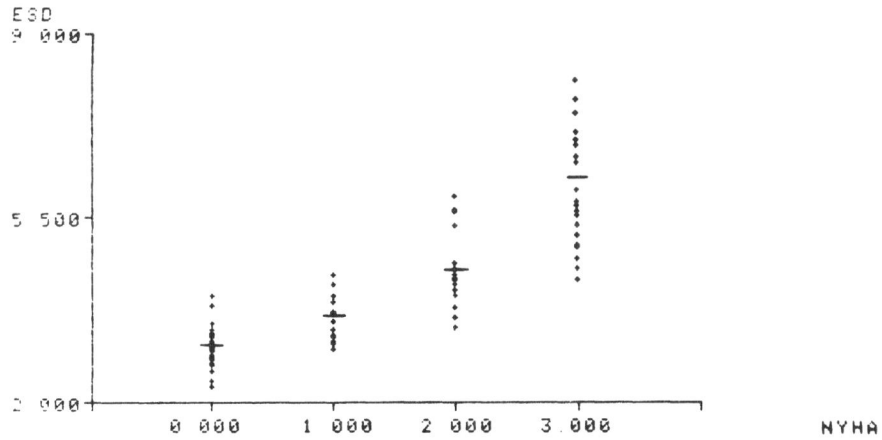

Fig. 1. Distribution of pre-op values in the control group and in
         patients with CAR in the different NYHA classes.

Figure 3, a large overlapping of values in the different groups of
patients was present. It is furthermore well known as these indexes
are generally overestimated in volume overloads.

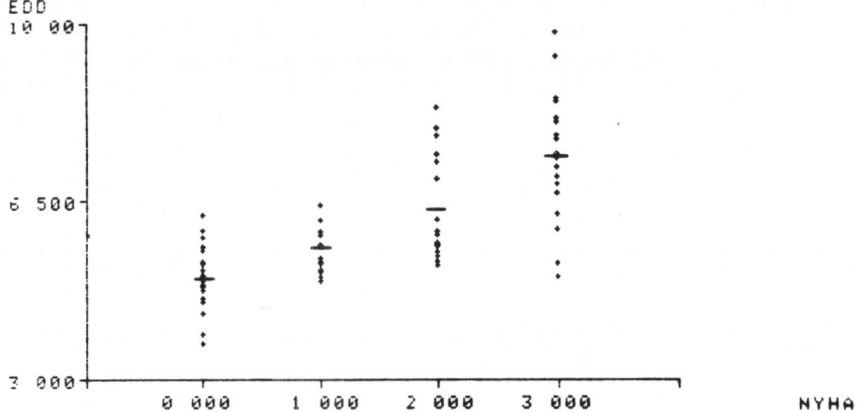

Fig. 2. Distribution of pre-op EDD values in the control group and
         in patients with CAR in the different NYHA classes.

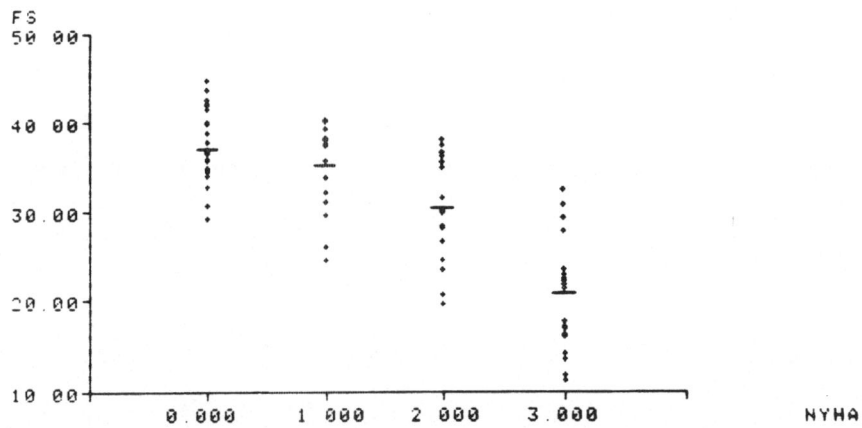

Fig. 3. Distribution of pre-op FS values in the control group and in
         patients with CAR in the different NYHA classes.

In relation to the P/D ratio (Figure 4) we observed a clear
cut-off value; below this value all patients in the Class III and 2
in the Class II (who had a clear compromise of ventricular contractility) were found. A similar distribution was observed in FS/
stress values, this ratio seemed to be more useful than the last one
in discriminating patients in Class I and II (Figure 5). A similar
behavior was observed in the stress values (Figure 6).

It is already known that the prognosis of aortic valve replacement for CAR is closely related to the reduction in left ventricular
dimensions. Recent studies have shown a prompt return of left ventricular end-diastolic dimension within 2 weeks after aortic valve
replacement.

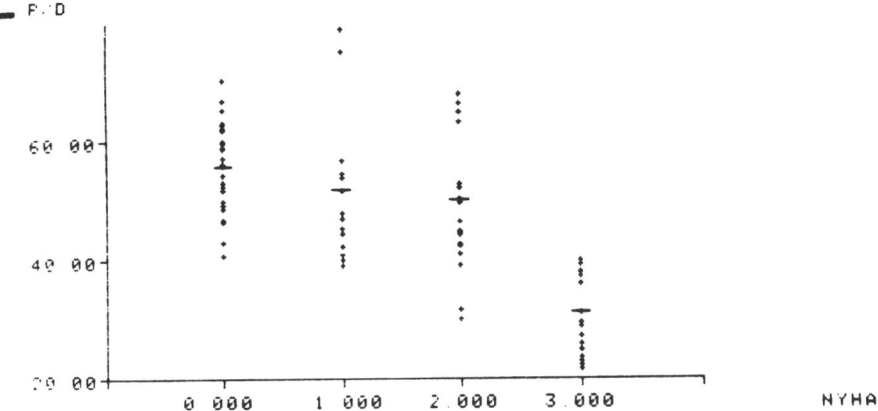

Fig. 4. Distribution of pre-op P/D values in the control group and
in patients with CAR in the different NYHA classes.

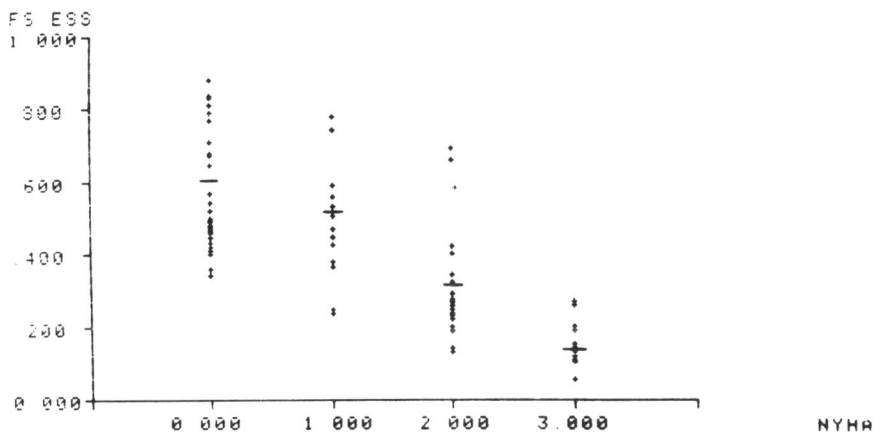

Fig. 5. Distribution of pre-op FS/stress values in the control group
and in patients with CAR in the different NYHA classes.

We have considered, therefore the correlations between post-op
EDD and some pre-op parameters (Table 2). We noted particularly as
significant correlations existed between post-op EDD and pre-op
ejection phase indexes, End-systolic stress and P/d ratio values.
The best correlation however was observed in comparing pre-op FS/
stress ratio values and post-op EDD.

DISCUSSION

Several reports in literature underline as long distance prognosis after aortic valve replacement for CAR is closely correlated
with left ventricular dimensions reduction[5,6].

Table 2.   Coefficient of correlation between post-EDD and pre-op
           indexes.   See Table 1 for abbreviations.   'p<0.05;
           ''p<0.001.

pre-op	EDD	ESD	FS	Vcf	EF	R/thD	P/D	ESS	FS/ESS
post-op									
EDD	0.26	0.48	0.67	0.69	0.68	0.36	0.48	0.54	0.83

Henry et al.[1] observed as post-op measurement of EDD more than
5.8 cm was a sign of bad prognosis, also Gaasch[5] observed that
post-op mortality was greater in patients with persistent left ventricular dilatation.

Many pre-op parameters have been proposed to evaluate the post-
op results. We tried to correlate directly some pre-op parameters
and the post-op EDD. In our experience we did not find good correlations with post-op EDD and pre-op echocardiographic left ventricular transverse dimensions. Bonow[5] defined the EDD as a sensitive parameter in identifying patients at high risk of late postoperative death. Kumpuris[6] also reported a good predictive value
for ESD, utilizing a cut-off value of 5 cm. In relation to ejection
phase indexes of left ventricular performance evaluated by echocardiography at rest, they were well correlated with EDD, but by
other Authors[13] they were shown not to be well correlated with the
same values obtained by angiography; Johnson[14], however, found that
echocardiography well correlated with angiographic EF. Kumpuris in
two separated studies[15,16] in patients with CAR showed significant
correlations between EF derived from echocardiographic dimensions and
angiographic EF, but found that echocardiographic data were consistently higher. It has been generally reported that these indexes are
less sensitive in detecting myocardial damage in conditions of volume
overload. However, a good predictive value has been reported
Kumpuris in CAR[4]. Many Authors[3,4,6] are in agreement in judging
the wall stress as a very sensitive prognostic index.

Our data also confirm these observations, although in our study
a poor predictive value was represented by R/Th ratio.

The study of the inotropic state of the left ventricle has
clearly resulted to be very important. The P/D end-systolic ratio as
well as FS/ESS ratio showed an optimal correlation with post operative left ventricular dimensions. The last parameter is the most
useful one in determining preoperatively the initial compromise of
contractile state and therefore the most useful in identifying the
ideal time for aortic valve replacement.

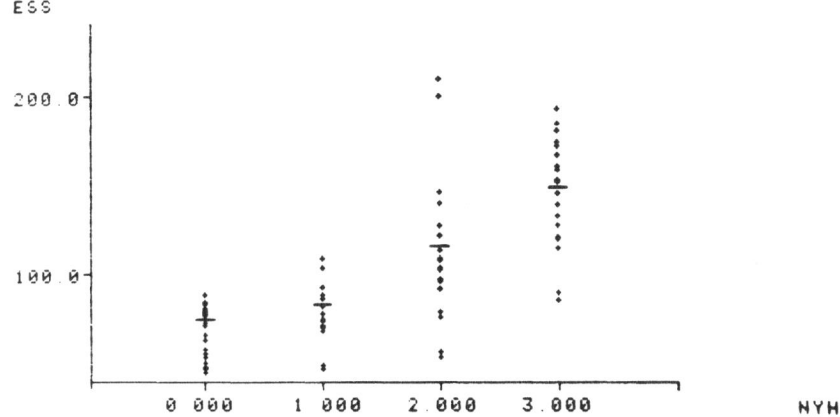

Fig. 6.   Distribution of pre-op ESS values in the control group and in patients with CAR in the different NYHA classes.

SUMMARY

        Controversy continues regarding the appropriate timing of oper-
ative intervention in patients with chronic aortic regurgitation
(CAR). We studied left ventricular (LV) contractility, pump function
and wall stress by echocardiogrphy in 52 pts affected by CAR, to
determine the clinical utility of some indexes to examine the relation with post-operative (post-op) persistent LV dilatation and
clinical status. It is already known that the prognosis of aortic
valve replacement for CAR is closely related to the reduction in LV
dimensions. We have considered, therefore, the correlations between
post-op EDD and some pre-op parameters. We noted that significant
correlations existed between post-op EDD and pre-op ESD ($r = 0.48$
$p<0.05$), FS ($r = 0.67$ $p<0.01$), Vcf ($r = 0.69$ $p<0.01$) and EF ($r = 0.68$
$p<0.01$), end-systolic stress (ESS) ($r = 0.54$ $p<0.05$), P/D end-
systolic ratio ($r = 0.48$ $p<0.05$).

        The best correlation however was observed in comparing FS/ESS
ratio values and post-op EDD ($r = 0.83$ $p<0.001$). The same indexes
have been shown to be also the most useful in discriminating an
initial status of myocardial damage in asymptomatic patients with
CAR.

REFERENCES

1. W. L. Henry, R. O. Bonow, J. S. Borer, J. H. Ware, K. M. Kent,
       D. R. Redwood, C. L. McIntosh, A. G. Morrow, and S. E.
       Epstein, Observations on the optimum time for operative
       intervention for aortic regurgitation, <u>Circulation</u>, 61:471
       (1980).

2. R. O. Bonow, D. R. Rosing, and C. L. McIntosh, The natural
    history of a symptomatic patients with aortic regurgitation
    and normal left ventricular function, Circulation, 68:509
    (1983).

3. R. O. Bonow, D. R. Rosing, K. M. Kent, and S. E. Epstein, Timing
    of operation for chronic aortic regurgitation, Am.J.Cardiol.,
    68:509 (1983).

4. J. Ross, Left ventricular function and the timing of surgical
    treatment in valvular heart disease, Am.Int.Med., 94:498
    (1981).

5. W. H. Gaasch, C. W. Andrias, and H. J. Levine, Chronic aortic
    regurgitation: the effect of aortic valve replacement on left
    ventricular volume, mass and function, Circulation, 58:825
    (1978).

6. A. G. Kumpuris, M. A. Quinones, A. D. Waggoner, D. J. Kanon,
    J. G. Nelson, and R. R. Miller, Importance of preoperative
    hypertrophy, wall stress and end-systolic dimension as echo-
    cardiographic predictors of normalization of left ventricular
    dilatation after valve replacement in chronic aortic insuf-
    ficiency, Am.J.Cardiol., 49:1091 (1982).

7. M. Osbakken, A. A. Bove, and J. F. Spann, Left ventricular
    function in chronic aortic regurgitation with reference to
    end-systolic pressure, volume and stress relations,
    Am.J.Cardiol., 47:193 (1981).

8. R. L. Popp, K. Filly, O. R. Brown, and D. C. Harrison, Effect of
    transducer placement on echocardiographic left ventricular
    dimensions, Am.J.Cardiol., 35:537 (1975).

9. J. D. Marsh, L. H. Green, J. Wynne, P. F. Chon, and W. Grossman,
    Left ventricular end-systolic pressure-dimension and stress
    length relations in normal human subjects, Am.J.Cardiol.,
    44:1311 (1977).

10. W. Grossman, D. Jones, and L. P. McLaurin, Wall stress and
    patterns of hypertrophy in the human left ventricle, J.Clin.
    Invest., 56:56 (1975).

11. K. W. Borow, L. H. Green, W. Grossman, and E. Braunwald, Left
    ventricular end-systolic stress-shortening and stress-length
    relations in humans, Am.J.Cardiol., 50:1301 (1982).

12. G. W. Burggraf and E. Graige, Echocardiographic studies on left
    ventricular wall motion and dimensions after valvular heart
    surgery, Am.J.Cardiol., 35:473 (1975).

13. A. M. Abdulla, M. J. Frank, M. I. Canedo, and M. A. Stefadouros,
    Limitations of echocardiography in the assessment of left
    ventricular size and function in aortic regurgitation,
    Circulation, 61:148 (1980).

14. A. D. Johnson, J. S. Alper, G. S. Frances, V. R. Vieweg, I.
    OcKone, and A. D. Hagan, Assessment of left ventricular
    function in severe aortic regurgitation, Circulation, 54:975
    (1976).

15. M. A. Quinones, E. Pickering, and J. K. Alexander, Percentage of
    shortening of the echocardiographic left ventricular

dimension: its use in determining ejection fraction and stroke volume, Chest, 74:50 (1978).

16. M. A. Quinones, W. H. Gaasch, J. S. Cole, and J. R. Alexander, Echocardiographic determination of left ventricular stress-velocity relations in man: with reference to the effects of loading and contractility, Circulation, 51:689 (1975).

SYSTOLIC TIME INTERVALS IN CORONARY ARTERY DISEASE

A. M. Weissler and *H. Boudoulas

Chairmen, Department of Medicine, Rose Medical Center
Professor of Medicine, University of Colorado
*Professor of Medicine, The Ohio State University
College of Medicine, Columbus, Ohio

It is well over a century since A. H. Garrod noted that the duration of left ventricular ejection in man bears an inverse relationship to heart rate[1]. Indeed, the measurement of the duration of systole, the first method applied in the evaluation of left ventricular function in humans preceded by at least two decades the inventions of the sphygmomanometer, the electrocardiogram and the chest roentgenogram. Despite their early introduction, clinical use of the systolic time intervals (STI) has lagged appreciably with respect to the more commonly applied clinical measures of stroke and minute output of the heart, ventricular diastolic and systolic pressure, and ejection fraction. Among these measures of left ventricular performance, the STI remain unique in that they offer insight into the timing of the left ventricular contraction cycle, a dimension which is not encompassed in other measures.

Early studies from our laboratory focused on the relationships between the STI and heart rate[2,5]. Simplified expressions for the pre-ejection period (PEP), the left ventricular ejection time (LVET) and the duration of electromechanical systole (QS2) corrected for heart rate were defined and termed the STI indices[6]. These indices yielded quantitative expressions of the extent of deviation in the individual STI from normal levels for any level of heart rate. Having defined the normal linear relationships between heart rate and STI our studies directed at the effects of left ventricular global dysfunction on the sequential phases of the cardiac cycle. These studies defined the presence of a highly reproducible and sustained alteration in the STI in left ventricular decompensation consisting of prolongation of the PEP and abbreviation in the LVET while QS2 remained within normal limits[4,5]. During these studies it was

appreciated that in the range of 50 to 100 beats per minute the ratio
of PEP/LVET in normal individuals was relatively uninfluenced by
heart rate (normal mean .34, S.D. .04). The PEP/LVET hence yielded a
convenient means for delineating the changes in STI associated with
left ventricular decompensation. Other studies were directed at the
influence on the STI of ventricular loading, inotropic state and
pharmacologic agents[7,17].

Additional investigations demonstrated a close correlation
between alterations in PEP/LVET and the left ventricular ejection
fraction in patients with chronic disease states[18,19]. On the
basis of these studies it was apparent that the STI, a noninvasively
derived measure, projected an accurate reflection of cardiac function
which was complementary to other contractile indices.

An issue of primary concern to the clinician who is applying any
measure of ventricular performance, be it invasive or noninvasive,
is whether abnormalities so detected have true clinical relevance.
It is for this reason that our more recent studies have focused on
the clinical meaning of alterations in the STI. It will be my
purpose in this presentation to summarize these studies, emphasizing
in particular the significance of STI in patients with chronic
coronary artery disease, the commonest cause of left ventricular
decompensation in modern practice.

Principles of Systolic Time Intervals

The STI, which constitute the sequential phases of left ventricular systole, are determined from simultaneous high-speed recordings
of the electrocardiogram, the phonocardiogram, and the carotid arterial pulse tracing. In practice, three primary measurements of the
phases of systole are determined. The QS2 is the interval that spans
the entire systolic period from the onset of the QRS complex on the
electrocardiogram to the closure of the aortic valve as denoted in
the second heart sound. The LVET is the phase of systole during
which the left ventricle ejects into the arterial system. The PEP,
the interval from the onset of ventricular depolarization to the
beginning of ejection, is derived by subtracting the LVET from the
QS2 interval. Calculation of the PEP in this manner discounts the
error resulting from the delay in transmission of the arterial pulse
from the proximal aorta to the point of its detection over the carotid artery. (The PEP can be derived alternatively by measuring the
interval from the onset on ventricular depolarization to the beginning of the upstroke of the carotid arterial pulse, from which the
arterial pulse transmission time [the A2-incisural interval] is subtracted).

To make these measurements, one must apply the same meticulous
methodological approach that is in use in the cardiac catheterizatior

laboratory.   In modern practice the standard or precordial lead which
elicits the earliest onset of electrical depolarization is selected.
This is best determined from simultaneous recordings of electrocardiographic leads which encompass three planes of depolarization
forces (e.g. leads I, aVF and V1).   The phonocardiogram should be
recorded over the upper precordium in the frequency range 100–500 Hz.
The carotid arterial pulse must be recorded with high-fidelity
instrumentation, preferably a strain gauge transducer that permits
flat frequency and phase response between 0.1 and 30.0 Hz and at
least a two-second time constant.   For best definition of the STI,
one must be certain that the recordings delineate a sharp inscription
of the initial high-frequency vibrations of the aortic component of
the second heart sound on the phonocardiogram, and a clearly discernible upstroke and incisural notch on the carotid arterial pulse
tracing.   Photographic recordings are preferred, and a minimum paper
of 100 mm/sec is deemed necessary for accurate determination of the
STI.   Calculations of the STI are derived from the mean of at least
10 consecutive beats obtained while the patient is supine during
quiet respiration.   The validation of the technical and physiologic
principles underlying the measurement of the STI have been summarized
by Lewis and co-workers[20,21].

    During the past decade, the STI have been determined from simul-
taneous recordings of the M-mode echocardiogram of the aortic valve
and the electrocardiogram employing the measurement of the interval
from the onset of ventricular depolarization to the opening of the
aortic valve (PEP) and that between the opening and closure of the
aortic valve (LVET).   This method is most useful in children in whom
the temporal landmarks are best delineated by the echocardiogram.
The measurements so obtained are identical to those derived by the
graphic methods described above[22].   Our preference remains with the
method employing simultaneous recordings of the electrocardiogram,
phonocardiogram and carotid arterial pulse since it is less costly
and because the echocardiographic visualization of the openings and
closing of the aortic valve, which is essential to accurate measurement of the STI, is present in only 80–85% adult patients.   The two
approaches are clearly complementary so that when the temporal landmarks for the STI are not definable by one method they may be readily
detected by the alternate approach.

    In simple terms, the STI delineate the duration of the two major
components of the systolic cycle of the left ventricle, namely the
pre-ejection and ejection phases.   The duration of these two sequential phases comprise total electromechanical systole.   Their clinical
application is rooted in the fact that, like the heart rate, cardiac
output and arterial pressure, the STI are well regulated under normal
circumstances, so that QS2, PEP, and LVET corrected for sex and heart
rate fall within narrow and well-defined physiological limits.   This
narrow distribution of the STI relative to heart rate permits easy
detection of deviations from the established normal relationship.

Physiological Determinants of the Temporal Dynamics of Left
Ventricular Chamber Contraction

    The earliest studies on the determinants of the duration of
the phases of systole in the intact heart employing modern physiologic recording techniques originated in the laboratory of Carl J.
Wiggers[23].   Later observations by Braunwald, Sarnoff and
Stainsby[24] and Wallace, Mitchell and Skinner[25] led to the view
that the duration of the phases of systole were influenced not only
by heart rate but by preload, afterload and the intrinsic contractile
state of the myocardium.   In a contemporary re-evaluation of these
physiological relationships, to be published later this year,
Nakamura, Wiegner Gaasch and Bing[26] employed an elegant computer-
assisted isolated rat papillary or trabecular left ventricular muscle
preparation, to study the determinants of the duration of the pre-
shortening period (PSP, the analogue of PEP), the isotonic contraction time (ICTC, the analogue of LVET) and the duration of electromechanical systole (EMS).   These studies confirmed in the isolated
myocardial segment, the independent contribution of loading and
intrinsic myocardial function to the duration of the sequential
phases of ventricular systole.   Of special interest in their findings
is the fact that the duration of the phases of systole in the isolated muscle strip reponded consistently with respect to loading.
Thus, increased preload caused an abbreviation in the PSP and an
increase in the ICTC with lengthening of EMS. The ratio of the PSP to
the ICTC thus decreased with increased preload.   Increases in total
load induced opposite changes with little influence on the duration
of EMS.   The responses of their preparation to the positive inotropic
effects of isoproterenol and calcium differed somewhat suggesting
that differences in the mechanisms of inotropic stimulation may yield
varying responses in the STI.   The responses to ventricular loading
and positive inotropy observed in the isolated myocardium are remi-
niscent although not identical to those observed in man as summarized
in Figure 1.   Thus increased preload in general induces a decrease in
PEP, a lengthening of LVET and a slight prolongation of QS2 while
PEP/LVET decreases.   Increases in total load (afterload), in general,
induce opposite changes.   The influence of afterload on the LVET is
influenced by concomitant changes in preload.   Thus, pharmacologi-
cally induced increases in afterload, when accompanied by increases
in preload, may prolong both PEP and LVET[27].   Major increases in
afterload in man are usually accompanied by increased preload which
must be accounted for in the interpretation of the influence of
systolic after loading.   The changes induced by positive inotropic
agents in man consist of an abbreviation in PEP, a variable response
in LVET dependent on the magnitude of the inotropic response (LVET
increases with maximum positive inotropic effect and is reduced or
remains unchanged at lower levels of stimulation) and a consistent
decrease in QS2 and PEP/LVET [8-15,17].

    Clearly, the duration of the phases of the cardiac cycle are
governed by the same factors (heart rate, preload, systolic afterload

	PEP	LVET	PEP/LVET	QS$_2$
Increased Preload	▽	△	▽	sl △
Increased Total Load	△	▽	△	sl △
Positive Inotropy	▽	+/−	▽	▽

*Acute interventions

Fig. 1. Factors influencing left ventricular* systolic time intervals. Schematic representation of the effects of increased preload, increased total load (after load) and positive inotropic influences on the PEP, LVET, PEP/LVET and QS2 in humans.

and myocardial function) that have been demonstrated in the past to control left ventricular stroke volume, stroke work, ejection fraction, and the rate of rise in systolic pressure. The fact that the responses in the STI are easily discernible and that the intervals can be measured repetitively with no discomfort or intrusion of the vascular system allows for their determination in series. It is this convenience which readily permits testing of the influence of load reducing agents and drugs which affect the inotropic state in man. Furthermore, it is knowledge of the influence of ventricular load and inotropic influences which permits analysis of the significance of alterations in STI in chronic disease states.

Systolic Time Intervals in Chronic Coronary Artery Disease

Our earliest studies in patients with coronary artery disease focused on the question "Do the STI afford information that current clinical methods cannot provide?" As an approach to this question we sought to define the relative frequency of abnormality in STI among patients who had convalesced completely from a documented injury to the left ventricular myocardium. Thirty-seven patients, in the age range from 35-75 years, who had recovered from an acute transmural myocardial infarction which occurred 3-60 months before the STI determinations, were investigated[19,28]. Documentation of the acute myocardial infarction included a typical history of precordial chest discomfort, characteristic serum enzyme changes and sequential electrocardiographical alterations with persistence of pathologic Q-waves. Each patient was categorized according to the New York Heart Association classification for fatigability, dyspnea and angina pectoris. The cardiothoracic ratio (CTR) was calculated for each patient from a standard roentgenogram. The presence of an S3 or S4 gallop sound was determined by auscultation and was considered to be present only when confirmed by phonocardiography.

Of special relevance to the question that was approached above were our observations on twenty patients who were asymptomatic for dyspnea, fatigability and angina pectoris. Among these twenty asymptomatic patients, all of whom had convalesced to return to a normal life pattern, fourteen (70%) exhibited abnormality in left ventricular performance by the PEP/LVET measurement, (upper normal level 0.42). Neither the presence of cardiomegaly on chest x-ray nor an S3 gallop could accurately distinguish the patients with normal PEP/LVET from those with abnormal PEP/LVET. Thus, among the 14 patients who demonstrated an abnormal PEP/LVET only four were found to have a CTR>0.5, and only one demonstrated the presence of an S3 gallop. The frequency of the presence of an S4 gallop was equivalent in the patients with and without abnormal function evidencing the ineffectiveness of the S4 in discriminating normal from abnormal left ventricular performance.

These early investigations gave credence to the thesis that the STI indeed reflect the presence of residual abnormality in left ventricular performance among patients with previous myocardial infarction in whom symptoms and signs did not suggest such abnormality. However, the high frequency of the abnormalities in PEP/LVET created a new question: were the deviations in PEP/LVET an accurate measure of the residual defect in the functional state of the left ventricle following acute myocardial infarction?

With respect to this question, we launched a prospective analysis of the accuracy of the PEP/LVET in detecting and quantitating left ventricular dysfunction in coronary artery disease among 66 patients with a history of previous myocardial infarction and 48 patients with angina pectoris in whom there was no clinical evidence of previous myocardial infarction[29]. All of the patients had diagnostic coronary arteriography and left ventriculography. Forty-one patients who were evaluated for atypical chest pain and found to have normal coronary arteries by coronary arteriography and normal left ventricular performance on contrast ventriculography served as control subjects. Multiple methods of statistical analysis were employed to test the accuracy of the STI employing the left ventricular ejection fraction (LVEF) as a standard of reference. Analysis of variance revealed that the group of patients with previous myocardial infarction were significantly different in all of the measures (PEP, LVET, PEP/LVET LVEF) from the normal group and the group with coronary artery disease without previous myocardial infarction. The latter two groups did not differ from each other in any of the measures. The discriminating power for the PEP/LVET and LVEF in separating the group with previous myocardial infarction from the two other groups was virtually identical. The PEP and the LVET (corrected for heart rate) were less discriminating than either PEP/LVET or LVEF in separating the groups. Cumulative distribution plots for the three groups of patients demonstrated virtually identical distribution of PEP/LVET and LVEF in each group. For the entire series of patients the correlation between PEP/LVET and LVEF was 0.84.

   To approach further the question as to whether PEP/LVET could
reliably separate patients with normal and abnormal ventricular
function, the sensitivity (abnormals detected by PEP/LVET/all abnormals by LVEF) and specificity (normals detected by PEP/LVET/all
normals by LVEF) of PEP/LVET was calculated employing a lower limit
of LVEF of 0.52. (The normal LVEF in our laboratory is .68 with an
SD of 0.8). A plot of the sensitivity and specificity at varying
limits of PEP/LVET is illustrated in Figure 2. The sensitivity-
specificity plot so delineated allows definition of the relative
discriminating power of various assigned limits of PEP/LVET in
detecting the presence of normal and abnormal left ventricular global
function. As is apparent in Figure 2, an upper normal limit of
PEP/LVET of .42 permitted separation of normal and abnormal LVEF with
a sensitivity and specificity exceeding 90%. Further analysis of the
sensitivity-specificity plot reveals that all patients with a PEP/
LVET <0.34 retain a normal LVEF while all patients with PEP/LVET
>0.45 have an abnormal LVEF. Thus, the close parallelism between
PEP/LVET and LVEF in defining the presence of normal and abnormal
left ventricular global function in patients with established coronary artery disease was confirmed.

   Having defined the accuracy of PEP/LVET we turned once again to
the high incidence of left ventricular dysfunction among patients
with a previous myocardial infarction. Our studies first focused on
the relationship of the presence of left ventricular dysfunction by
the measurement of PEP/LVET to the extent of coronary arterial
disease and to the presence of previous myocardial infarction[30].
Among 48 consecutively studied patients with angina pectoris and one,
two and three-vessel coronary arterial occlusive disease in the
absence of previous myocardial infarction the relative incidence of
abnormality in left ventricular global function, as defined by the
PEP/LVET, was minimal at all levels of angiograpically defined

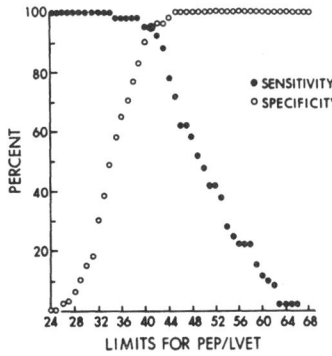

Fig. 2.   Sensitivity and specificity of various upper limits of
          normal for PEP/LVET. The sensitivity is denoted by closed
          circles and the specificity by open circles (Reprinted with
          permission from Am. J. Cardiol. ref. 29).

coronary arterial disease (Figure 3). Conversely, among 66 consecutively studied patients with angina pectoris and previous myocardial
infarction a high frequency of persistent left ventricular global
dysfunction, as defined by the PEP/LVET, was apparent at all levels
of coronary arterial occlusive disease (Figure 4). It is noteworthy
in this regard that patients with previous myocardial infarction and
three-vessel disease retained a distinctly higher incidence of left
ventricular dysfunction (72%) when compared to individuals with one-
and two-vessel disease in whom the incidence of abnormal left ventricular performance was in the range of 31-43%.

Collating all of the data among patients with and without previous myocardial infarction, the prevalence of left ventricular
dysfunction by PEP/LVET among patients without previous myocardial
infarction was quite low (4%), while that among patients with previously documented myocardial infarction was impressively high (58%).
These observations were confirmed employing the ventriculographically
determined LVEF[30]. From the practical point of view, the finding
of an abnormal PEP/LVET in the patient with established coronary
artery disease is associated with a 95% probability that the patient
has sustained a previous myocardial infarction.

It is notable that in the above studies comparing the PEP/LVET
and LVEF, only patients with sinus rhythm were included. Patients
with left bundle branch block (LBBB), or auscultatory evidence of
aortic or mitral valvular disease, each of which influences the

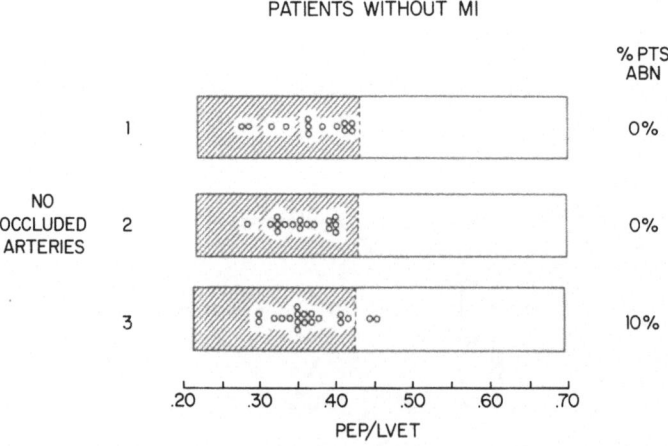

Fig. 3. PEP/LVET in coronary artery disease. Distribution of PEP/
         LVET among patients with angina pectoris in the absence of
         previous myocardial infarction. The data are grouped ac-
         cording to the presence of arterial occlusive disease in
         one, two or three coronary arteries. The shaded area re-
         presents the normal range of data. (Data obtained from
         ref. 30).

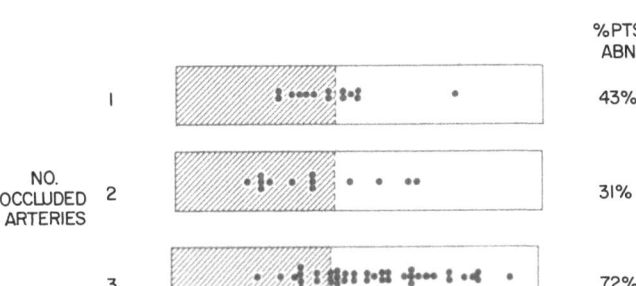

Fig. 4. PEP/LVET in coronary artery disease. Distribution of PEP/
LVET among patients with angina pectoris and previous myocardial infarction. The data are grouped according to the
presence of arterial occlusive disease in one, two or three
coronary arteries. The shaded area represents the normal
range of data. (Data obtained from ref. 30).

PEP/LVET independent of the effect of left ventricular dysfunction,
were excluded. No restrictions on drug therapy were made. It has
been commented by some investigators that the usefulness of STI in
chronic cardiac disease is limited by the fact that STI are influenced by ventricular loading and agents with influence inotropic
state. While it is true the the STI are so influenced, the high
degree of agreement between PEP/LVET and LVEF make it evident that
the LVEF is similarly affected. Thus, in the absence of LBBB and
mitral or aortic valvular disease the measurement of PEP/LVET or LVEF
reflects the combined effects of intrinsic left ventricular function,
loading conditions and inotropic drugs on left ventricular performance.

     Concomitant with observations on the STI among patients with
coronary artery disease we have obtained data on the incidence of
clinical descriptors which are conventionally applied in the detection of abnormalities in left ventricular function, namely the
presence of dyspnea (NYHA Class 3 or 4), the presence of S4 and S3
gallop on cardiac auscultation (verified phonocardiographically) and
the measurement of the cardiothoracic ratio on chest roentgenogram.
These data allowed an overall assessment of the accuracy of these
clinical descriptors in defining the presence of normal and abnormal
left ventricular function. Since left ventricular dysfunction was
prevalent only among the patients with previous myocardial infarction, the relative value of these clinical descriptors were compared
to PEP/LVET among patients with a previously documented myocardial
infarction. Calculations of the sensitivity, specificity, postive

predictive value and negative predictive value, as well as the overall accuracy in predicting normal and abnormal ventricular function
for these descriptors, among 136 consecutive patients with previous
myocardial infarction are summarized in Table 1. (Analysis of data
taken from reference 31). It is clear from this analysis, that none
of the time-honored clinical indicators have sufficient accuracy to
be practically useful in discriminating the presence of left ventricular dysfunction among patients with previous myocardial infarction.

Mechanisms of Disparities Between PEP/LVET and LVEF

It has been noted by Boudoulas and coworkers[32] that the agreement between PEP/LVET and LVEF is not perfect (89% overall agreement,
11% discrepancies for PEP/LVET vs LVEF). In attempting to better
understand the reasons for such disparities in patients with chronic
cardiovascular disease they theorized that the two factors which
can produce such disagreements are a diminution in the pressure
developed during isovolumic systole (aortic diastolic minus left
ventricular end-diastolic pressure <45 mm Hg) and the occurrence of
large segmental contraction abnormalities of the left ventricular
chamber. The presence of a diminished isovolumic pressure, which is
induced largely through major elevations in end-diastolic pressure
may result in a lesser increase in PEP and PEP/LVET for any level of
left ventricular contractile dysfunction. The presence of major
segmental contraction abnormalities, in distorting the geometry of
the left ventricular chamber, may induce an error in the calculated
left ventricular ejection fraction for any given level of global
dysfunction as measured in the PEP/LVET. These two theoretic expla-
nations for discrepancies between PEP/LVET and LVEF were tested among
453 patients, including 273 patients with coronary artery disease, 64
with primary myocardial disease, 42 with hypertensive heart disease,
4 with constrictive pericarditis, 2 with amyloid heart disease and
68 with no demonstrable cardiac abnormality. Patients with intra-

Table 1. Accuracy of Clinical Indicators of LV Dysfunction in CAD*

	Sensitivity	Specificity	Pos pred value	Neg pred value	Accuracy
S₄ Gallop	49%	46%	48%	47%	48%
S₃ Gallop	24%	95%	83%	56%	60%
Dyspnea Class III-IV	32%	95%	86%	58%	64%
CTR >0.5	37%	91%	80%	59%	64%

*Data derived from 136 patients with previous myocardial infarction
 (31). Accuracy is % accurate predictions/all predictions. Standard
 of reference for LV dysfunction is PEP/LVET.

ventricular conduction disturbances, valvular disease and atrial
fibrillation were excluded. The sensitivity, specificity and accuracy (overall agreement) between PEP/LVET and LVEF among three
groups of patients, those with a diminished isovolumic pressure,
those with an obvious localized segmental contraction abnormality
on cineventriculography and those with neither of the above abnormalities, is summarized in Table 2. Most notable in this analysis was
finding of 96% sensitivity, specificity and accuracy among the 310
patients in whom neither low isovolumic pressure nor segmental contraction abnormality was noted. The lowered sensitivity, specificity
and accuracy of PEP/LVET relative to LVEF in the presence of low
isovolumic and segmental contraction abnormality is readily apparent.

It is the view of Boudoulas et al. that the discrepancies
between PEP/LVET and LVEF in the presence of major segmental contraction abnormalities (in the absence of diminished isovolumic
pressure) reflects an error in the ejection fraction determination
by the area length method. The determination PEP/LVET, which is
independent of geometric considerations, is hence felt to be a more
accurate expression of left ventricular global performance in the
presence of segmental contraction abnormalities. The presence of
low isovolumic pressure induces an error in the estimation of left
ventricular global performance by the PEP/LVET when compared to LVEF.
The occurrence of a low isovolumic pressure was most commonly caused
by markedly elevated left ventricular end-diastolic pressure. This
was observed to occur among patients with overt congestive heart
failure, a circumstance in which PEP/LVET is rarely needed for the
detection of left ventricular dysfunction.

The Physiologic Significance of the Agreement Between PEP/LVET and
LVEF

In concluding, we should like to bring attention to the physiologic significance of the high degree of agreement between the LVEF

Table 2. Sensitivity, Specificity and Accuracy of PEP/LVET Relative
to LVEF in Presence and Absence of Disorders of LV Chamber
Function

	N	Sensitivity %	Specificity %	Accuracy %
Decreased isovolumic pressure	46	16%	86%	39%
Segmental contraction abnormality	97	75%	82%	78%
Neither of the above	310	97%	97%	97%

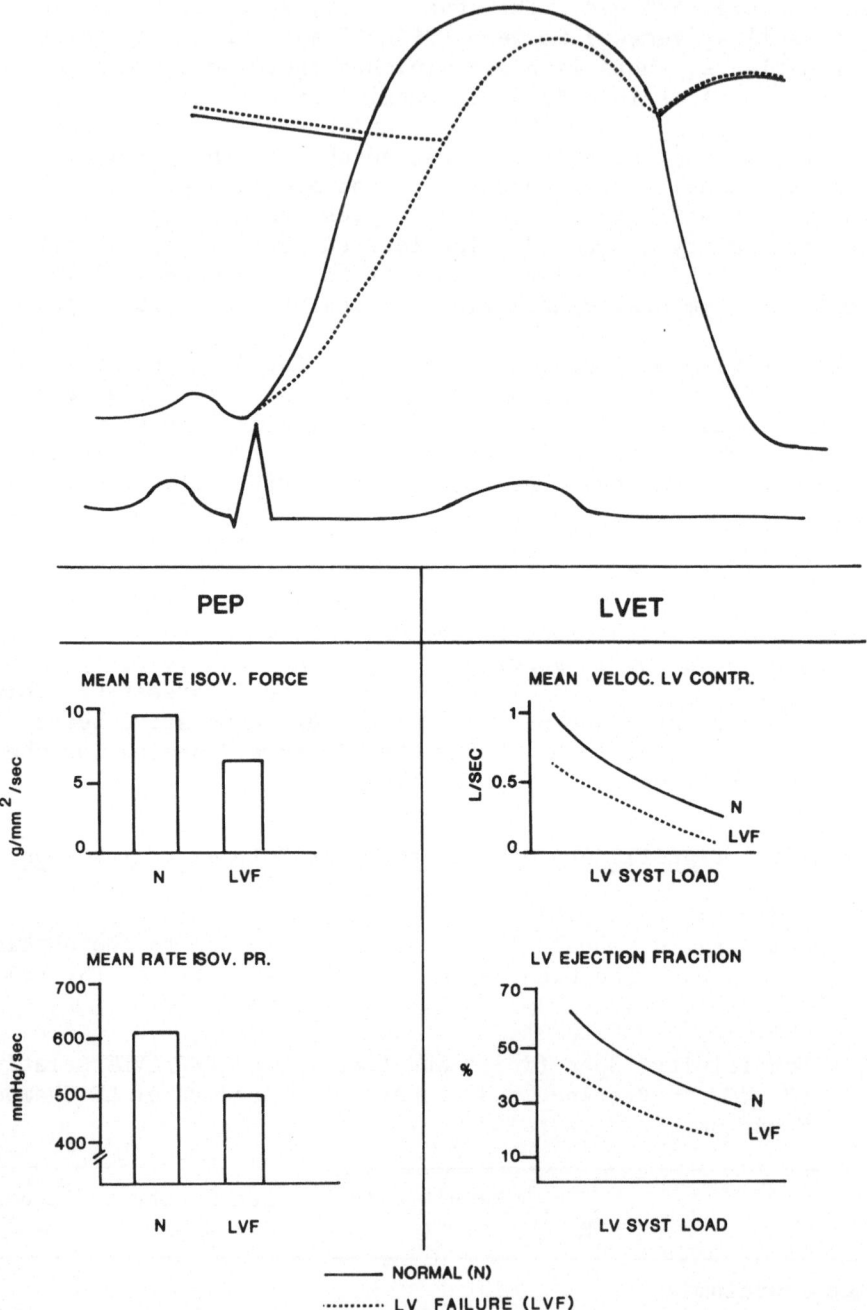

Fig. 5.   Theoretic representation of the myocardial and hemodynamic
          factors which account for the close relationship between
          PEP/LVET and LVEF in left ventricular decompensation.

and the PEP/LVET, in particular among patients in whom left ventricular global dysfunction occurs in the absence of segmental contraction abnormalities or diminished isovolumic pressure. It is our
thesis that this concordance reflects a fundamental disorder in left
ventricular muscle dynamics in left ventricular dysfunction. As
delineated several years ago in the studies of Spann, Buccino,
Sonnenblick and Braunwald[33], myocardial function in heart failure
is characterized by a decrease in the rate of force developed per
unite load during isovolumic systole and downward shift in the force/
velocity characteristics of the myocardium during isotonic systole
while the total duration of systole did not change. Translated into
left ventricular chamber dynamics (Figure 5), for any level of systolic load, a diminished rate of left ventricular isovolumic force
development induces a diminished rate of left ventricular pressure
rise during isovolumic systole with a consequent prolongation of the
PEP. The resultant delay in onset of ejection at a time when the
duration of total electromechanical systole is unchanged, results in
a relative shortening of the LVET. PEP/LVET is hence increased.
During the abbreviated ejection phase the diminished velocity of
contraction of the decompensated myocardium per unit systolic load is
translated into a decrease in the extent of contraction, with a
resultant reduction in ejection fraction. The interdependent relationships of these hemodynamic variables thus accounts for the
close correlation of PEP/LVET and LVEF among patients with left
ventricular decompensation.

In summary, the STI offer an accurate temporal description of
the sequential phases of the cardiac cycle which are influenced
physiologically by virtually the same variables that affect other
measures of global left ventricular performance. The STI reflect a
dimension of ventricular function which is unique when compared to
measures based on ventricular volume and pressure. The STI hence
offer a measure of ventricular function which augments other measures
of left ventricular performance and amplifies our understanding of
the altered contractile events with accompany left ventricular decompensation.

REFERENCES

1. A. H. Garrod, On some points connected with the circulation of
    the blood arrived at from a study of the sphygomograph-trace.
    Proc.Roy.Soc.London, 23:140, 1874-1875.
2. A. M. Weissler, R. G. Peeler, and W. H. Roehll Jr., Relationship
    between left ventricular ejection time, stroke volume, and
    heart rate in normal individuals and patients with cardio-
    vascular disease, Am.Heart J., 62:367 (1961).
3. A. M. Weissler, L. C. Harris, and G. D. White, Left ventricular
    ejection time index in man, J.Appl.Physiol., 18:919 (1963).
4. A. M. Weissler, W. S. Harris, and C. D. Schoenfeld, Systolic

time intervals in heart failure in man, Circulation, 37:149 (1968).

5. A. M. Weissler, W. S. Harris, and C. D. Schoenfeld, Bedside technics for the evaluation of ventricular function in man, Am.J.Cardiol., 23:577 (1969).

6. A. M. Weissler, R. P. Lewis, and R. F. Leighton, The systolic time intervals as a measure of left ventricular performance in man, in: "Progress in Cardiology," P. N. Yeu and J. F. Goodwin, eds., Philadelphia, Leea and Febiger, p.155 (1971).

7. R. W. Stafford, W. S. Harris, and A. M. Weissler, Left ventricular systolic time intervals as indices of postural circulatory stress in man, Circulation, 41:485 (1970).

8. P. T. Shiner, W. S. Harris, and A. M. Weissler, Effects of acute changes in serum calcium levels on the systolic time intervals in man, Am.J.Cardiol., 24:42 (1969).

9. S. Cohen, A. M. Weissler, and C. D. Schoenfeld, Antagonism of the contractile effect of digitalis by EDTA in the normal human ventricle, Am.Heart J., 69:502 (1965).

10. A. M. Weissler, A. R. Kamen, R. S. Bornstein, C. D. Schoenfeld, and S. Cohen, Effect of deslanoside on the duration of the phases of ventricular systole in man, Am.J.Cardiol., 15:153 (1965).

11. A. M. Weissler and C. D. Schoenfeld, The effect of digitalis on systolic time intervals in heart failure, Am.J.Med.Sci., 259:4 (1970).

12. W. F. Forester, R. P. Lewis, A. M. Weissler, and T. A. Wilke, The onset and magnitude of the contractile response to commonly used digitalis glycosides in normal subjects, Circulation, 49:517 (1974).

13. W. S. Harris, C. D. Schoenfeld, R. H. Brooks, and A. M. Weissler, Effect of beta adrenergic blockade on the hemodynamic responses of epinephrine in man, Am.J.Cardiol., 17:484 (1966).

14. W. S. Harris, C. D. Schoenfeld, and A. M. Weissler, Effects of adrenergic receptor activation and blockade on the systolic preejection period, heart rate and arterial pressure in man, J.Clin.Invest., 46:1704 (1967).

15. H. Boudoulas, R. P. Lewis, R. E. Kates, and G. Dalamangas, Hypersensitivity to adrenergic stimulation after propranolol withdrawal in normal subjects, Ann.Intern.Med., 87:433 (1977).

16. H. Boudoulas, S. F. Schaal, R. P. Lewis, T. G. Welch, P. Green, and R. E. Kates, Negative inotropic effect of lidocaine in patients with coronary artery disease and normal subjects, Chest, 71:170 (1977).

17. C. V. Leier, C. E. Desch, R. D. Magorien, D. W. Triffon, D. V. Unverferth, H. Boudoulas, and R. P. Lewis, Positive inotropic effects of hydralazine in human subjects: comparison with prazosin in the setting of congestive heart failure, Am.J.Cardiol., 46:1039 (1980).

18. C. L. Garrard, A. M. Weissler, and H. T. Dodge, The relationship of alterations in systolic time intervals to ejection fraction in patients with cardiac disease, Circulation, 42:455 (1970).

19. A. M. Weissler, R. S. Stack, Y. S. Sohn, and D. Schaffer, Systolic time intervals, in: "Heart Failure," A. P. Fishman, ed., Hemisphere Publishing, Washington DC, p.203 (1978).

20. R. P. Lewis, S. E. Rittgers, W. F. Forester, and H. Boudoulas, A critical review of time intervals, Circulation, 56:146 (1977).

21. R. P. Lewis, R. F. Leighton, W. F. Forester, and A. M. Weissler, Systolic time intervals, in: "Non-invasive Cardiology," A. M. Weissler, ed., Grune and Stratton, New York, p.301 (1974).

22. S. Hirschfeld, R. Meyer, D. C. Schwartz, J. Korfhagen, and S. Kaplan, Measurement of right and left ventricular systolic time intervals by echocardiography, Circulation, 51:304 (1975).

23. C. J. Wiggers, Studies on the consecutive phases of the cardiac cycle. I. The duration of the consecutive phases of the cardiac cycle and the criteria for their precise determination, Am.J.Physiol., 56:415 (1921).

24. E. Braunwald, S. J. Sarnoff, and W. N. Stainsby, Determinants of duration and mean rate of ventricular ejection, Circ.Res., 6:319 (1958).

25. A. G. Wallace, J. H. Mitchell, N. S. Skinner, and S. J. Sarnoff, Duration of the phases of left ventricular systole, Circ.Res., 12:611 (1963).

26. Y. Nakamura, A. W. Wiegner, W. H. Gaasch, and O. H. L. Bing, Systolic Time Intervals: Assessment by Isolated Cardiac Muscle Studies, to be published in JACC.

27. J. A. Shaver, F. W. Kroetz, J. J. Leonard, and H. W. Paley, The effect of steady state increases in systemic arterial pressure on the duration of left ventricular ejection time, J.Clin.Invest., 47:217 (1968).

28. R. S. Stack, C. C. Lee, B. P. Reddy, M. L. Taylor, and A. M. Weissler, Left ventricular performance in coronary artery disease evaluated with systolic time intervals and echocardiography, Am.J.Cardiol., 37:331 (1976).

29. R. S. Stack, Y. H. Sohn, and A. M. Weissler, Accuracy of the systolic time intervals in detecting abnormal left ventricular performance in coronary artery disease, Am.J.Cardiol., 47:603 (1981).

30. A. M. Weissler, R. S. Stack, and Y. H. Sohn, Global left ventricular performance in chronic stable angina pectoris: the critical role of myocardial infarction, Coeur Med. Interne, 19:143 (1980).

31. A. M. Weissler, W. W. O'Neill, Y. H. Sohn, R. S. Stack, P. C. Chew, and A. H. Reed, Prognostic Significance of Systolic Time Intervals After Recovery from Myocardial Infarction, Am.J.Cardiol., 48:994 (1981).

32. H. Boudoulas, P. Geleris, C. A. Bush, R. P. Lewis, P. K.
      Fulkerson, A. K. Kolibash, and A. M. Weissler, Assessment of
      ventricular function by combined noninvasive measures:
      Factors accounting for methodologic disparities,
      Int.J.Cardiol., 2:493 (1983).
33. J. F. Spann, Jr., R. S. Buccino, E. H. Sonnenblick, and E.
      Braunwald, Contractile state of cardiac muscle obtained from
      cats with experimentally produced ventricular hypertrophy and
      heart failure, Circ.Res., 21:341 (1967)

EFFECT OF PROPRANOLOL AND LABETALOL ON LEFT VENTRICLE SYSTOLIC TIME INTERVALS AND M-MODE ECHOCARDIOGRAPHIC DIMENSIONS AT REST AND DURING EXERCISE IN PATIENTS SUFFERING FROM ANGINA PECTORIS

A. Cherchi and C. Lai

Cattedra di Cardiologia
Université di Cagliari
Cagliari, Italy

INTRODUCTION

The effect of adrenergic blocking agents on left ventricular function remains under investigation.

This study was designed to investigate the influence of propranolol and labetalol, respectively, beta and alpha beta blocking agents, on echocardiogram and systolic time intervals (STI) at rest during exercise.

MATERIAL AND METHODS

The study was undertaken on six patients, male mean age 58 (32-67) years, suffering from stable angina pectoris of effort, with a normal electrocardiogram at rest and a typical depression of ST during exercise. The echocardiogram was normal at rest. No significant variations of wall motion were found in the explored sections of the heart, i.e. part of the interventricular septum and posterior wall of the left ventricle.

The experiment was carried out in the morning, at least one hour after a light breakfast, in relation to i.v. injection of propranolol (0.15 mg/kg), labetalol (1.5 mg/kg) and placebo (saline solution). The experiment was balanced with latine squares and double blind.

The patients, after giving consent, were studied with a record of the blood pressure by the cuff method, STI according to Weissler and M-mode echocardiogram. After 30 minutes of bed rest, the patients received the drugs. The records were taken after 10 minutes of rest

115

supine after three minutes of standing and during a progressive
exercise on the bicycle with 10 watt/min each two minutes and at the
end of exercise (Figure 1).

From the records we obtained the next parameters: heart rate
(HR), systolic blood pressure (SBP), HR x SBP product, end-diastolic
(EDD) and end-systolic (ESD) dimensions, the fractional shortening
(FS%) of the LV, the SBP x EDD product, derived from the formula of
Gaasch et al.[1], as an approximation of LV systolic wall stress.

PEP, LVET, P/L and QS2 were expressed as a percent of normal
values at rest (Weissler)[2] or during exercise (Cherchi)[3].

RESULTS

Heart Rate

Heart rate, in comparison with placebo, was decreased after
propranolol at rest supine and sitting, after labetalol at rest
supine and during exercise (Figure 2).

Systolic Blood Pressure

In relation to placebo, SBP was found decreased during exercise
after propranolol and also at rest supine and standing after
labetalol (Figure 3).

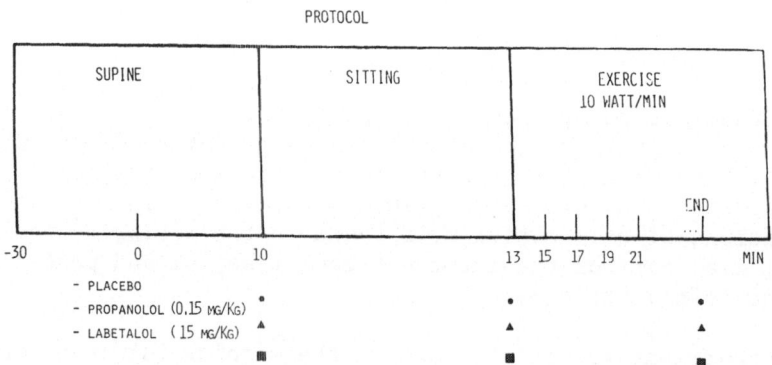

Fig. 1. Protocol of the trial. The patients have been studied after
         ten minutes of bed rest in supine position, three minutes of
         sitting on the bycicle and during exercise, 10/watt min on
         the bycicle, in upright position. STI – systolic time
         intervals.

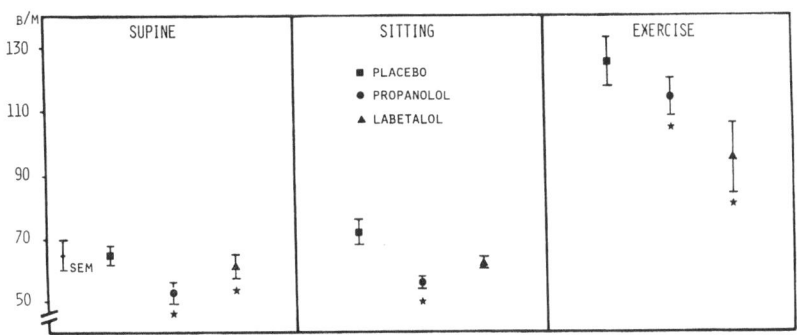

Fig. 2.   The heart rate was decreased after propranolol at rest in supine and sitting position, and during maximal exercise common to all tests.   The heart rate, after labetalol, has been found reduced in supine position and during exercise.

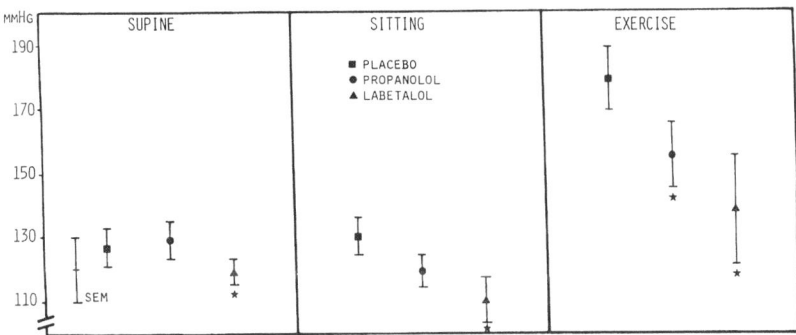

Fig. 3.   The systolic blood pressure was decreased after propranolol during exercise and, after labetalol, at rest, in supine and sitting position and during exercise.

HR x SBP Product

The HR x SBP was reduced, in comparison with placebo, after propranolol and labetalol at rest supine and sitting and during exercise (Figure 4).

End Diastolic Dimension

In relation to placebo, end-diastolic dimension was increased at rest supine and sitting and during exercise.   No significant variations were found after labetalol (Figure 5).

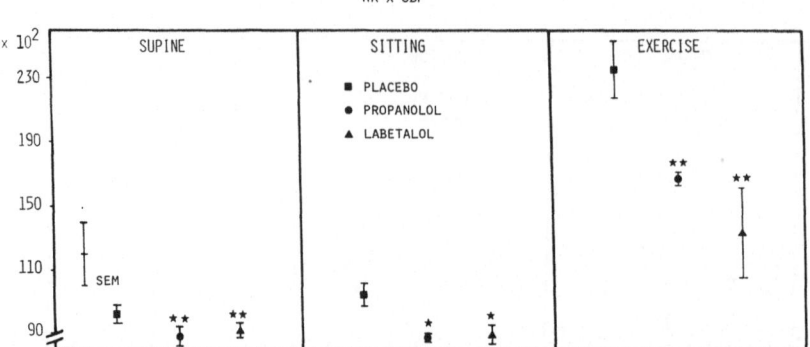

Fig. 4.   The heart rate x systolic blood pressure product (HR x SPB)
          was significantly decreased after propranolol and labetalol,
          at rest, supine and sitting, and during exercise.

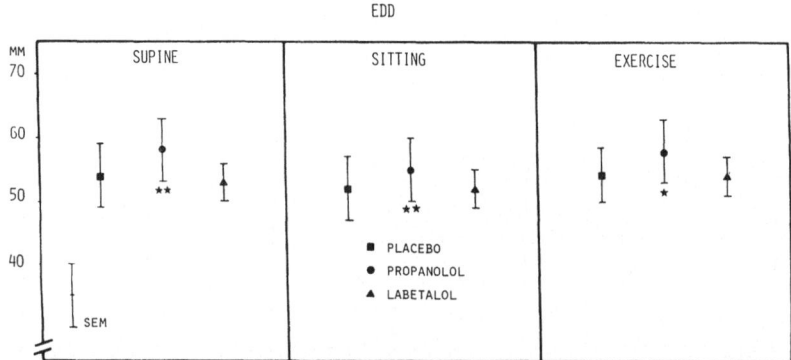

Fig. 5.   The end diastolic dimension (EDD) was increased after
          propranolol at rest, in supine and sitting position, and
          during exercise.

End-Systolic Dimension

     Similarly to end-diastolic dimension, ESD was found increased
after propranolol, in relation to placebo, at rest and during exercise. No significant variations were observed after labetalol
(Figure 6).

Shortening Fraction

     The shortening fraction was not significantly modified after
propranolol and labetalol, related to placebo (Figure 7).

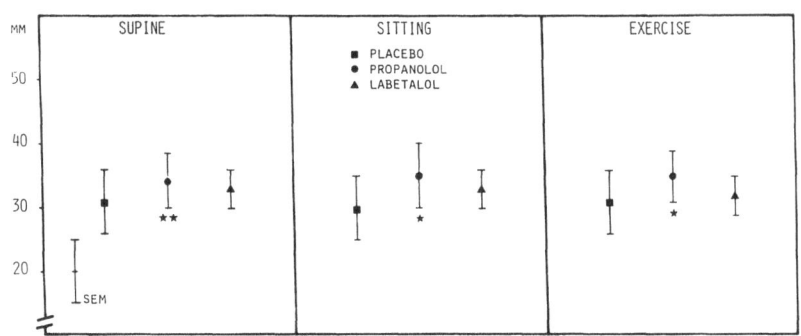

Fig. 6.   The end systolic dimension (ESD) was increased at rest in
          supine and sitting position, and during exercise.

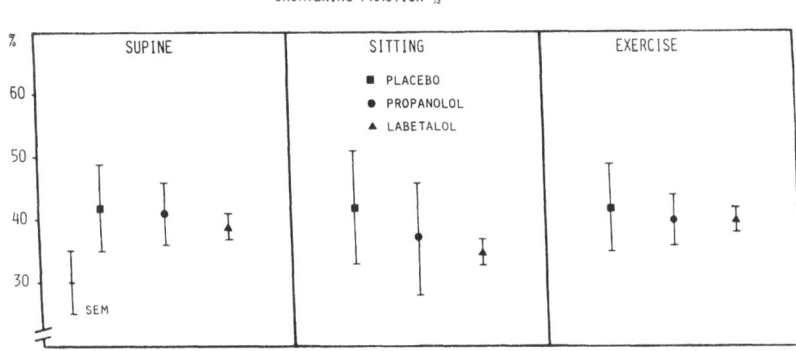

Fig. 7.   The shortening fraction was not significantly changed after
          propranolol and labetalol.

SBP x EDD Product

The SBP x EDD, an approximate of LV systolic wall stress was
decreased during exercise after labetalol in relation to placebo
(Figure 8).

PEP %

PEP % was decreased at rest supine after propranolol (Figure 9).

LVET %

LVET % was significantly increased after propranolol and
labetalol during exercise (Figure 10).

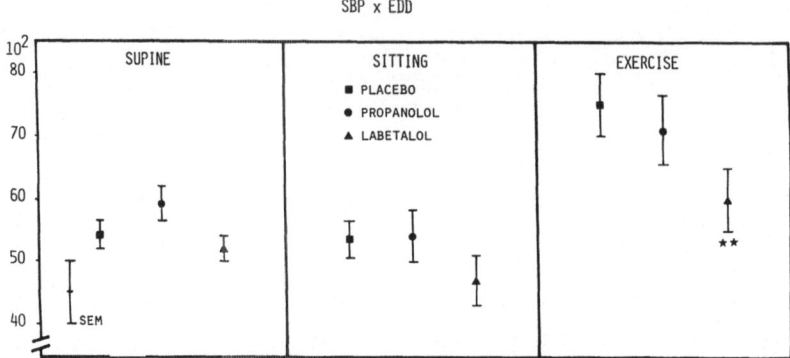

Fig. 8. The product systolic blood pressure xend dyastolic
         dimension, an expression of afterload, was decreased after
         labetalol, during exercise.

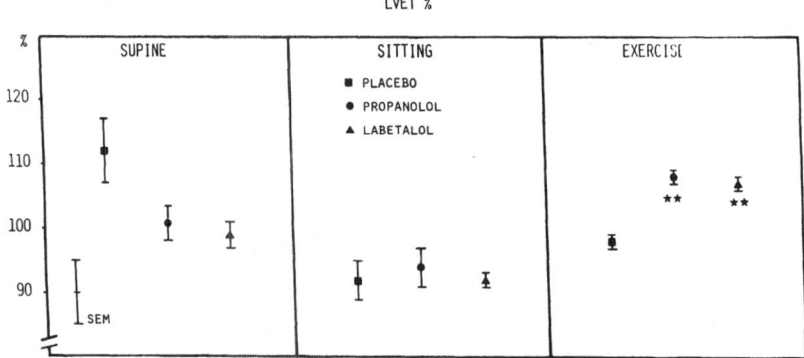

Fig. 9. The pre-espulsive period (PEP %) was decreased after
         propranolol in supine position.

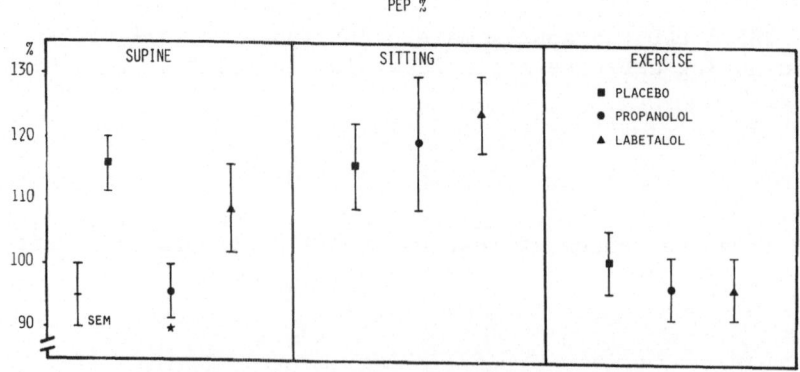

Fig. 10. The left ventricular ejection time (LEVT %) was increased
          after propranolol and labetalol, during exercise.

QS2 %

QS2 % was found significantly augmented after propranolol, sitting and during exercise and labetalol, during exercise (Figure 11).

PEP/LVET %

In relation to placebo, P/L was found decreased after propranolol, at rest supine and during exercise, and after labetalol, during exercise (Figure 12).

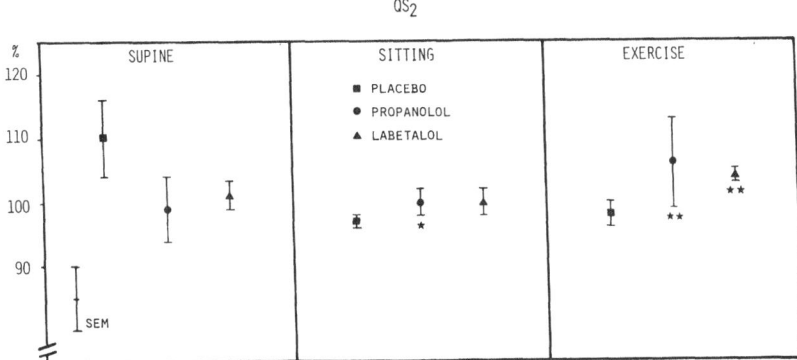

Fig. 11.   The QS2 interval was increased after propranolol in sitting positions and during exercise and after labetalol during exercise.

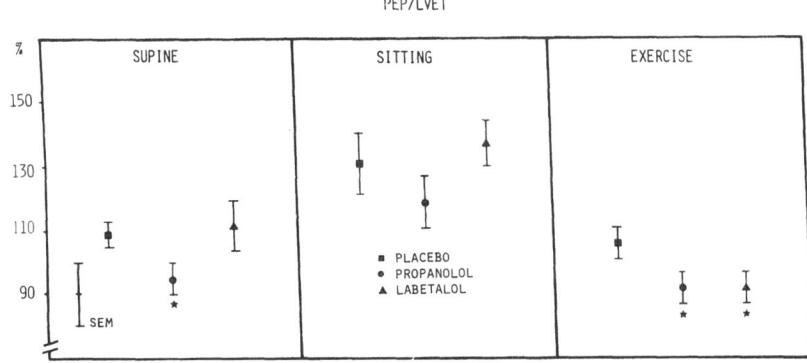

Fig. 12.   The PEP/ET ratio was found decreased after propranolol at rest in supine position, and during exercise and after labetalol, during exercise.

DISCUSSION

    After propranolol, there was an increase of end-diastolic dimen-
sion, an expression of preload, without significant change of SBP x
EDD, an approximation of systolic wall stress, i.e. of the afterload.
The fractional shortening, an expression of LV performance, was
unchanged after propranolol.

    The increase of QS2 after propranolol is difficult to explain;
but as a decrease of QS2 is related to an increase of contractility,
one can speculate on a decrease of systolic contractility.

    PEP was found increased at rest supine in relation to the in-
crease of preload.

    The increase of LVET during exercise is probably related to
increase of stroke volume.

    The reduction of PEP/LVET at rest is attributable to an increase
of LV performance, in relation to the increase of preload according
to Frank-Starling law.

    After labetalol there was not an increase of end-diastolic
dimension, differently from propranolol. In fact whereas propranolol
causes venous constriction, blocking beta receptors and letting alpha
receptors free.

    The SBP x EDD, an expression of afterload, was found decreased
in relation to the vasodilatator action of the arteries. The fractional shortening was unchanged, in relation to placebo, at rest and
during exercise.

    Similarly to propranolol, QS2 was significantly increased after
labetalol.

    LVET was increased during exercise, probably in relation to the
increase of stroke volume and the decrease of afterload.

    Finally P/L was decreased during exercise in relation to an
increase of LV performance.

    In conclusion, the i.v. propranolol, blocking beta 1 and beta 2
receptors, letting alpha receptors free, cause an increase of preload, probably a decrease of contractility, no significant variations
of afterload.

    Labetalol essentially reduced afterload, without changing pre-
load, possibly decreasing contractility.

    After the two drugs the left ventricular performance is not
impaired.

REFERENCES

1. W. H. Gaasch, Left ventricular radius to wall thickness ratio, Am.J.Cardiol., 43:1189 (1979).
2. A. M. Weissler, W. S. Harris, and C. D. Schoenfeld, Systolic time intervals in heart failure in man, Circulation, 37:149 (1971).
3. A. Cherchi, P. L. Montaldo, and F. Sau, Sulle variazioni degli intervalli di tempo sistolici nel corso dell'esevcizio muscolare in rapporto all'età ed al sesso, Boll.Soc.It. Cardiol., 23:1 (1978).

SEGMENTAL MYOCARDIAL WALL MOTION AND MYOCARDIAL ISCHEMIA

J. Forrester, R. A. Silverberg, D. Tzvoni
G. A. Diamond and R. Vas

Cedars-Sinai Medical Center University of
California, USA

Segmental wall motion abnormalities reliably occur with acute
ischemia and can be detected during angina pectoris, with myocardial
infarction, and during stress in patients with significant coronary
artery disease. Some chronic contraction abnormalities can be reversed by successful revascularization surgery[1], as well as by
physiological interventions that reduce ischemia by improving the
myocardial oxygen sypply-demand ratio[2-5]. Demonstration of reversible asynergy during intervention ventriculography can aid in selecting low-risk patients who will benefit from aortocoronary bypass
surgery, whereas irreversible asynergy signifies infarction and is
associated with a poorer prognosis after surgery[2]. Furthermore,
the extent and location of segmental contraction abnormalities determines the level of global ventricular pump function which is a major
determinant of morbidity and mortality after myocardial infarction[6,7]. The chapter will describe the effects of acute ischemia
upon the mechanical behavior of myocardial segments, and will review
the use of cardiokymography to detect and record cardiac motion in
man.

NORMAL AND ISCHEMIC SEGMENTAL WALL MOTION

Since the normal left ventricular contraction pattern is basic-
ally symmetrical, the movement of any single segment closely resembles the left ventricular volume curve. Figure 1 illustrates
normal segmental wall motion throughout the cardiac cycle on the
control state. There is an initial increase in length during
isovolumic systole. Shortening of the myocardial segment begins near
the peak of left ventricular (LV) pressure, rather than with the
opening of the aortic valve, and terminates when LV pressure begins
to fall.

125

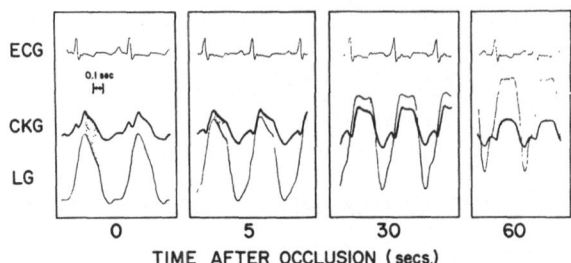

Fig. 1.   Identical changes in ischemic segment length after coronary
          artery occlusion as measured by a mercury-in-Silastic length
          gauge sutured to the epicardium.   See text for explanation.

This movement is sharply inward during systole.   There is little
change in length during isovolumic relaxation, and segment length
increases passively with filling during diastole.

    There is a specific and predictable sequence of changes in the
pattern of segmental wall motion induced by sudden coronary occlusion[8-11] or graded reduction in coronary flow, also shown in
Figure 1.   The earliest detectable change, observed within 5 second
of occlusion, is a lengthening in isovolumic relaxation, while the
amount of systolic shortening remains unchanged.   The ventriculographic correlate of dissimilar segment motion with preservation of
the magnitude of contraction was described by Herman and associates
as "asyneresis"[7].

    This lengthening rapidly in magnitude and duration, encroaching
on late systole, then progressing to ultimately involve all of systole with a reduction of systolic shortening.   The ventriculor-
graphic correlate of this intermediate phase with a reduction in
force development is "hypokinesis".   Concomitant with the decrease in
systolic shortening is an increase in end-diastolic length and
progressive lengthening during isovolumic systole associated with an
absence of systolic shortening and subsequent holosystolic expansion
by 1 to 3 minutes.   The ventriculographic correlate of the absence of
shortening in systole is "akinesis", and the equivalent of paradoxical systolic expansion is "dyskinesis".

    We performed serial left ventricular cineangiograms after coron-
ary occlusion in dogs.   Frame-by-frame analysis of individual segments demonstrates the same sequence of ischemic abnormalities
recorded directly from myocardial gauges, beginning at 5 to 10 second
after occlusion.   However, the ventriculogram demonstrates akinesis
after 1 to 2 minutes as might occur if a bulging segment is carried
inward by adjacent contracting nonischemic segments, thus balancing
the over-all movement and appearing as an apparent absence of movement.   These changes in segmental wall motion are essentially
completed by 1 minute following coronary occlusion.   With early

reperfusion following occlusion of short duration, the return of
segment motion to normal invariably procedes through the same
predictable sequence of changes, in reverse order.

Pathophysiology of Abnormal Segmental Wall Motion

A physiological explanation for these changes can be derived
from in vitro studies of isolated papillary muscle function. Hypoxia
causes a decrease in the duration of contraction, associated with a
decrease in the rate of onset and in the magnitude of developed
tension. When hypoxic muscle is placed in tandem with normal muscle
segments, the earliest effect is a dissociation in the time of onset
and duration of contraction occurring between the segments, resulting
in passive lengthening of the hypoxic muscle by the normally contracting muscle in the early and late phase of force development.
The hypoxic muscle is still able to develop force when it is allowed
to contract isometrically. In the intact heart, this could result in
a pattern of lengthening in isovolumic systole and relaxation, with
preservation of shortening during ejection expressed as dysynchrony
or dysineresis. As hypoxia progresses there is substantial reduction
in force development in the isolated papillary muscle, which would
cause a decrease in the magnitude of shortening during ejection in
the intact heart, or hypokinesis. Profound hypoxia leads to an
absence of force development which in the intact heart would be
manifested as an absence of contraction in systole, or akinesis. If
the noncontracting papillary muscles are subjected to repetitive
passive stretch by the normally contracting muscle, there is as
increase in the tension-length relationship, which can cause an
increase in segment length and paradoxical systolic expansion in the
intact heart. These changes represent a progressive decrease in the
capacity of the ischemic myocardial segment to shorten during systole
against the force developed by the simultaneous contraction of con-
tiguous normally perfused segments of the ventricle. The increased
end-diastolic length with paradoxical systolic expansion probably
reflects the repetitive passive stretching of the noncontracting
ischemic muscle by ajacent normal fibers, resulting in an early
increased compliance.

The dynamic nature of this segmental dysfunction is also highly
specific as a marker for myocardial ischemia, particularly when the
abnormalities are induced by stress and disappear with rest or with
interventions designed to reduce ischemia. Thus, the demonstration
by Franklin et al.[12] of these same abnormalities of segmental wall
motion being induced in conscious dogs with partial coronary occlusions when subjected to exercise is analogous to the precipitation
of ischemic signs and symptoms in exercising patients with coronary
artery disease.

In the early stages of a complete occlusion, the akinetic,
apparently noncontractile, muscle remains responsive to post-extra-

systolic potentiation[4], nitroglycerin, propranolol, intracoronary
injections of calcium and isoproterenol, particularly in the marginal
zone near the central ischemic area. Intracoronary isoproterenol can
reduce ischemic myocardium to contract while it simultaneously
"worsens" ST-segment elevation. Therefore, improved contraction may
not always signify ultimate "improvement" and recovery, nor does
resting segmental dysfunction always signify irreversible damages.
However, if occlusion is prolonged, permanent changes occur in the
central ischemic area and normal segmental function can no longer be
restored by reperfusion or by effective stimuli. This unresponsive-
ness, or irreversible asynergy, represents the physiological and
functional evidence of nonviable myocardium.

Segment wall motion changes are, therefore, one of the most
sensitive indices of acute myocardial ischemia. These changes can be
detected by serial noninvasive techniques, one of which is cardiokymography.

VALIDATION OF THE CKG AS A METHOD FOR ASSESSING SEGMENTAL WALL MOTION

Two types of studies have been used to determine the validity of
CKG as a method for assessing segmental wall motion: simultaneous
direct measurement of segmental wall motion in animals and comparison
to other established methods of assessing segmental wall motion in
man[13,14].

Figure 2 shows a normally contracting myocardial segment of an
intact human heart. As in dogs, a normally contracting segment shows
a continuous inward or downward movement throughout the ejection
phase. In some cases, however, a small but insignificant outward
movement can be seen. Segmental lengthening of the isovolumetric
relaxation period usually starts just before the second heart sound.

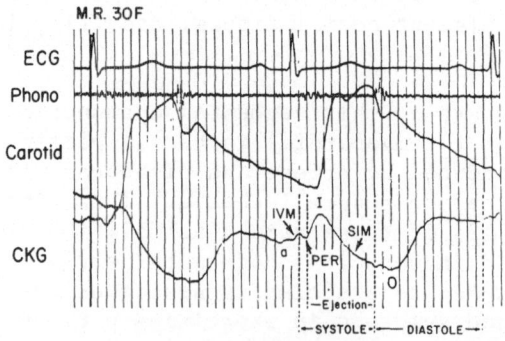

Fig. 2. The normal pattern of segmental wall motion in man. Dotted
         lines backet isovolumic systole (IS) ejection (E), iso-
         volumic relaxation (IVR).

The common feature to all abnormally contracting segments is a
large systolic outward movement or a bulge occurring during mid- or
late-systole. In some cases, a holosystolic bulge can be seen. In
others this bulge can be smaller or shorter in duration. In comparison, a normal tracing has a continuous downsloping movement
during the ejection phase.

In 70 patients studied by CKG and left ventriculography, ninety-
seven percent of anterior and lateral wall and 95% of posterior wall
abnormalities were depicted by the CKG[14]. Figure 3 shows a comparison of wall motion as depicted by the CKG and the same segment as
constructed from a frame analysis of ventriculograms in the same
patient during the same cardiac cycle.

Figure 4 shows an example of a normal CKG and electrocardiographic treadmill response: the most apparent change in the post-
exercise tracing is accentuation of pre-ejection movements. Inward
movement throughout the systolic ejection period, however, persists.
The treadmill electrocardiographic response and the coronary angiogram were also normal. Figure 5 shows an abnormal electrocardiographic and CKG response to exercise. A normal control CKG tracing
was obtained at rest. Following exercise, a holosystolic bulge was
recorded, which decreased in magnitude by 2 minutes, and by 5 minutes, returned to control morphology. The illustration also exhibits
a strongly positive ischemic electrocardiographic response, manifest
by 2 mm ST depression. At coronary angiography, the patient was
found to have a subtotal occlusion of the left main coronary artery.

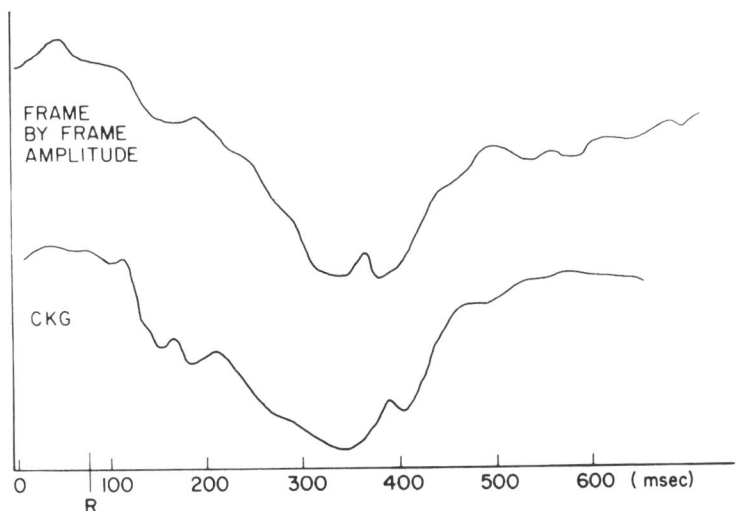

Fig. 3. Comparison of CKG and by frame analysis of the anterior
         wall motion from a left ventricular angiogram in the same
         patient.

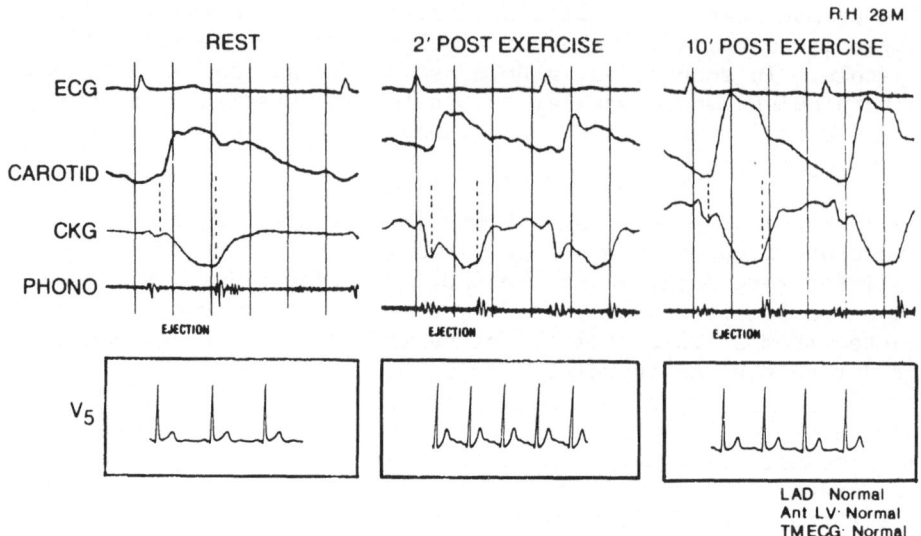

Fig. 4.   The normal CKG response to exercise. Dotted lines indicate the period of systolic ejection.

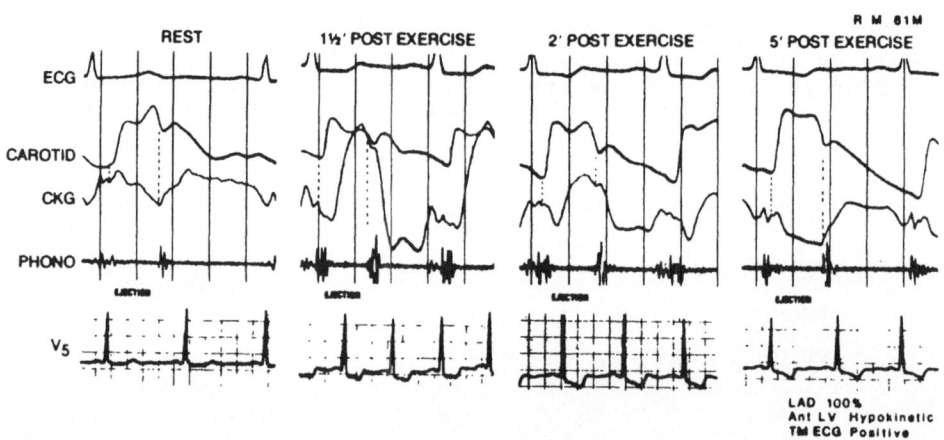

Fig. 5.   A positive ECG and CKG stress test in a patient with coronary disease as documented by angiography.

Figure 6 demonstrates a patient with coronary artery disease which was not detected by treadmill ECG but was detected by exercise CKG. As with the previous patient, he developed a systolic bulge immediately after exercise which returned to control later. The ECG remained negative, and the angiographic study showed a 70% left anterior descending coronary artery lesion. The angiographic study revealed no evidence of coronary artery disease.

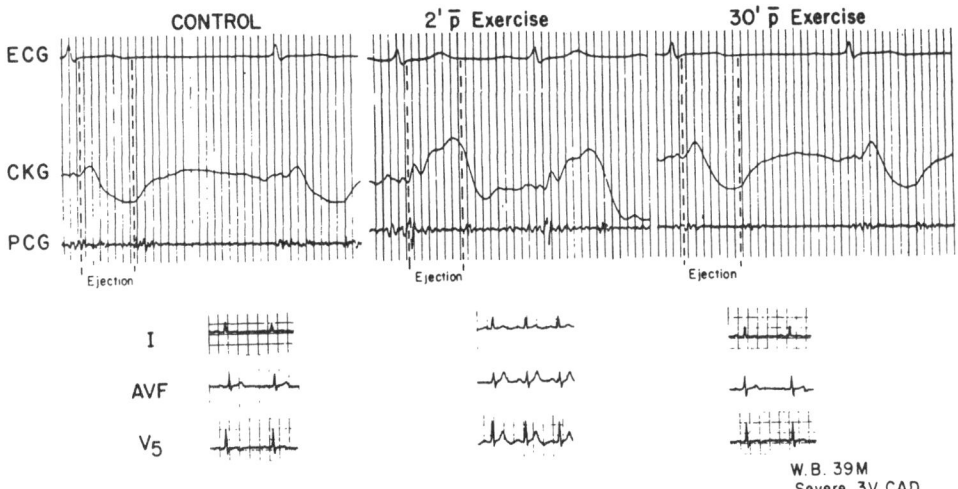

Fig. 6. A false negative ECG, and true positive CKG test in a
         patient with a typical angina, and coronary disease at
         angiography.

Table 1 summarizes the sensitivity of the CKG stress test in
patients studied by coronary angiography. The sensitivity of the CKG
is approximately 65% and the specificity is approximately 90%, values
comparable to both ECGs, thallium, and technetium stress testing.

The important limitation in cardiokymography is the potential
for misinterpretation created by superimposition of total cardiac
movement upon the segmental wall motion. The CKG signal is the sum
of segmental movement total heart movement and chest wall movement.
The magnitude of contribution of each movement to the signal undoubtedly varies with each individual, with his physiological state,
and with the location on the chest wall over which the recording is
obtained. We know little about these variations.

Nevertheless, segmental wall motion is an acceptably sensitive
and specific marker of coronary artery disease. The CKG, as low cost
device used as an adjunct to ECG stress testing, can provide valuable
additional information in properly selected populations.

REFERENCES

1. K. Chatterjee, H. J. C. Swan, W. W. Parmley et al., Depression
      of left ventricular function due to acute myocardial ischemia
      and its reversal after aortocoronary saphenous vein bypass,
      N.Engl.J.Med., 288:1117 (1972).
2. R. H. Helfant, R. Pine, and S. G. Meister, Nitroglycerin to un-
      mask reversible asynergy. Correlation with post coronary

bypass ventriculography, Circulation, 50:108 (1974).

3. S. H. Dyke, P. F. Cohn, R. Gorlin et al., Detection of residual
    myocardial function in coronary artery disease using post-
    extrasystolic potentiation, Circulation, 50:694 (1974).

4. H. R. Horn, L. E. Teichholz, P. F. Cohn et al., Augmentation of
    left ventricular contraction pattern in coronary artery
    disease by an inotropic catecholamine. The epinephrine
    ventriculogram, Circulation, 49:1063 (1974).

5. V. S. Banka, M. M. Bodenheimer, R. Shah et al., Intervention
    ventriculography: Comparative value of nitroglycerin, post-
    extrasystolic potentiation and nitroglycerin plus post-extra-
    systolic potentiation, Circulation, 53:632 (1976).

6. M. V. Herman and R. Gorlin, Implications of left ventricular
    asynergy, Am.J.Cardiol., 23:538 (1969).

7. H. J. C. Swan, J. S. Forrester, G. A. Diamond, K. Chatterjee,
    and W. W. Parmley, Hemodynamic spectrum of myocardial infarc-
    tion and cardiogenic shock. A conceptual model, Circulation,
    45:1097 (1972).

8. J. V. Tyberg, L. A. Yeatman, W. W. Parmley, C. W. Urschel, and
    E. H. Sonnenblick, Effect of hypoxia on mechanics of cardiac
    contraction, Am.J.Physiol., 281:1780 (1970).

9. J. S. Forrester, H. L. Wyatt, J. V. Tyberg, S. Goldner, W. W.
    Parmley, and H. J. C. Swan, The pressure-length loop: A new
    method for simultaneous measurement of segmental and total
    cardiac function, J.Appl.Physiol., 37:771 (1974).

10. J. S. Forrester, H. C. Wyatt, J. V. Tyberg, P. L. DaLuz, G. A.
    Diamond, and H. J. C. Swan, Functional significance of
    regional ischemic contraction abnormalities, Circulation,
    54:64 (1976).

11. H. L. Wyatt, P. L. DaLuz, J. S. Forrester, G. A. Diamond, R.
    Chagrasulis, and H. J. C. Swan, Functional abnormalities in
    nonoccluded regions of myocardium after experimental coronary
    occlusion, Am.J.Cardiol., 37:366 (1976).

12. D. Franklin, H. Tomoike, D. McKown, S. Kemper, M. Guberek, B.
    Crozatier, and J. Ross Jr., Exercise induced regional myo-
    cardial dyskinesia in dogs with limited coronary flow,
    Circulation, 54(Supp.II):II-69 (1976).

13. T. G. Gay, R. Bas, D. E. Pittman, and C. R. Jayner, The dis-
    placement cardiograph: A non-invasive technique for recording
    myocardial wall motion, Circulation, 53:139 (1976).

14. J. Detwiler, M. Crawford, R. Vas, H. Henning, and R. O'Rourke,
    Comparative effectiveness of videotracking (VT) and displace-
    ment cardiography (DCG) in detecting wall motion abnormal-
    ities, Circulation, 51(Suppl II):195 (1975).

HEMODYNAMIC EFFECTS OF VENTRICULAR PACING

A. Raineri, P. Assennato, B. Candela, G. L. Piraino,
G. Mercurio and M. Traina

Cattedra di Fisiopatologia Cardiovascolare
Università di Palermo, Policlinico, Palermo, Italy

INTRODUCTION

Fixed rate ventricular pacing is used in most cases of implanted pacemakers. It does not usually create considerable hemodynamic problems, as the heart, by its adaptation mechanism, is capable of varying its pump performance to meet the metabolic needs[1,3].

During strenuous physical activity, or when the cardiac performance is reduced, stroke volume becomes fixed so that the adaptation mainly results from heart rate variation[4]. Considering this, pacemakers which can follow the patients' hemodynamic needs, with reference not only to his condition at the moment of implant, but also for future conditions must be implanted. Among these atrial involvement in pacing is to be cited, because the normal atrio-ventricular sequence of contractility is said to determine a contribution to cardiac output ranging from zero to as much as 30% [5,10].

Non invasive techniques of investigation must be set out and they should be efficient enough to show the patient's hemodynamic situation. The left ventricle ejection fraction (LVEF) has become a standard measure of ventricular performance both at rest and during exercise. This value is important in many pathological circumstances because both cardiac output and stroke volume are not capable of discovering patients with impaired ventricular function[11]. In this report we have studied LVEF and left ventricle end-diastolic volume (LVEDV) at various rates of pacing both at rest and during exercise in patients with programmable pacemakers with the aim of evaluating the reciprocal relationship that exists at various conditions of examination. The same study was carried out in a patient with a

133

physiological pacemaker where variations of ventricular rate were
guided by the trend of the atrial rate.

STUDY OF PATIENTS AND METHODS

    The study has been carried out on 16 patients, 14 males and
2 females ranging in age from 34-64 years, with complete A-V block,
3 patients had congenital A-V block. In 15 patients a programmable
on demand and in 1 patient a physiological VDD (RS4-CPI) pacemakers
were implanted. VDD pacemaker has 2 modalities for ventricular
pacing that can be selected. The first is stimulation in VVI, with
ventricular pacing on demand inhibited by ventricular sensing. The
other modality is ventricular too, but, by atrial sensing, the ventricular rate is variable following atrial rate. The patients
studied were selected from a group of 60 patients with programmable
pacemakers. The choice was made on the basis of the following
criteria: 1) being able to be exercised, 2) maintaining pacemaker
rhythm during exercise. LVEF and LVEDF were detected both at rest
and during exercise, at the heart rates of 70,95,115 bpm (beats per
minute). The evaluation was carried out using the "Nuclear
Stethoscope", which is a non image scintillator probe, elaborating
cardiac volumes[12,13]. The exercise protocol included a first
cycloergometer test carried out in the upright position to determine
the maximum work load and to ascertain the maintenance of pacemaker
rhythm. A second test was performed in the supine position for each
of the 3 programmed heart rates. This exercise test was performed at
a level of 40% of the maximum effort reached at the first evaluation
(moderate level) for a period of 5 minutes at each of the 3 programmed heart rates. A strenous exercise which lasted 5 minutes
equal to 90% of the maximum work load was subsequently carried out at
115 bpm. Between each exercise step there was a 10 minute interval
for recovery. The hemodynamic parameters were evaluated at the 4th
minute of every work load by radionuclide techniques with the injection of 20 mCi of technitium pyrophosphate detected by the Nuclear
Stethoscope. The LVEF is an absolute value, whereas LVEDV is a
relative one.

    On the basis of studies carried out on subjects free of cardio-
pulmonary disease the left ventricle is considered normal when EF
increases at least 5% during exercise. The statistical analysis was
carried out with the "t" Student for paired data.

RESULTS

Ejection Fraction at Rest and During Exercise at Different
Ventricular Pacing Rates

    In 13 patients out of 15 with programmable pacemakers the LVEF
with ventricular pacing at the initial rate of 70 bpm was on average

58±4.8% at rest. With an increase in the rate of pacing the LVEF
decreased, as at 95 bpm it was 54.5±5% and at 115 bpm it reached
49±8% (Table 1 and Figure 1). The variation of LVEF from 70 to
95 bpm was not statistically significant instead it became statistically significant from 95 to 115 bpm (p<0.05) and from 70 bpm
to 115 bpm (p<0.001). During moderate physical exercise, the LVEF
reached the mean value of 64±5.5% at the initial rate of 70 bpm with
a percentage increase of +9.4 (range 5±20). The LVEF increased
further, reaching the mean value of 72.5±4% at 95 bpm, with a percentage increase of +20.8 (range 10-30). This increase was significant (p<0.001). When the pacing rate was 115 bpm, the LVEF
reached the mean value of 67.5±6.2% with a percentage increase of
+26.8 (range 12-40). These values compared with the ones reached at
70 and 95 bpm were not significant. A strenuous exercise carried out
at 115 bpm determined an increase in the EF to 75±6.3% with a percentage increase of +34.6 (range 13.4-51). The comparison between
the LVEF bpm and 95 bpm during moderate exercise, and the LVEF at
115 bpm during strenuous exercise was significant (p<0.001).

In 2 patients the LVEF underwent a different behavior. In the
first (Figure 2A) at rest and at a pacing rate of 70 bpm the LVEF was
68% and at 95 the LVEF was 75%. At 115 bpm the LVEF was 59%. When
the patients underwent moderate exercise the LVEF was lowered at all
the pacing rates with a value of 56% at 70 bpm, 49% at 95 bpm, and
47% at 115 bpm.

In the second patient (Figure 2B) at rest and at 70 bpm the EF
was 61%, at 95 bpm it was 57%, and at 115 bpm it was 59%. During
moderate physical exercise, at 70 bpm the EF was 64% at 95 bpm 70%,
and at 115 bpm 59%.

Left Ventricle End Diastolic Volume at Rest and During Exercise at Various Ventricular Pacing Rates

Ventricular pacing at programmed rates of 70, 90, and 115 bpm
determined a decrease of LVEDV at rest, starting from a value of
1.41±0.4 going to 1.37±0.4 and ending with 1.33±0.7. The indicated
variations were not statistically significant (Figure 3). During
moderate exercise LVEDF increased significantly from 70 bpm to 95 bpm
(1.47±0.6; 1.56±0.6; 1.51±0.7 respectively). From 95 to 115 bpm
LVEDV was not significant, as from 70 to 115 bpm. Strenuous exercise
determined a LVEDV increase which reached the value of 1.61±0.8 at
115 bpm. The statistical comparison with the respective values at
the various rates during moderate exercise show statistical differences which are significant (p<0.001).

In the two patients reported separately due to the different
behavior of the LVEF exercise, the values of LVEDV found in the
various conditions were: in the first (Figure 2A) at rest and 70 bpm

Table 1. Left Ventricle Ejection Fraction (LVEF) and End-diastolic
          Volume (LVEDV) at Rest and During Moderate or Strenuous
          Exercise at Different Heart Rates (HR). No. pts = 13. Mean
          Values.

	HR (beats/min)	LVEF (%)	LVEDV (relative)
		moderate exercise	
rest	70	58+4.8	1.41+0.4
exercise	70	64+5.5	1.47+0.6
Δ % (range)		9.4 (5-20)	4 (2-9.3)
rest	95	54.5+5	1.37+0.4
exercise	95	72.5+4	1.56+0.6
Δ % (range)		20.8 (10-30)	12.1 (5.2-20)
rest	115	49+8	1.33+0.7
exercise	115	67.5+6.2	1.51+0.7
Δ % (range)		26.8 (12-40)	11.9 (4.2-24)
		strenuous exercise	
rest	115	49+7	1.32+0.6
exercise	115	75+6.3	1.61+0.8
Δ % (range)		34.6 (13.4-51)	18(7.2-25

the LVEDV was 1.52; at 95 bpm it was 1.60; at 115 bpm it was 1.43.
During physical exercise the LVEDV decreased progressively with the
values from 1.39 to 1.31.

In the other patient (Figure 2B), at rest and 70 bpm, LVEDV was
1.44; at 95 bpm it was 1.40, at 115 bpm it was 1.42. The physical
exercise at 70 bpm determined an LVEDV increase to 1.47. This was
greater at 95 bpm (LVEDV=1.53), while at 115 it was not modified in
respect to the value at rest (1.42).

Ejection Fraction at Rest and During Exercise in a Patient with a Physiological Pacemaker

The physiological pacemaker determined the variations of LVEF
which were valued in relation to the modality of stimulation. During
VVI mode (ventricular pacing and ventricular sensing) the LVEF was
56% at rest, with fixed rate of 72 bpm. During moderate exercise the
LVEF was 64% with a percentage increase of +14. During strenuous
exercise the LVEF reached the value of 66% with a percentage increase

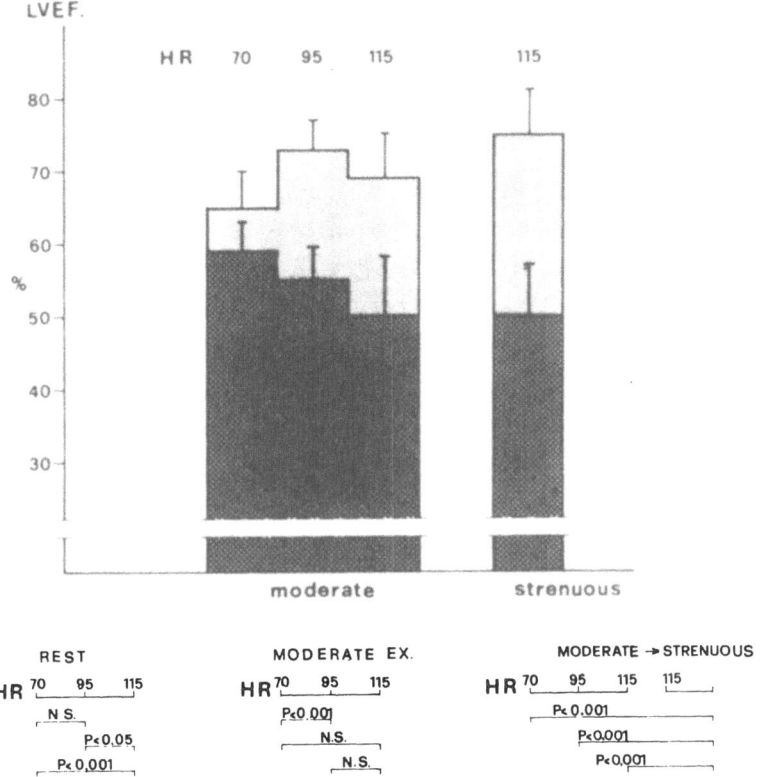

Fig. 1.   Left ventricle ejection fraction at rest and during exercise
           at different ventricular pacing rates.

of +17 (Table 2, Figure 4). When the pacemaker was switched over the
physiological (VDD) the rate at rest was 68 bpm and the LVEF was 61%.
During moderate exercise the ventricular rate was 90 bpm, the LVEF
was 76% with a percentage increase of +24.5. During the VVI mode
with a fixed rate of 72 bpm, the LVEDV was 1.38 at rest. During
moderate exercise the LVDEV was 1.47 with a percentage increase of
6.5. During strenuous exercise the LVEDV reached 1.49 (%+7.9). With
the pacemaker switched over to physiological (VDD) the rate at rest
was 68 bpm and the LVEDV was 1.44. During exercise at 90 bpm the
LVEDV was 1.61 with a percentage increase of +11.8. The patient
underwent a spiroergometrical evaluation. The O_2 consumption in the
VVI mode was 950 ml/min and in the VDD mode 1100 ml/min (Figure 5).

A. RAINERI ET AL.

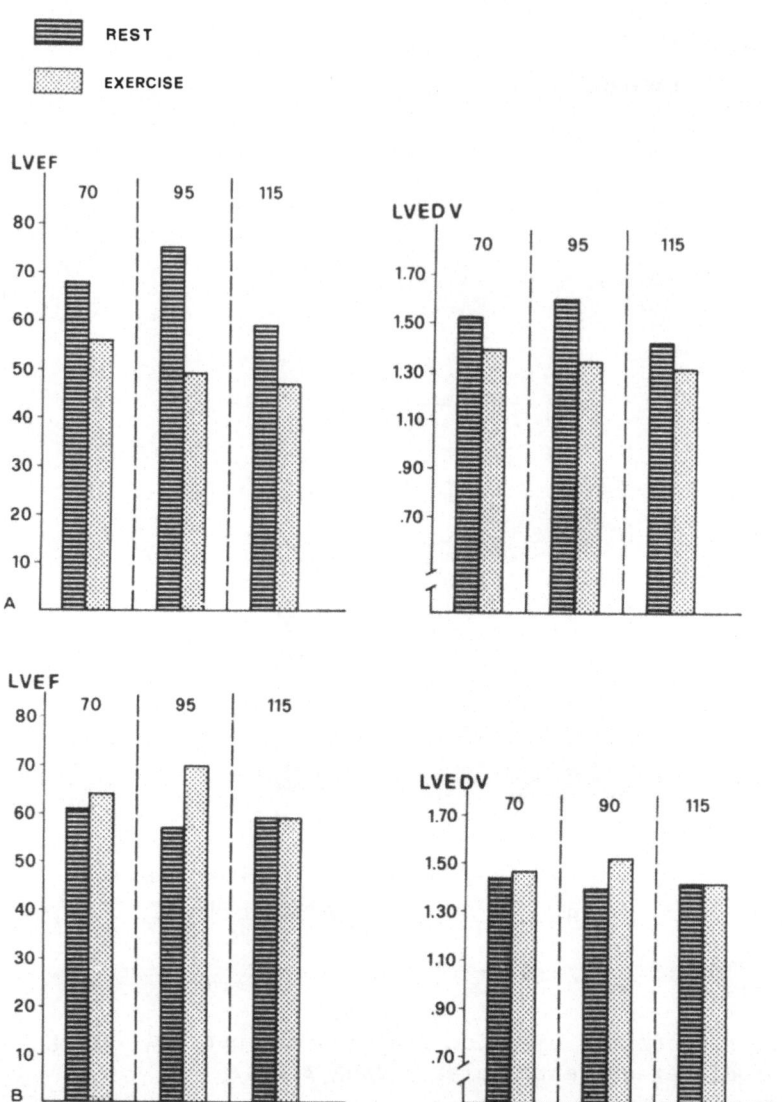

Fig. 2. Different behavior of LVEF and LVEDV in 2 patients with
         reduced L.V. performance.

The equivalent of ventilation in VVI precociously exceeded the VDD
for similar O_2 consumption (Figure 6).

DUSCUSSION

     Our results obtained with nuclear techniques show that the
increase of heart rate obtained by ventricular pacing at rest determines the reduction of the ejection fraction.

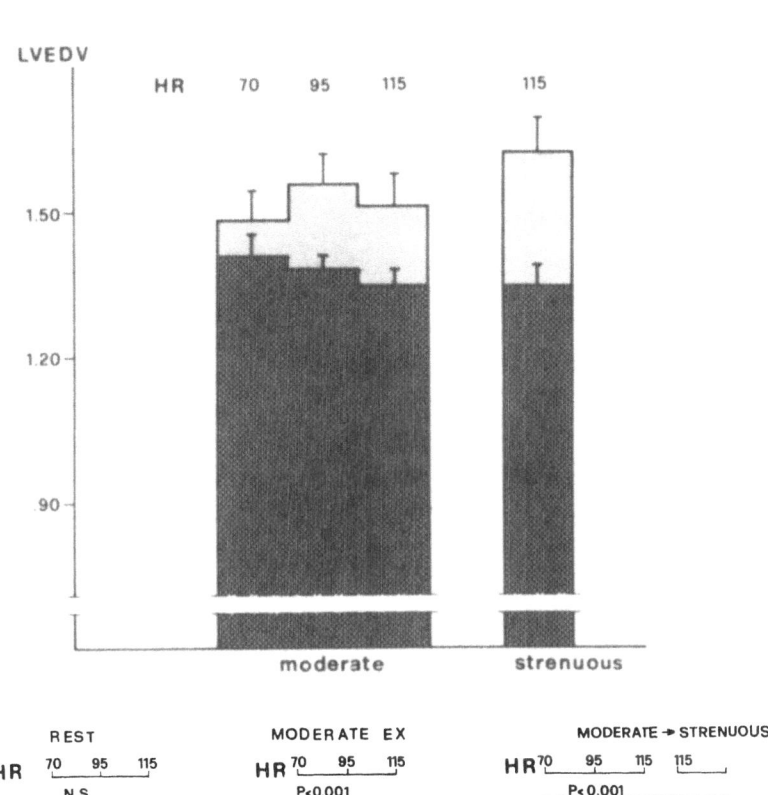

Fig. 3.   Left ventricle end diastolic volume at rest and during
          exercise at different ventricle pacing rates.

The inverse relationship at rest between ejection fraction and
heart rate both in animals and human beings has been demonstrated[14-16].

Some authors have found that for increases of 30 bpm such a
difference reaches the limits of statistical significance[17]. The
LVEDV decrease was another outcome which was found in our results,
but its reduction was not significant at rest.   The decrease in
end-diastolic volume as the heart rate is increased as reported in
previous data[14,15] shows the dependance of the ejection fraction
on the preload[18,19].   Nevertheless other factors affecting
response to tachycardia can interact.   The increased inotropic effect

Table 2. Left Ventricle Ejection Fraction (LVEF) and End–diastolic
          Volume (LVEDV) at Rest and During Moderate and Strenuous
          Exercise with Pacing in VVI and VDD at Different Heart
          Rates

	HR (beats/min)	LVEF (%)	LVEDV (relative)
moderate exercise in VVI			
rest	72	56	1.38
exercise	72	64	1.47
Δ %		14	6.5
strenuous exercise in VVI			
rest	72	56	1.38
exercise	72	66	1.49
Δ %		17	7.9
moderate exercise in VDD			
rest	68	61	1.44
exercise	90	76	1.61
Δ %		24.5	11.8

associated with tachycardia affects stroke volume, and thus end-
systolic volume[20]. Another factor which may influence the results
is the impedance to ventricular ejection: mean aortic pressure tends
to increase with pacing induced tachycardia and decreases ejection
fraction[21].

     The increase of the ejection fraction during physical exercise
is a well known event. On the basis of studies in subjects without
evidence of cardiopulmonary disease, the normal response to exercise
is at least a 5% increase of the ejection fraction[22,23].

     At the beginning of physical activity the normal heart responds
with an increase of stroke volume[24]. After this initial increase,
the stroke volume remains relatively constant, and the increase of
cardiac output is directly proportional to the increase of the heart
rate[25]. When the rate rises it is limited, such as in heart block,
the stroke volume plays a much more important part. The increase of
stroke volume during exercise is characteristic of pacemaker patients
with a good myocardial function and in this group the cardiac output

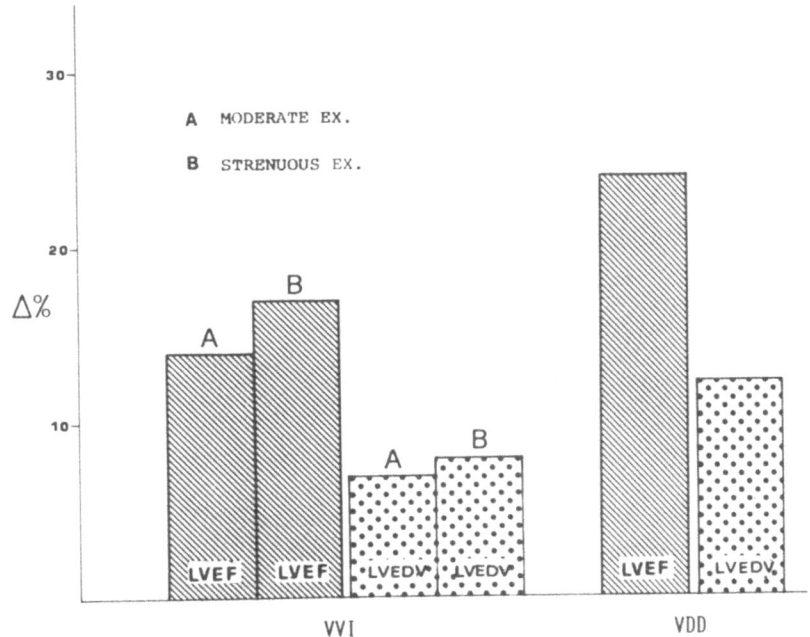

Fig. 4.   Percentage increase (Δ%) of left ventricle ejection fraction
          and left ventricle end diastolic volume during moderate (A)
          and strenuous (B) exercise in a patient during ventricular
          pacing and ventricular sensing (VVI) and during physio-
          logical pacing (VDD).

rate is mostly independent of heart rate and it is only influenced by
the exercise level[1-3].   On the other hand, patients with a poor
myocardial function are incapable of increasing the stroke volume, so
in this condition with an artificial pacemaker the cardiac output is
dependent on the heart rate even during exercise[1-4].   In the preliminary reports we found that the changes in LVEF induced by physical activity are rate dependent in paced patients[26].   The results
reached in this study confirm that the ejection fraction during
moderate exercise increases.   This tendency is increased further by
the increase of the heart rate from 70 bpm to 95 bpm.   The percentage
increase of the ejection fraction in comparison to the condition at
rest is high at 95 bpm, moreover, the ejection fraction at rest is at
its lowest level.   When this variation is present at 115 bpm it
certainly becomes an unfavorable data, as the value reached by the
ejection fraction at this pacing rate shows a reduction.   The variations of end-diastolic volume which develop parallelly to the LVEF,
clearly show that the variations of the LVEF in some way depend on
preload.   Therefore our research allows us to consider the variations
f heart rate valid in the group of patients with good performance
oo.   This observation is confirmed by the behavior of the LVEF in

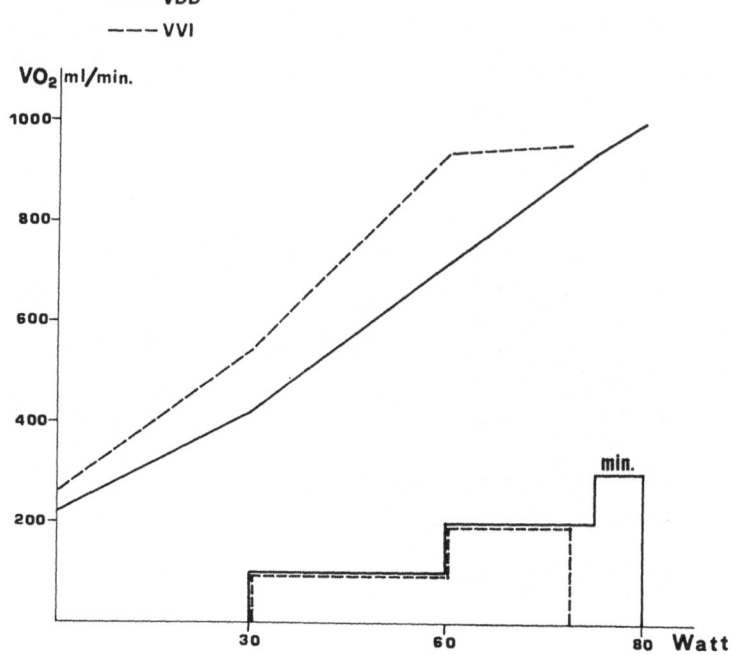

Fig. 5. Work capacity (watt min) and aerobic capacity (VO$_2$) at
         different ventricular pacing (VVI-VDD).

the patient with a physiological pacemaker. In this patient in fact
at the fixed rate of 72 bpm and demand mode the LVEF during moderate
exercise had a considerable increase, but however it was clearly
inferior to the one reached by physiological mode, when the pacemaker
was sensed by the atrial rate and paced the ventricle at 90 bpm. In
this way the LVEF reached was higher than that obtained during strenuous exercise. The relationship between ventilation and oxygen
consumption in this patient was calculated by a further demonstration
of the optimum heart rate which is reached when the pacemaker is
switched such as to sense atrial rate. In this case the VE/VO$_2$
relationship is lower than on demand. This result was found in our
previous experience, where during exercise paced patients in respect
to other patients who recovered sinus rhythm or their own rhythm,
they showed a greater slope of ventilation - oxygen consumption
curve[27]. The different behavior of the patients' hemodynamic
variables suggest the following consideration.

First of all ejection fraction is a very sensitive index of
cardiac performance in subjects that clinically might be considered
normal. In Figure 2 the reduction which the ejection fraction undergoes during exercise in patient A is very significant in this consideration. The variations in patient B show that the levels of the
pacing rate are important to show the degree of functional reserve.

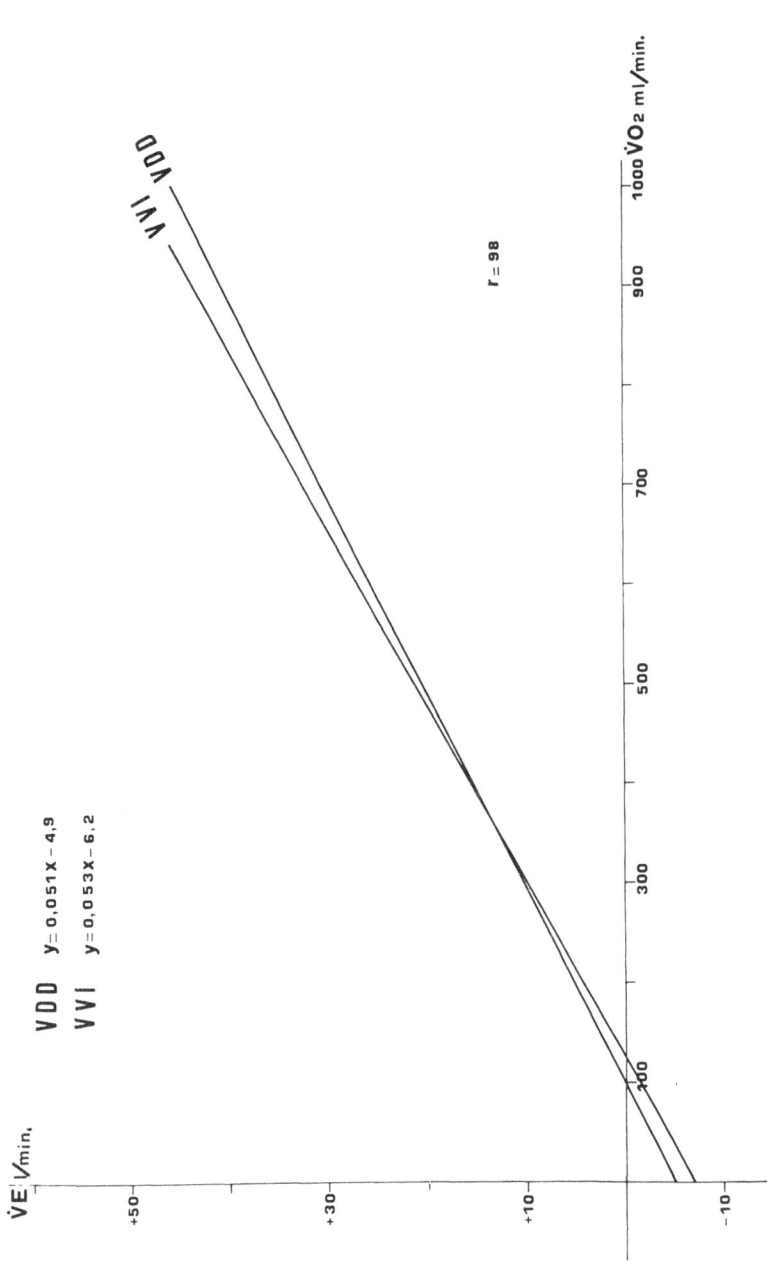

Fig. 6. VE/VO$_2$ relationship during physical work at different ventricular pacing (VVI-VDD).

In conclusion the course of the LVEF at various pacing rates does not show the univocal tendency found in patients with a good performance. Seeing the difference of behavior during physical exercise where patient B had a better performance in respect to patient A, it might be possible to give a predictive value of behavior at rest. In fact, the LVEF increased from 70 to 95 bpm at rest only in patient A with the worse performance. Moreover, as already reported by others, our results show that the heart rate at which the LVEF is determined must be taken into great consideration to get an exact valuation of the cardiac performance[17].

In conclusion we could therefore affirm, that a relationship exists between heart rate and hemodynamic parameters so it would be useful to have pacemakers capable of varying the rate in relation to the needs imposed by metabolic necessities and the functional adaptability of the pump when it becomes more critically dependent on heart rate. This seems possible by using a physiological pacemaker in patients whose atrial activity is present.

SUMMARY

In order to evaluate the hemodynamic effects of ventricular pacing, LVEF and LVEDV were detected both at rest and during exercise in 15 patients with programmable pacemakers at heart rates of 70, 95, 115 bpm and in one patient with a physiological pacemaker. The test was performed on a cycloergometer in the supine position with a submaximal single work load. The evaluation was carried out using the "Nuclear Stethoscope" which is a non imaging nuclear detector, elaborating cardiac volume. The conclusion are:

1) In patients at rest an increase of the heart rate over 70 bpm may decrease the LVEF.
2) If the variation of the heart rate is coupled with physical exercise there is an increase in the LVEF which is proportional to some extent to the heart rate.
3) The study carried out by us is capable of giving useful information regarding the left ventricle performance.
4) The physiological pacemaker is capable of improving the hemodynamic data of ventricular function varying the rate in relation to the needs imposed by the metabolic necessity.

REFERENCES

1. M. McGregor and G. A. Klasser, Observations on the effect of heart rate on cardiac output in patients with complete heart block at rest and during exercise, Circulation. (Suppl.2) 215:14-15 (1964).
2. E. Sowton, Haemodynamic studies in patients with artificial pacemakers, Brit.Heart J., 26:737 (1964).

3. S. Bervegard, B. Jonsson, I. Karlof, H. Lagergren, and E. Sowton, Effect of changes in ventricular rate on cardiac output and central pressures at rest and during exercise in patients with artificial pacemakers, Cardiovasc.Res., 1:21 (1967).

4. A. Benchimol, Y-B. Li, E. G. Dimond, R. B. Voth, and A. S. Roland, Effect of hear rate, exercise, and nytroglycerin on the cardiac dynamics in complete heart block, Circulation, 28:510 (1963).

5. R. A. Gesell, Cardiodynamics in heart block as effected by auricular systole, auricular fibrillation and stimulation of vagus nerve, Am.J.Physiol., 40:267 (1961).

6. P. Samet, W. Bernstein, and S. Levine, Significance of atrial contribution of ventricular filling, Am.J.Cardiol., 15:195 (1965).

7. P. Samet, W. H. Bernstein, S. Lavine, and A. Lopez, Hemodynamic effects of tachycardias produced by atrial and ventricular pacing, Am.J.Med., 39:905 (1965).

8. P. Samet, C. Castillo, and W. H. Bernestein, Hemodynamic sequelae of artial ventricular and sequential atrioventricular pacing in cardiac patients, Am.Heart J., 72:725 (1966).

9. S. Furman, Physiologic Pacing, Pace, 3:639 (1980).

10. B. N. Goldreyer, Physiologic Pacing: The role of AV Synchronomy, Pace, 5:613 (1982).

11. S. S. Young, L. G. Bentivoglio, V. Maranhäo, and H. Goldberg, "From Cardiac Catheterization Data to Hemodynamic Parameters," FIA, Davis Co., Philadelphia, page 160 (1972).

12. H. N. Wagner, P. Rigo, R. H. Baxter, P. O. Alderson, K. H. Douglass, and F. D. Housholder, Monitoring ventricular function at rest and during exercise with a nonimaging nuclear detector, Amer.J.Cardiol., 43:975 (1979).

13. H. J. Berger, R. A. Davies, W. P. Batsford, P. B. Hoffer, A. Gottschalk, and B. L. Zaret, Beat to beat left ventricular performance assessed from the equilibrium cardiac blood pool using a computerized nuclear probe, Circulation, 63:133 (1981).

14. J. D. Bristow, R. E. Fergusson, F. Mintz, and E. Rapaport, Influence of heart rate on left ventricular volume in dogs, J.Clin.Invest., 42:649 (1963).

15. G. Glick, J. R. Williams, Jr., D. C. Harrison, A. G. Morrow, and E. Braunwald, Cardiac dimensions in intact unanesthetized man. VI. Effects of changes in heart rate, J.Appl.Physiol., 21:947 (1966).

16. A. G. Trakiris, D. E. Donald, R. E. Sturm, and E. H. Wood, Volume, ejection fraction, and internal dimensions of left ventricle determined by biplane videometry, Fed.Proc., 28:1358 (1969).

17. D. R. Ricci, A. E. Orlick, E. L. Alderman, N. B. Ingels, G. T. Daughters, and E. B. Stinson, Influence of heart rate on left ventricular ejection fraction in human beings, Am.J.Cardiol., 44:447 (1979).

18. H. P. Krayenbuhl, W. D. Bussman, M. Turina, and E. Lütley, Is
     the ejection fraction an index of myocardial contractility?
     Cardiology, 53:1 (1968).
19. R. D. Gentzler, A. S. Hunter, and J. H. Gault, Preload depen-
     dence of ejection fraction (abstr.) Am.J.Cardiol., 33:139
     (1974).
20. D. R. Ricci, A. E. Orlick, E. L. Alderman, N. B. Ingles, Jr., G.
     T. Daughters II, C. A. Kusnick, B. A. Reitz, and E. B.
     Stinson, Role of tachycardia as an inotropic stimulus in man,
     J.Clin. Invest., 63:695 (1979).
21. F. H. Mahler, J. Ross, Jr., R. A. O'Rourke, and J. W. Covell,
     Effects of changes in preload, afterload and inotropic state
     on ejection and isovolumic phase measures of contractility in
     the conscious dog, Am.J. Cardiol., 35:626 (1975).
22. S. K. Rerych, P. M. Scholz, G. E. Newman, D. C. Sabiston, Jr.,
     and R. H. Jones, Cardiac function at rest and during exercise
     in normals and in patients with coronary heart disease:
     evaluation by radionuclide angiography, Ann.Surg., 187:449
     (1978).
23. H. J. Berger, L. A. Reduto, D. E. Johnstone et al., Global and
     regional left ventricular response to bicycle exercise in
     coronary artery disease: assessment by quantitative radio-
     nuclide angiography, Am.J.Med., 66:13 (1979).
24. S. Bevegard, Studies on the regulation of the circulation in
     man, Acta Physiol.Scand., 57:(Suppl.200) (1962).
25. A. Holmgren, B. Jonsson, and T. Sjöstrand, Circulatory data in
     normal subjects at rest and during exercise in recumbent
     position, with special reference to the stroke volume at
     different work intensities, Acta Physiol.Scand., 49:343
     (1960).
26. A. Raineri, Hemodynamic effects of ventricular pacing, Impulse,
     A special 10th anniversary issue from CPI, April 1982, page
     19.
27. A. Raineri, G. Mercurio, A. M. Milito, and P. Assennato, Func-
     tional evaluation of patients with implanted pacemakers, In:
     "Selected Topics in Exercise Cardiology and Rehabilitation,"
     Plenum Press, New York and London (1980).

TIMING OF OPERATION FOR CHRONIC

SEVERE AORTIC REGURGITATION*

E. Geraci

Division of Cardiology
"V. Cervello" Hospital
Palermo, Italy

The indication for aortic valve replacement in chronic severe aortic regurgitation poses different problems. In symptomatic patients the operation offers better results than medical treatment alone[1], so that in general there are no doubts about surgical indication once a patient develops more than trivial symptoms. Rather, in these cases the problem is of a prognostic nature: which patients will probably have an excellent postoperative recovery and which others might have a worse outcome? Today this is not a very difficult question, since we possess quite accurate predictive criteria based on both invasive and noninvasive evaluation of left ventricular function[1-6]. In asymptomatic (or midly symptomatic) patients, on the contrary the problem is of decisional nature: if and when aortic valve replacement is indicated. This is a more difficult task and will be the main theme of this paper.

If we consider the natural history of chronic severe aortic incompetence under the presence or not of symptoms and of left ventricular functional status**, we can schematically identify three models of evolution[7,8]:

- left ventricular function becomes impaired yet the patient remains asymptomatic for a certain period of time (Figure 1A);
- symptoms appear in coincidence with left ventricular dysfunction (Figure 1B);
- symptoms appear yet left ventricular function remains normal for

---

*Pre-arranged intervention on the paper "Evaluation of Left Ventricular Function as Indication of Aortic Valve Surgery" presented by G. Mattioli et al.

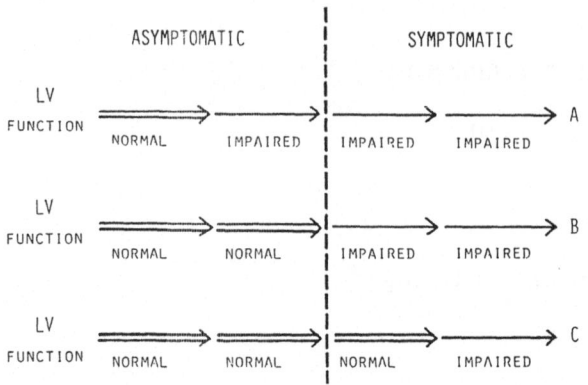

Fig. 1. Chronic severe aortic regurgitation. Schematic evolution in
         terms of symptoms and left ventricular (LV) function: three
         possible models.

a certain period of time (Figure 1C). So one would argue that
symptoms are angina or syncope rather than dyspnea or fatigue
(see footnote).

As stated before, there is a general agreement as to recommend
surgery for all patients who develop significant symptoms, even when
left ventricular function remains normal[8]. Some doubts are raised
only for patients at the opposite end of the spectrum, those with extremely severe left ventricular dysfunction, for whom the operation
might be useless or even harmful[1].

On the contrary, there are contrasting attitudes towards
asymptomatic patients. The most aggressive is that of a minority of
investigators who recommend "prophylactic" aortic valve replacement
for all patients with relevant aortic regurgitation, including the
asymptomatic ones with normal left ventricular function[9]. This
drastic attitude has been recently well confuted[10] and will not be
discussed further here. The real controversy is the one among those

---

**One point that appears to be overlooked in the pertinent medical
literature is that in these patients significant symptoms are not
necessarily in strict connection with left ventricular dysfunction
in terms of depressed contractility. In fact, under the label
"symptomatic" also patients with angina or syncope, are included
which could have different pathogenesis from that of dyspnea and
fatigue. This point, however, is not very relevant to the present
discussion, concerning essentially asymptomatic patients.

authors who maintain that surgery is not indicated unless significant
symptoms appear[11,12] and the ones for whom the operation is warranted once left ventricular dysfunction develops even in patients
who are asymptomatic[6,8,13-18].

Within latter strategy different tactics have been proposed:
to operate if there is proof of left ventricular dysfunction; to
operate when a worsening is found between two successive measurements; to operate beyond a definite threshold of functional impairment. Several methods of evaluation (often the same used for prognostic purposes in symptomatic patients) and many levels of left
ventricular dysfunction have been advocated as indicators of the
optimum time for the operation[13-21].

On the whole, early surgical treatment for patients with chronic
severe aortic regurgitation who develop left ventricular dysfunction
but remain asymptomatic, appears the preferred policy. Today this is
based upon two main assumptions: 1) the time elapsed between left
ventricular dysfunction and occurrence of symptoms is short; 2) once
symptomatic, patients with depressed left ventricular function will
undergo the operation with high risk of irreversible myocardial
impairment and a consequent unfavorable outcome.

Both these assumptions, are supported mainly by NIH group of
Bethesda[8,19], are confutable: 1) it has been observed that many
asymptomatic patients with chronic severe aortic regurgitation and
impaired (not too severely, of course) left ventricular function can
remain stable for long periods of time[22]; 2) the progress achieved
in the past years has greatly enhanced the chances of good surgical
results in such patients even if the operation is delayed until
symptoms develop[23,24].

If the rationale for early aortic replacement is the prevention
of irreversible left ventricular deterioration[8], the rational for
delaying, the operation is that the patient will have a shorter
period of being at risk for valve-related complication and will
benefit from further refinements of the operative technique, myocardial protection measures and prosthetic design[12].

As a matter of fact, the analysis of medical literature offers
no definite proof that the detection of left ventricular dysfunction
in asymptomatic patients with chronic severe aortic regurgitation is
per se an urgent reason for valve replacement[25-28]. Therefore the
aggressive attitude of recommending surgery in these patients at the
first sign of impaired left ventricular performance does not appear
warranted now. A better approach would be to wait for evidence of a
worsening between two successive functional measurements and/or to
adopt a proper threshold of left ventricular dysfunction beyond which
the operation would represent a convenient option. This threshold
cannot be unequivocally indicated: based upon a critical review of

the literature and the surgeons' own experience. One must choose
among a multitude of criteria, also the locally available diagnostic
facilities, must be considered.

REFERENCES

1. G. Fasoli, R. Scognamiglio, R. Chioin, and S. Dalla Volta, Indi-
    cazioni alla terapia chirurgica delle cardiopatie valvolari
    acquisite: insufficienza aortica. Cardiologia, (Boll.Soc.
    Ital.Cardiol.), Suppl.Atti XLIII Congr.Naz.Soc.Ital.Cardiol.,
    (Roma, 16-18 Dec. 1982) 43-52 (1982).
2. W. L. Henry, R. O. Bonow, J. S. Borer, J. H. Ware, K. M. Kent,
    D. R. Redwood, C. L. McIntosh, A. G. Morrow, and S. E.
    Epstein, Observations on the optimum time for operative
    intervention for aortic regurgitation. I. Evaluation of the
    results of aortic valve replacement in symptomatic patients.
    Circulation, 61:471-483 (1980).
3. J. Greves, S. H. Rahimtoola, J. H. McAnulty, H. de Mots, D. G.
    Clark, B. Greenberg, and A. Starr, Preoperative criteria
    predictive of late survival following valve replacement for
    severe aortic regurgitation, Am.Heart J., 101:300-308 (1981)
4. A. G. Kumpuris, M. A. Quinones, A. D. Waggoner, D. J. Kanon,
    J. G. Nelson, and R. R. Miller, Importance of preoperative
    hypertrophy, wall stress and end-systolic dimension as echo-
    cardiographic predictors of normalization of left ventricula
    dilatation after valve replacement in chronic aortic insuf-
    ficiency, Am.J.Cardiol., 49:1091-1100 (1982).
5. C. A. Peter and R. H. Jones, Cardiac response to exercise in
    patients with chronic aortic regurgitation, Am.Heart J.,
    104:85-91 (1982).
6. H. J. Levine and W. H. Gaasch, Ratio of regurgitant volume to
    end-diastolic volume: a major determinant of ventricular
    response to surgical correction of chronic volume overload,
    Am.J.Cardiol., 52:406-410 (1983).
7. S. H. Rahimtoola, Valve replacement should not be performed in
    all asymptomatic patients with severe aortic incompetence,
    J.Thorac.Cardiovasc.Surg., 79:163-172 (1980).
8. R. O. Bonow, D. R. Rosing, K. M. Kent, and S. E. Epstein, Timir
    of operation for chronic regurgitation, Am.J.Cardiol.,
    50:325-336 (1982).
9. H. J. Smith, J. M. Neutze, A. H. G. Roche, T. M. Agnew, B. G.
    Barratt-Boyes, The natural history of rheumatic aortic re-
    gurgitation and the indications for surgery, Brit.Heart J.,
    38:147-154 (1976).
10. R. O. Bonow, D. R. Rosing, C. L. McIntosh, M. Jones, B. J.
    Maron, K. K. G. Lan, E. Lakatos, S. L. Bacharach, M. V.
    Green, and S. E. Epstein, The natural history of asymptomat:
    patients with aortic regurgitation and normal left ventric-
    ular function, Circulation, 68:509-517 (1983).

11. A. Selzer, Cardiac valve replacement: an unanswered question, Am.J.Cardiol., 37:322-324 (1976).
12. J. W. Hirshfeld, Jr., Valve replacement for chronic severe aortic regurgitation: when should it be done? Intern.J. Cardiol., 3:243-247 (1983).
13. H. P. Krayenbuehl, M. Turina, O. M. Hess, M. Rothlin, and A. Senning. Pre- and postoperative left ventricular contractile function in patients with aortic valve disease, Brit.Heart J., 41:204-213 (1979).
14. D. A. Samuels, G. D. Curfman, A. L. Friedlich, M. J. Buckley, and W. G. Austen, Valve replacement for aortic regurgitation: long-term follow-up with factors influencing the results, Circulation, 60:647-654 (1979).
15. R. A. O'Rourke and M. H. Crawford (Editorial), Timing of valve replacement in patients with chronic aortic regurgitation, Circulation, 61:493-495 (1980).
16. R. Forman, B. G. Firth, and M. S. Barnard, Prognostic significance of preoperative left ventricular ejection fraction and valve lesion in patients with aortic valve replacement, Am.J.Cardiol., 45:1120-1125 (1980).
17. L. I. Bonchek, Current status of cardiac valve replacement: selection of a prosthesis and indications for operation, Am.Heart J., 101:96-106 (1981).
18. W. Paulsen, D. R. Boughner, J. Persaud, and L. Devries, Aortic regurgitation. Detection of left ventricular dysfunction by exercise echocardiography, Br.Heart J., 46:380-388 (1981).
19. W. L. Henry, R. O. Bonow, D. R. Rosing, and S. E. Epstein, Observations on the optimum time for operative intervention for aortic regurgitation. II. Serial echocardiographic evaluation of asymptomatic patients, Circulation, 61:484-492 (1980).
20. J. Ross, Jr., Left ventricular function and the timing of surgical treatment in valvular heart disease, Ann.Int.Med., 94:498-504 (1981).
21. K. M. Borow, L. H. Green, W. Grossman, and E. Braunwald, Left ventricular end-systolic stress-shortening and stress-length relations in humans. Normal values and sensitivity to inotropic state, Am.J.Cardiol., 50:1301-1308 (1982).
22. N. Goldschlager, J. Pfeiffer, K. Cohn, R. Popper, and A. Selzer, The natural history of aortic regurgitation. A clinic and hemodynamic study, Am.J.Med., 54:577-588 (1973).
23. F. Schwarz, W. Flameng, F. Langebartels, M. Sesto, P. Walter, and M. Schlepper, Impaired left ventricular function in chronic aortic valve disease: survival and function after replacement by Björk-Shiley prosthesis, Circulation, 60:48-58 (1979).
24. P. Fioretti, J. Roelandt, R. J. Bos, R. S. Meltzer, D. van Hoogenhuijze, P. W. Serruys, J. Nauta, and P. G. Hugenholtz, Echocardiography in chronic aortic insufficiency. Is valve replacement too late when left ventricular end-systolic

dimension reaches 55 mm? <u>Circulation</u>, 67:216-221 (1983).

25. G. J. Dehmer, B. G. Firth, L. D. Hillis, J. R. Corbett, S. E. Lewis, R. W. Parkey, J. T. Willerson, Alterations in left ventricular volumes and ejection fraction at rest and during exercise in patients with aortic regurgitation, <u>Am.J.Cardiol.</u>, 48:17-27 (1981).

26. I. Mirsky, C. Henschke, O. M. Hess, and H. P. Krayenbuehl, Prediction of postoperative performance in aortic valve disease, <u>Am.J.Cardiol.</u>, 48:295-303 (1981).

27. G. Schuler, K. Von Olshausen, F. Schwarz, H. Mehmel, M. Hofmann, H.-J. Hermann, D. Lange, and W. Kübler, Noninvasive assessment of myocardial contractility in asymptomatic patients with severe aortic regurgitation and normal left ventricular ejection fraction at rest, <u>Am.J.Cardiol.</u>, 50:45-52 (1982).

28. R. L. Huxley, F. A. Gaffney, J. R. Corbett, B. G. Firth, R. Peshock, P. Nicod, J. S. Rellas, G. Curry, S. E. Lewis, and J. T. Willerson, Early detection of left ventricular dysfunction in chronic aortic regurgitation as assessed by contrast angiography, echocardiography, and rest and exercise scintigraphy, <u>Am.J.Cardiol.</u>, 51:1542-1550 (1983).

PRE-ARRANGED INTERVENTION ON THE PAPER

"HEMODYNAMIC EFFECTS OF CARDIAC PACING"

A. Galassi and R. Russo

Cardiology Department
Garibaldi Hospital
Catania, Italy

The study of hemodynamic effects of a pacemaker implantation, as Prof. A. Raineri has shown, is the fundamental problem that characterizes modern cardiac pacing.

Cardiac pacing presently can't only be intended as a rhythmic electric heart stimulation, but it represents an essential thera-peutic device for the vital function of this organ, that is the pumping function with involvement of all anatomo-functional compon-ents of this system and adapting pacing rate to the changing meta-bolic conditions. In this way we can say that we have obtained nowadays nearly a personalized cardiostimulation and we aren't far from a physiological one.

For this reason cardiologists must investigate those indexes that can better reveal the cardiac performance and reach the physio-logical stimulation, considering above all the kinetic importance of pacing rate and the maintenance of temporal relationship between the atrial and ventricular contractions, even if for the latter, opinions are often conflicting.

More than to privilege a certain modality or a pure index - even if on the basis of valid physiopathology conditions - we agree on the need to look to a latent poor cardiac performance, prior to implant. This because either clinic experience and recent physiopathologic studies have shown that heart failure can appear also several times after an implantation, with an evident relationship with a pacemaker implantation.

Noninvasive diagnostic methods are to be preferred for their possibility of continuous monitoring of cardiac performance,

when their reliability is proved at least the same as invasive
techniques.

The pursuit of optimal pacing rate with a VVI pacemaker technique could be substituted by A-V sequential pacing mode when the
sinoatrial node actively is good enough or by a DDD pacing when the
sinoatrial node activity is poor, having previously chosen the
optimal pacing rate by a temporary A-V sequential pacing.

Unfortunately the ratio risk/benefit and cost/benefit for DDD
mode and for A-V sequential mode is higher than for the VVI pacing
mode, and for this reason at present there is higher percentage of
ventricular inhibited rate-adjustable pacemakers implanted. Highly
sophisticated technology can induce more risks of damage; programing
methods are more complex and their batteries have a minor life; it is
possible the arising of arrhythmias pacemaker-dependent; at last
particular conditions don't benefit of A-V sequential pacing.

We hope that the cost of these prostheses will diminish in the
next future, because if their cost is justified for the single
patients, this is not true for the community.

REFERENCES

1. C. Alicandri, F. M. Fouad, R. C. Tarazi, L. Castle, and V.
    Morant, Three cases of hypotension and syncope with pacing:
    possible role of atrial reflexes, Am.J.Cardiol., 137:42
    (1978).
2. D. A. Bognolo, R. R. Vijayanagar, and P. F. Epstein, Atrial and
    atrioventricular sequential pacing. Rationale and clinical
    experience. J.Florida Med.Assoc., 1028:66 (1979).
3. L. Fananapariz, D. H. Bennett, and P. Monks, Atrial synchronized
    ventricular pacing: contribution of the chronotropic response
    to improved exercise performance, PACE, 601:6(part.I) (1983).
4. H. D. Funke, 18 mois d'experience clinique avec un stilumateur
    implantable séquential optimisé (OSS). VI Symposium Mondial
    sur le Stimolation Cardiaque, Montreal 2-5 October (1979).
5. I. Kruse, K. Arnman, T. B. Conradson, and L Rydén, A comparison
    of the acute and long-term hemodynamic effects of ventricular
    inhibited and atrial synchronous ventricular inhibited
    pacing, Circulation, 846:65(5) (1982).
6. R. M. Luceri, A. V. Ramirez, A. Castellanos, L. Zaman, R. J.
    Thurer, and R. J. Myerburg, Ventricular tachycardia produced
    by a normally functioning AV sequential demand (DVI) pace-
    maker with "committed" ventricular stimulation, J.Am.
    Coll.Cardiol., 1177:1(4) (1983).
7. G. P. Marinoni, Comunicazione personale.
8. K. A. Narahara and M. L. Blettel, Effect of rate on left
    ventricular volumes and ejection fraction during chronic

ventricular pacing, Circulation, 323:67(2) (1983).

 9. A. F. Rickards and R. M. Donaldson, Rate responsive pacing,
     Clin.Prog.Pacing and Electrophysiol., 12:1(1) (1983).

10. J. W. Rubin, M. J. Frank, J. P. Boineau, and R. G. Ellison,
     Current physiologic pacemakers: a serious problem with a new
     device, Am.J.Cardiol., 88:52 (1983).

11. P. Samet, W. H. Bernstein, D. A. Nathan, and A. Lopez, Atrial
     contribution to cardiac ouput in complete heart block, Am.J.
     Cardiol., 1:16 (1965).

EFFECTS OF ADRENERGIC STIMULATION ON SYSTOLIC TIME INTERVALS

IN PATIENTS WITH ARTIFICIAL PACEMAKER

P. P. Campa

Cattedra di Cardiologia
University, L'Aquila
Italy

The aim of this research was to contribute for a better knowledge of the physiological determinants of the temporal dynamics of
the left ventricular chamber contraction. We have studied the effects of adrenergic stimulation on systolic time intervals in
patients with artificial pacemaker (PM). These subjects are of
considerable pathophysiological interest representing an useful
experimental model; compared with subjects with spontaneous electrical activity, patients with PM present: 1) a fixed heart rate or
only slightly modified heart rate; 2) frequent occurrence, within
short time intervals, of systoles with different electrogenesis; 3)
abnormal intraventricular conduction. In previous papers[1-5] we
have employed the study of systolic time intervals in the
pathophysiological investigation of patients with PM; the present
report deals with results of a study on the effect of microdoses of
epinephrine on systolic time intervals in patients with demand
pacemaker presenting, at the time of the study, systoles of different
electrogenesis. The microdoses of epinephrine employed may be
considered "metabolic", not unlike the amounts secreted in
physiological conditions and without hypertensive effect upon the
systemic circulation[6-10].

MATERIALS AND METHODS

The study was performed on 11 patients with "demand" PM (VVI),
selected on account of the presence, at the time of the study, of at
least two of the following types of systoles: 1) spontaneous sinusal
or ventricular systoles (SS); 2) PM systoles preceded within 0,2 sec
by P wave (P-PMK); 3) PM systoles not preceded within 0,2 sec by P
wave (PMK). We consider separately P-PMK and PMK systoles depending
upon the presence or not or atrial activity in normal time on account

of the relationship existing between atrial activity and ventricular performance[1]. Patients included in the study had not received any form of treatment for at least 3 days prior to the investigation; 2 of the patients were males and 9 females; mean age of 68 years (53-86). The reason for the PM implantation was in 6 a complete heart block, in 4 a sick sinus syndrome and in 1 a bradiarrhythmia.

The investigation was carried out with the patient in the clino-static position, in warm comfortable surroundings, and after at least 30 minutes of complete relaxation. A 6-channel Battaglia-Rangoni polygraph with photographic recording and speed of 100 mm/sec was employed. The study included EKG tracing, two phonocardiographic tracings with 70 and 140 Hz filters and a carotid pulse tracing. The recordings was made during mid expiratory apnoea, and repeated every 15 minutes, the first two times during infusion of a saline solution, the following i.v. infusion of 0.04, 0.08 and 0.12 µg/Kg/min epinephrine and finally during infusion of saline solution 15 and 30 min after stopping epinephrine infusion. The investigation lasted for a total of 105 min. The following indices were calculated at every record: 1. Electromechanical interval (EMI); 2. Total electro-mechanical systole (TEMS); 3. Left ventricular ejection time (LVET); 4. Left ventricular ejection time corrected for heart according to Meiner's nomogram (LVETc); 5. Pre-ejection period (PEP); 6. PEP/LVET ratio; 7. Isovolumetric contraction (IVC), calculated as PEP - EMI. Student's "t" test for paired data was used in the statistical analysis.

RESULTS

Nevertheless the differences in the basal values between the spontaneous and PM systoles, as already well known[1,2], the behavior during epinephrine infusion of EMI, TEMS and LVET was similar. As can be seen in Figure 1, correcting the LVET for heart rate, the behavior of this parameter becomes less uniform: for spontaneous systoles a significant increase is observed with the 0.04 µg/Kg/min infusion; for the PMK and P-PMK systoles a significant increase is observed only after infusion of 0.08 µg/Kg/min. After stopping epinephrine infusion, a significant difference with respect to basal values persists for 15 min for spontaneous and for P-PMK systoles, whereas the values are quickly close to those basals for PMK systoles.

Also the behavior of PEP during epinephrine infusion (Figure 2) and after stopping is different for SS, P-PMK and PMK systoles, a significant difference being observed with respect to basal values earlier and more persistent for SS; while significant differences for P-PMK and PMK systoles are observed only with 0.08 µg/Kg/min epinephrine infusion, and after stopping a difference remains for only 15 min for P-PMK systoles, with a rapid return to basal levels once

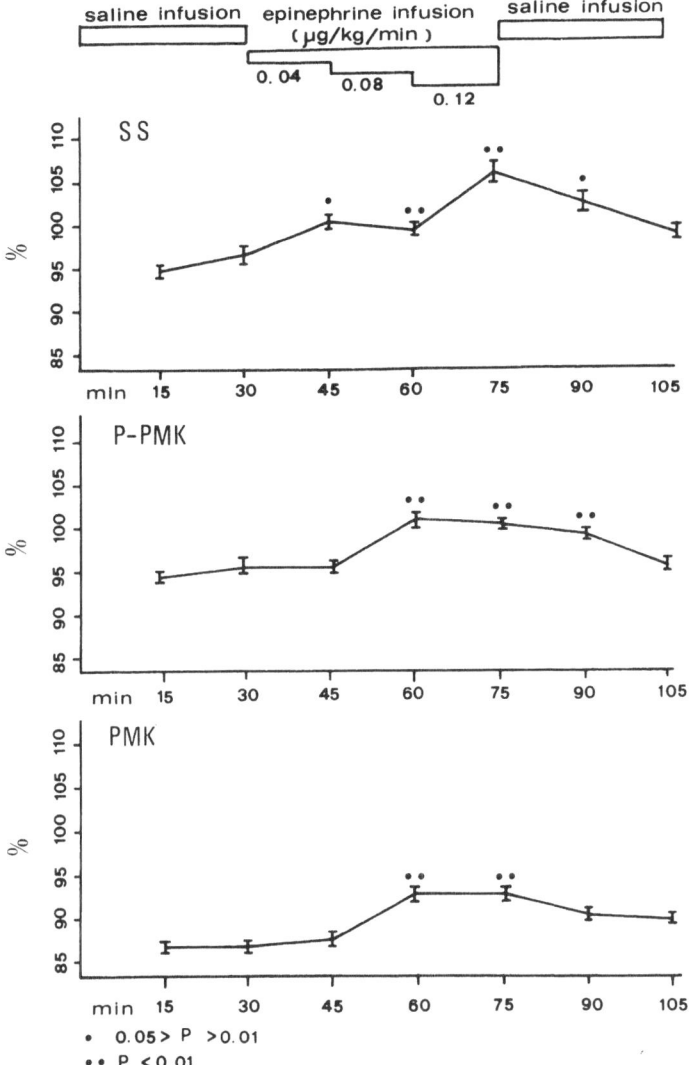

Fig. 1. Behavior of LVETc (rate corrected left ventricular ejection
time) before, during and after epinephrine infusion.

15 min after stopping epinephrine for the PMK systoles. Epinphrine
appears to induce maximum variations of decreasing entity from SS to
P-PMK and to PMK systoles (Figure 3).

PEP/LVET ratio (Figure 4). A significant decrease during epinephrine infusion in respect to basal values was observed for SS,
P-PMK and PMK systoles with a return to basal levels after stopping
infusion more rapid for P-PMK and PMK than for SS.

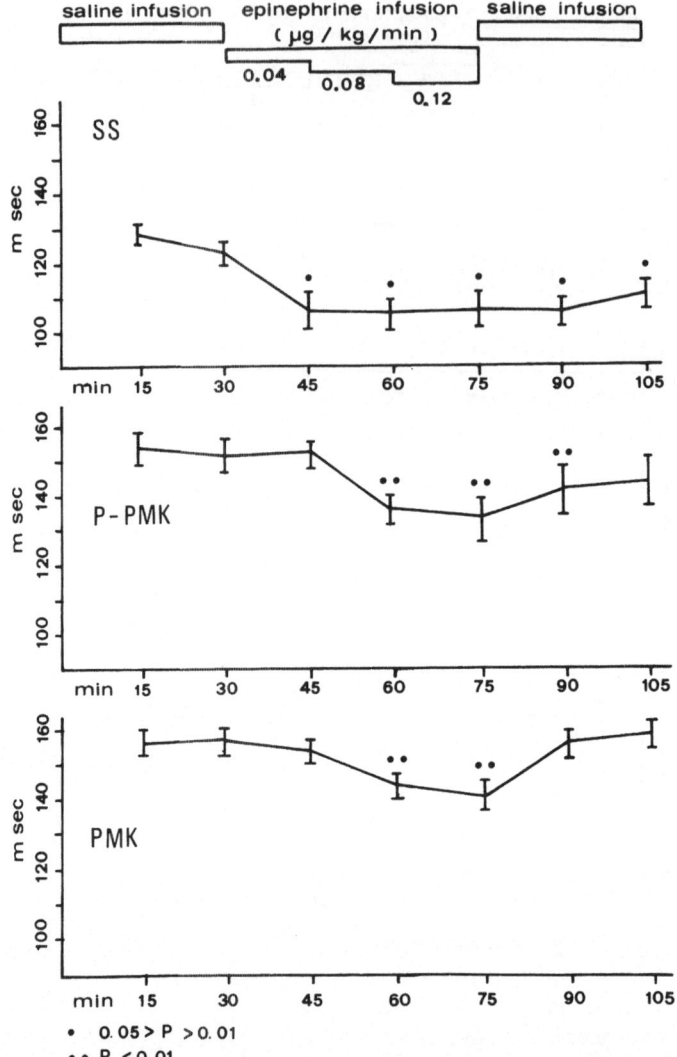

Fig. 2. Behavior of PEP (pre-ejection period) before, during and
         after epinephrine infusion.

Isovolumetric contraction (IVC, Figure 5). The various types of
systoles show a similar behavior, but with different degree of significance. A significant reduction was observed for SS even with
0.04 µg/Kg/min epinephrine infusion, which persist for at least 15
min after stopping epinephrine infusion; for the P-PMK and PMK
systoles the decreases in IVC values are significant only with the
higher doses (0.08 and 0.12 µg/Kg/min) with a rapid return to basal
values after withdrawal of epinphrine, main for the PMK systoles.

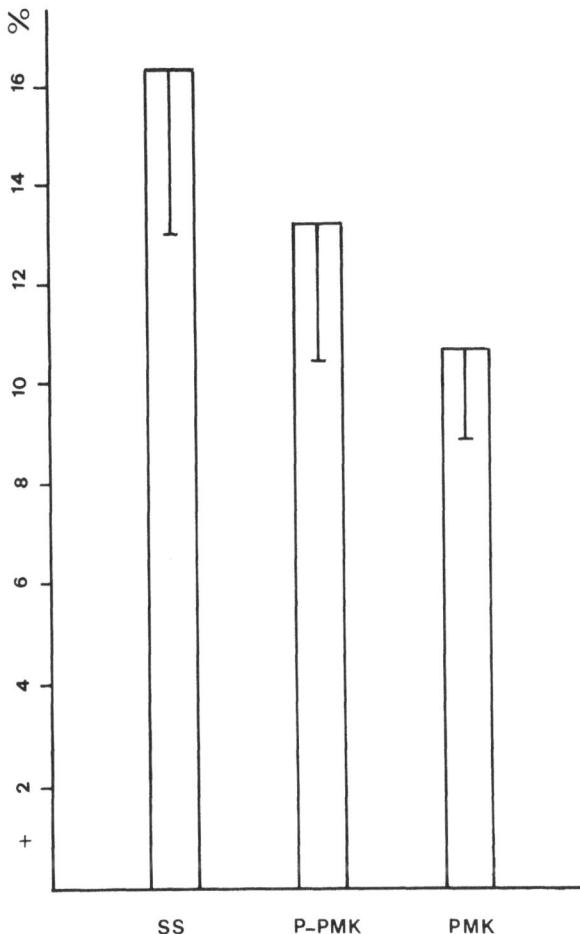

Fig. 3. Epinephrine induced maximum variations in PEP (pre-ejection
         period) with respect to basal values.

DISCUSSION

    Epinephrine infusion induced modifications in the systolic time
intervals, both in spontaneous and PM systoles. Nevertheless differences are present for the various types of systoles, namely: 1)
significant variations in regard to basal levels are visible for SS
even with 0.04 µg/Kg/min epinephine, but are visible only at higher
doses for PM systoles; 2) the entity of variations observed was
usually greater for spontaneous than for PM systoles; 3) after with-
draw of epinephrine infusion the systolic time intervals remain
significantly different from basal values for 15 or 30 min for spontaneous systoles while PM systoles show a rapid return to basal

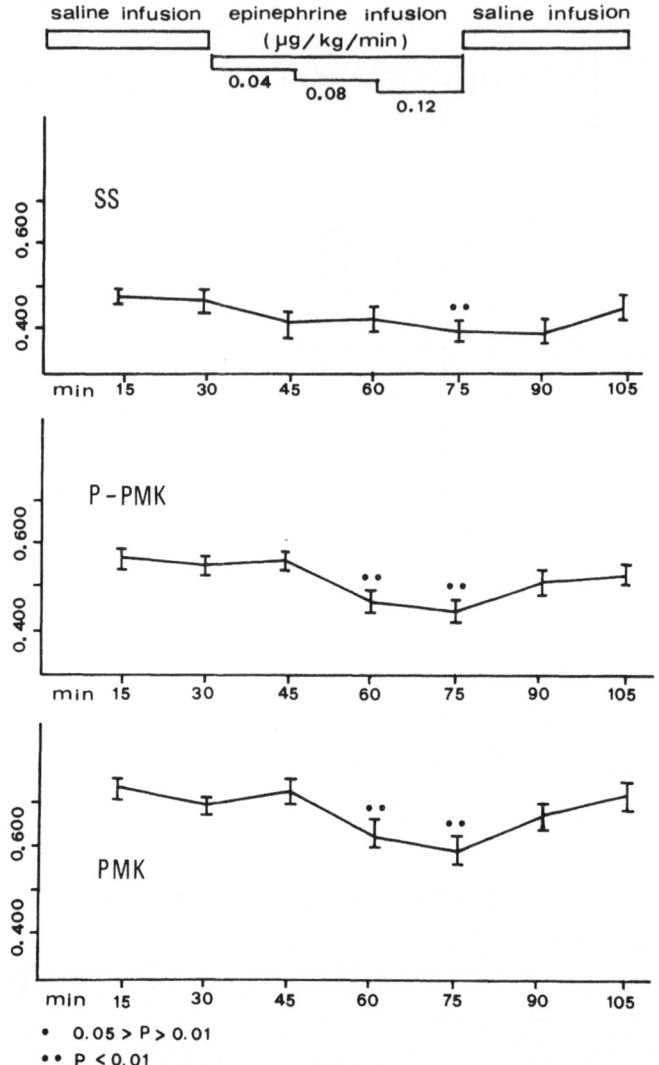

Fig. 4.   Behavior of PEP/LVET ratio before, during and after
          epinephrine infusion.

levels; 4) the variations in the systolic time interval observed for
P-PMK systoles were smaller than for spontaneous systoles, but
greater than for PM systoles not preceded by P wave (PMK systoles).

     Therefore PM systoles like spontaneous systoles are sensitive to
the effect of adrenergic agents at physiological concentrations.
Nevertheless, with respect to spontaneous systoles, PM systoles
present: a) a less marked response; b) a higher threshold limit of

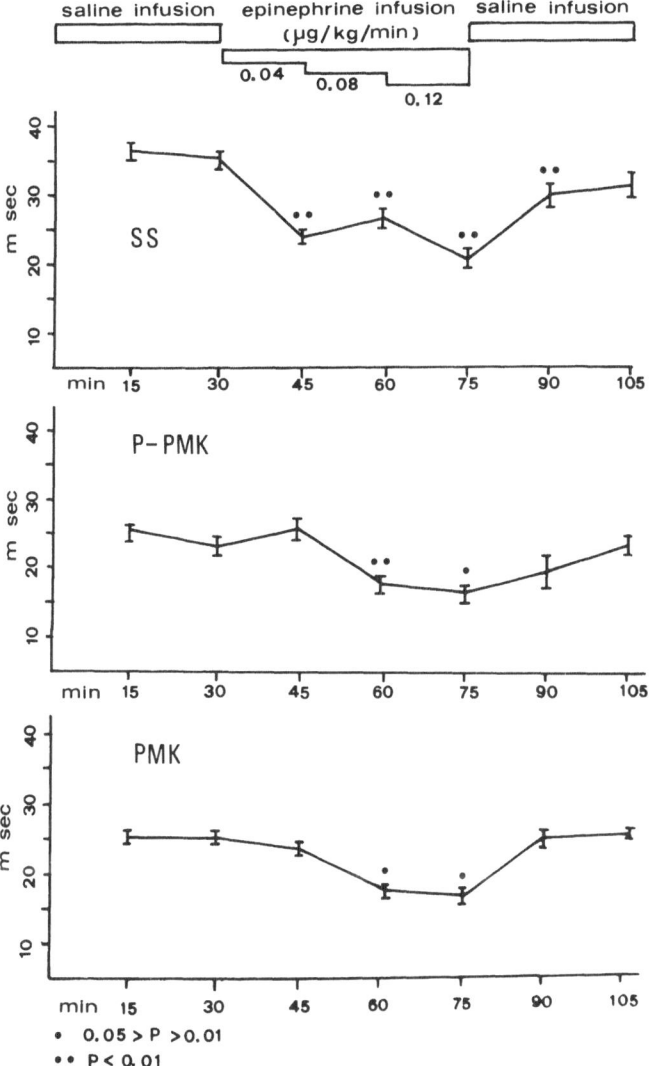

Fig. 5. Behavior of IVC (isovolumetric contraction) before, during
and after epinephrine infusion.

adrenergic stimulation; c) a duration of the effect strictly limited
to the period of epinephrine infusion. The differences founded in
behavior between spontaneous systoles and PM systoles preceded within
0.2 sec by P wave demonstrate that not only the lack of atrial contraction but also the modality with which intraventricular conduction
occurs is a determinant of a lesser degree of the sensitivity to the
adrenergic of agents of PM systoles.

REFERENCES

1. P. P. Campa, P. Tomassini, G. Speca, P. Di Sabatino, and G. De
   Curtis, Indagini policardiografiche in portatori di
   segnapassi artificiale: Effetto dell'attività atriale sulla
   contrazione ventricolare, Boll.Soc.It.Cardiol., 22:1632
   (1977).
2. P. P. Campa, G. De Curtis, G. Speca, G. Marcellini, and P.
   Tomassini, Effetto della -metildigossina su alcuni indici
   policardiografici in portatori di segnapassi artificiale,
   Boll.Soc.It.Cardiol., 22:1646 (1977).
3. P. P. Campa, G. Speca, G. De Curtis, and P. Cingoli, Indagini
   sul controllo neuroumorale dell'attività ventricolare da
   pacemaker, Giorn.It.Cardiol., 8(Suppl.3):156 (1978).
4. P. P. Campa, Adrenergic regulation of ventricular myocardial
   activity in patients with artificial pacemaker, Bol.Soc.
   Portuguesa Cardiol., 17(Suppl.1):35 (1979).
5. G. Speca, G. Marcellini, P. Pecce, F. Prosperi, C. Napolitano,
   G. De Curtis, and P. P. Campa, Riflessi atriali e
   retroconduzione ventricolo-atriale nella malattia del nodo
   del seno durante elettrostimolazione, Clin.Ter.Cardiovasc.,
   3:211 (1982).
6. L. Lundholm, The mechanism of the vasodilatator effect of
   adrenaline. I. Effect on skeletal muscle vessel, Acta
   Physiol.Scand., 39(Supp.1):133 (1956).
7. P. P. Campa, G. Bellisario, F. Pugliese, A. Santucci, and G.
   Filocamo Jr., Ricerche sull' effetto dell'infusione di
   adrenalina sull'attività fosforilasica delle emazie e sulla
   lattacidemia nelle pecore, Boll.Soc.It.Biol.Sper., 42:437
   (1966).
8. P. P. Campa, A. Santucci, G. Bellisario, S. Rizzo, and G.
   Calcagnini, Sugli effetti vascolari e metabolici
   dell'infusione di microdosi di adrenelina in soggetti
   normali, Progr.Med., 22:351 (1966).
9. P. P. Campa, A. Santucci, G. Bellisario, S. Rizzo, and A.
   Parrinello, Ricerche sui rapporti tra attività metabolica ed
   effetto vasoattivo di infusioni di microdosi di adrenalina in
   soggetti con splenomegalia congestizia, Progr.Med., 22:396
   (1966).
10. L. Ceremuzynski, K. Herbaczynska-Cedro, B. Broniszewska Ardelt,
    J. Nauman, A. Nauman, B. Wozniewikz, and J. Lawecki, Evidence
    for the detrimental effect of adrenaline infused to healthy
    dogs in doses imitating spontaneous secretion after coronary
    occlusion, Cardiovasc.Res., 12:179 (1978).

CHAPTER II:

DISCUSSION

DENOLIN

We have to try to connect the problem of physical capacity with left ventricular function, which is the topic of our course. We should use in principle dynamic exercise and only in a few cases we could use static exercise; as the level of exercise we should use a maximal possible level of exercise. The most important variable is the load achieved of the maximum oxygen consumption achieved, but I would like to remind you that oxygen consumption is the expression of cardiac output, that is the stroke volume and the heart rate and the artero-venous oxygen difference. If we need more details of the cardiac function we have to use other methods like cardiac output, ejection fraction, left diastolic pressure and so on. But to be practical, to remain at the clinical level, we can know exactly the physical capacity of a patient and we can prescribe a program of training only on the base of the load achieved or the maximal oxygen consumption achieved.

KELLERMANN

I should like to make a comment: in angina pectoris patients, an excellent prognostic parameters, the level of the angina rate, has been shown by my coworker, Dr Hayad. I think that, in patients without angina, it seems to be of much more importance the blood pressure response to exercise. Forty five per cent of the patients who died in our group of 211 patients the blood pressure response in the last test was flat or decreased. Most of the patients with flat blood pressures or decreased blood pressures had no angina.

OPASICH

Often patients with myocardial infarction and reduced left ventricle function have a low work capacity. How can we decide on training? There is a training at 85% of this work capacity? Is it still efficient?

KELLERMANN

As I have shown in the slides with a training of 16 weeks, we can start with people who have a very low capacity if they have normal exercise testing responses. Of course exercise capacity in stable angina is low and there is a normal physiological response to exercise, this is no contraindication.

ROTHLIN

In some patients with coronary artery disease and very bad left ventricular function it is amazing how good their exercise performance is. And I wonder whether Dr Niederberger has correlated ejection fraction or other parameters of bad ventricular function in his cardiomyopathy patients, with the extent of exercise these parameters could perform.

NIEDERBERGER

We did not do any significant measurement of ejection fraction. At rest the ejection fraction was more reduced in the more impaired patients, but there is a weak correlation. So, we have some patients, especially in young patients, with ejection fraction below 20%. They relatively could exercise in spite of low capacity.

WEISSLER

What I begin to derive from the two studies, especially from Dr Kellermann's longer experience in the field, is that with rehabilitation conditioning one adapts the heart better to its circulation without necessarily improving heart functions or even long term prognosis, but that makes the patients much more healthy during the period they live.

KELLERMANN

Dr Weissler, we know from literature that there are two possibilities to assess the results of certain procedures. One may be induction, the second may be deduction. So allow me to use deduction here as a reply to your question. Our mortality is 0.8% per year which means the same mortality we have in a normal healthy population in our country. Mean age group at admission in all the 3 groups was

52.3 to 53.8 years. So the explanation would be that despite the
fact that we are not improving function and that there is a question
about the central mechanism of exercise response on a long term basis
in these patients, the fact is they live longer. This may be biased
by environment, by psychology, by placebo, by many other things, but
we are not using exercise as a single parameter, we have comprehensive care programs including beta blocking agents, if indicated,
calcium antagonists, if indicated, and other drugs, which means there
is no single ad hoc surgery, there is no single procedure available
today. Exercise per se could be bad medicine.

WEISSLER

       I would like to direct this to Dr Niederberger. I am somewhat
concerned about the conclusion that what pressure fall is not related
to deterioration in left ventricular function. I'm interested in
remarking the measurement of cardiac output during a steady state, at
the time that the blood pressure fell, because if you would not,
there is a good chance that your cardiac output determination would
not reflect true stroke volume, because you are taking a heart rate
which is averaged over some period of time, and taking a cardiac
output determination that is determined over quite a few beats, and
so your calculations of stroke volume may not reflect what the true
deterioration of the ventricular function might be, that's one point.

       Secondly in my experience in doing a large number of pacing
stress test, patients that are in high risk subsets or coronary heart
diseases, are the ones that demonstrate very significant deterioration of blood ventricular function after a post-pacing period, after
angina has been induced.

       So I am worried about that conclusion.

NEIDERBERGER

       I did not say that left ventricular function is not involved in
the fall of blood pressure, certainly it is, but it is not the only
factor according to these data, but we have also a contribution of
peripheral vasculature. Concerning the measurement of stroke volume,
we did it by thermodilution, it was not during steady state, but it
was in the second minute after a work load that had been increased by
25 watts. So we are aware it is not a steady state, but we took
always 3 measurement by thermodilution and heart rate was recorded at
the time of injection.

I said that we had a significant difference in stroke volume response between the 3 subgroups: those with a systolic blood pressure fall had a fall in stroke volume, but at the same time they had - in our opinion - an amazing further fall in peripheral vascular resistance.

We should have expected as levelling of peripheral resistance, which was not the case, though it is a derived parameter, of course.

And let me add that I am also convinced that systolic blood pressure fall indicates severely impaired people, severe coronary disease. For instance, a similar thing has been said before of those patients who had coronary angiography and had this abnormal blood pressure response, 80% were subsequently submitted to bypass surgery.

As a further effect of training, one of my coworkers many years ago demonstrated that in coronary patients the training can improve the cardiac output and the stroke volume in patients with relatively good physical conditions, at the beginning. In those who have a poor condition, the benefit is only related to an increase in the arterovenous oxygen difference.

ASSENNATO

Dr Forrester, at the current level of technology, what is the medium anticipated cost in dollars of your new digital angiography equipment?

FORRESTER

The projected cost for the imaging device in the United States is right now in a range of 250,000 not as expensive as nuclear magnetic resonance, but not very cheap, either. There is an argument that this could save money by moving coronary angiography from the in-patient to the out-patient area. But it has not yet been demonstrated that this can be done.

ASSENNATO

Can you remove artifacts so well that you can analyze the left ventricle during exercise?

FORRESTER

The answer is "yes, it has been done". It has been done both with exercise and with pacing. However my opinion is that the quality of the nuclear technetium image is sufficient for that particular application, it is likely that technetium ventricularography will continue to be used even though it is possible to do it as you suggest.

PETERSON

When you showed those wash out rates, if I saw them correctly, you probably were looking at an area that had density in three dimensions, not just 2. And you might have superimposition of the posterior wall with the anterior; are you sure that this is reflecting regional wash out?

FORRESTER

We do not as yet have sufficient experience to know the answer to that question. But theoretically it seems, because the anterior wall and the posterior wall can be superimposed, that there will be error in the calculation wash out rates by Videodensitometry.

RAZZOLINI

Dr Zardini, I'd like only to ask an information about the data of your experience on mitral regurgitation, since in Padua we had some light different results as for left ventricular mass and the proportion of the mass-volume relationship, which is always in our experience lower than 1, i.e. about 0.8.

What was the kind of mitral regurgitation in your cases? Because I think that the difference is right here.

ZARDINI

I think that it is extremely useful this specification, that is there isn't always a compensation between mass and end-volume - I mean - in mitral regurgitation. It is easier to find it in aortic insufficiency than in mitral regurgitation. Our cases had post-reumatic mitral regurgitation in which the angiographic score assessment gave a high degree of regurgitation and in which the calculation of thickness in relation to dilatation of ventricular chamber was remarkably lower than those observed in aortic insufficiency.

DENOLIN

I would like to understand better what Dr Mattioli and coworkers are doing. I see there are a lot of indexes. What are the limits and when you decide that you are moving from normality to abnormality? We saw a large correlation, but I'd like to know what you are doing with the specific patient coming to your office, and what are the results of your treatment in this group of cases.

MATTIOLI

We have dedicated the first part of our paper to the problem of overlap of the indexes because it is very important.

The choice of an incorrect index is reflected in the diagnosis but above all on the prognostic evaluation. You have seen that ejection indexes remarkably overlap in normal and abnormal subjects. If I had to suggest a reliable index, that is an index with a auss bell like dispersion and a minimal area of overlap, I would choose wall stress and even better, end-systolic and end-diastolic diameters which differ greatly in statistical significance in normal subjects in respect to abnormal subjects affected by aortic insufficiency.

WEISSLER

The vary indexes that we measure reflect either symptoms, class 1, 2, 3, or 4 or reflect an immediate post operative result but not a long term effect on prognosis. And I was wondering whether you have any data or whether any preoperative indeces are sensitive indicators of which patients have longer longevity post operatively and which indeces indicate a poor post operative longevity result.

MATTIOLI

The assessment of these indexes to evaluate the survival and not the function needs a very long follow-up. End-diastolic diameter reduction is linked to the clinical outcome. Patients without a significant reduction of this parameter show particularly bad clinical results.

PETERSON

I would like to make a comment on Dr Geraci's intervention. One of the major problems that we have in the patient with volume overload is that you can have very significant hemodynamic embarrassment at the same time that the patient is asymptomatic. And also the question is how long can you wait during that period of embarrassment before you replace the valve? As Dr Bonow recently referred in Munich it is their opinion at the moment that the time span in chronic aortic regurgitation is relatively short before the patient moves from the asymptomatic state to the symptomatic state once he begins to show compromise of his exercise tolerance and once he begins to show a quite significant dilation of the ventricle.

If you look at all the old literature on mitral regurgitation it suggests that even more so than in aortic regurgitation these patients can go on for many years with significant regurgitation and significant regurgitation fraction, a very large dilated ventricle and can live reasonably normal lives without valve replacement. So I think it is an even more critical question in mitral regurgitation than it is in aortic regurgitation.

GERACI

Attention is directed on the end systolic dimension of the left ventricle as a prognostic and predictive index of ventricular function after surgery. This index may be deceptive. In fact end-systole dimension depends on afterload and contractility. Now we must verify our capacity to evaluate how much end-systolic dimension is due to either the loss of contractility which is not reversible after surgery or to the increase of afterload that surgery can modify.

PETERSON

I do think that in volume overload we have to be very careful about using end-systolic dimension as an index of depression of myocardial contractility and I'll show some animal data in the instrumental dog tomorrow where we have induced mitral regurgitation and looked what happens to the progression of end-systolic relationship, end-systole stress dimension relationship, end-systolic dimensions. All the different indeces that we can look at, and there is no doubt that what the imposition of an overload alone causes, is an immediate shift to the right of the curve, which does not necessarily indicate depression of myocardial contractility. So I think this is a very valid point and I don't think we know yet what the threshold is at which a certain end-systolic dimension or volume indicates depression of contractility.

LEACHMAN

I would like to ask if there is anyone here that is actually trying to decide when to operate on these patients on the basis of indexes like mean wall stress or shortening fraction. It seems to me that these parameters have not really helped me very much in making decisions for patients.

GAASCH

We can use a number of tests to predict who is going to do well following aortic valve replacement for chronic aortic regurgitation. We have particularly used the M-mode echocardiogram coupled with the two dimensional study in some patients, but I think nuclear studies and exercise studies might be helpful too.

When I say "do well" I mean develop normal left ventricular end-diastolic dimension following surgery. That means, the left ventricular end-diastolic dimension declines from ranges of 7, 8 centimeters to less than 5 and ½ centimeters. That reduction in dimension in most patients, but not all, is associated with a regression of hypertrophy, a marked reduction in the need for medications a reduction in symptoms but it is almost impossible to show it always results in a reduction of death or in an increase of survival.

The question is whether you can use that sort of information to decide when to replace an aortic valve in a patient, that's another question! And I don't know if we can do that. I think we can identify who is going to do pretty well, and who is going to do poorly after the operation, fairly well.

LEACHMAN

Would you currently use the echocardiogram to follow the patient and decide to do an operation in an asymptomatic state on the basis of the beginning of a increase in end-systolic volume?

GAASCH

Now we come down to the important question.

If we look at the extremes, there is no question what we would do! The problem is the borderline patient, the patient who falls on the border between all these indeces we keep talking about.

That patient I don't think if he is asymptomatic should be operated on. I have the feeling that the patients are not going to develop irreversible LV dysfunction if they have borderline measurements and we are following them twice a year. That is not the man on whom we would operate now, we may be wrong. Finally I would like to make one suggestion for students and lecturers here. If you should ask what would be the best way physiologically, theoretically to follow ventricular functions in patients with abnormal valves. I would say the best way would be to calculate mid wall stress, not meridional stress because that's 90° to the shortening, not circumferential stress because we don't know the long axis. If we have an echocardiogram, calculate mid wall stress using a simple cylinder model, plot that against mid wall shortening and following that relation over time.

KELLERMANN

I have two or maybe three very short questions to Dr Weissler. First, I wonder whether he can give us some information about timing, when the test was done, after myocardial infarction, moreover whether there was any classification done as to the time past myocardial infarction and the testing, if the testing was repeated and if so, how was the reproducibility.

Question number two: in some slides he showed that there was a very low rate of abnormality in non infarcted patients with 1, 2 and 3 vessel occlusions. In 3 vessel occlusions you have 10% abnormality, with infarction you had 72%. But then in 2 vessel diseases, you had 31%, while in 1 vessel disease you had 43%. I wonder why we have lower rate of abnormality in 2 vessel diseases than in 1 vessel disease.

And then, my third remark - I believe - well, I shall remember probably later when I have recieved the replies to these questions.

WEISSLER

The studies are made 2 or 16 months after infarction, with an average of 12 month. So these were patients who had recovered from infarction and what we have found is a marked reproducibility in the data.

The point that the first slide showed, however, was that among individuals in whom one could not detect left ventricular dysfunction by clinical methodology, either by symptoms, signs or chest X-rays, the systolic intervals demonstrated a remarkable incidence of abnormality, of abnormal functions. When I look at coronary heart disease, as one progresses from 1 to 2 to 3, especially from 2 to 3, there is a large rhythmic increase in the consequences of coronary heart disease in myocardial infarction hence in ventricular dysfunction. The point of the slide is that when one in fact has an infarct, then the extent of the disease is an independent contributor to the extent of left ventricular functions. Having an infarction in a 3 vessel disease imparts a higher frequency of dysfunction, than having an infarction with 1 or 2 vessel disease, probably because the patients with a 3 vessel disease who have an infarct have a more extensive infarct than those with 1 or 2, but that last statement is speculation.

KELLERMANN

Now my third question, I remembered. I believe all these tests were carried out at rest and not during exercise and so the ejection fraction was also taken at rest only, so we have no exercise figures for both methods.

WEISSLER

I must say that as a measure of distinct changes in systolic function with exercise, systolic intervals don't seem to be highly sensitive. In exercise the coexistence of opposite changes as of preload, afterload etc. actually neutralize the effect on systolic time intervals. Moreover the exercise in coronary heart disease shows the effects on the ventricle not only of the intrinsic abnormality that there was at rest, but of the superimposed distortion in ventricular function due to ischemia, and it is very hard to interpret it. Exercise is a very good test for perfusion disorders; I don't think it is a very good test for intrinsic contractile reserve.

A short technical question to Prof. Cherchi. In what percentage of cases are you able to obtain a good echo during exercise with a heart rate of 180? I asked this question because in my department we are able to obtain a good echo, only in a very small number of cases, even in the lying position.

CHERCHI

It is much easier to do physical exercise in the sitting position rather than in the lying one. I have done both. Maximal or submaximal workload in 18-20 year patients can be carried out on 8, 9 out of 10 patients. In patients between 40-50 years in 6 out of 10. I must add that we have been carrying out bidimensional echoes for the past year, in the sitting position, it is extremely easy to register the left ventricle in 2 or 4 chambers, it is usually very easy in 7 or 8 out of 10 patients.

LEACHMAN

I would like to make a comment to Prof. Raineri's paper. I think it is important to put into perspective that pace making at least relates to two specific problems: one arrhythmic problem that occur in the patient with a normal heart and the other of the arrhythmic problem that occur in a patient who has abnormal heart. Does every disease of the ventricle have an optimal heart rate, and if there is an optimal heart rate for each disease of the ventricle, can we predict what that heart rate should be to implant a pacemaker?

WEISSLER

"I don't know!" but I would comment very briefly on the
philosophy of the question. That is how does one know when pacing is
optimal physiologically. I do not feel that acute changes in various
dynamic measures such as ejection fraction tell us that the ventricle
is doing well for the body, this is not necessarily the way to approach it. Indeed I worry about such indeces as ejection fraction
which are very good in the chronic disease state but aren't true
indicators of integrity of ventricular chamber performance when we
use it, as a functional measure, in doing acute changes.

The many aspects of what is the optimal heart rate: I consider
the point in which the patient does the most with his heart, with the
least bit of dyspnea and congestion. I think that probably one
measure that we should be doing in most of the patients that we test
pacemakers in, is to obtain a measure of left atrial pressure or
pulmonary capillary wedge pressure. Secondly I consider the delivery
of flow in terms of cardiac output as probably being important. Is
there an optimum heart rate at which these measures are maximal? I
don't know if these data are available, I would suspect we would have
a broad curve in most individuals.

LEACHMAN

Docter Raineri, do you have any comment to make on what you
heard here, is there something you would like to say?

RAINERI

Certainly, I haven't tried to solve the problems, but I would
like to point out 2 of the problems: one is the necessity to have a
procedure to assess the patient and in this respect I think that the
procedure must be simple and non invasive, and I have to say that
certainly ejection fraction is a good marker. Perhaps end-systolic
volume can be another good marker. In fact, we have some preliminary
results in patients with atrial fibrillation. We selected these
patients to avoid the problem of atrial contribution. In these
particular patients we found an increase of end-systolic volume with
higher rate ventricular pacing, which can lead to left ventricle
enlargement after a period of time. This could be on reason to set
heart rate not too high. But to interpret our present results we are
convinced that the indications we have from exercise are the best
answer. In fact, the ejection fraction increases during exercise and
this increment is more important with the higher rates, but this has
a limit. It means that there is optimal rate for a cardiac pacing
and the EF is a good parameter to detect it. Moreover I would like
to add that in general atrial activity is an indicator of the rate at

which heart should beat in patients with compete AV block. The RS4 pacemaker exploits this condition in raising the particular paced rate, and we demonstrated, just through the EF analysis, that the higher beat rate is useful to improve the cardiac performance. As for as cardiac output is concerned I can remember the studies of Benchimol, Sowton, Samet and other. They found that cardiac output may improve in more of the patients, as the heart rate is set between 70 and 90 beats per minute. But I am not sure whether the cardiac output is a good parameter, or, at least, it is good to judge the circulatory sufficiency, but it isn't to judge the cardiac performance; in fact in many circumstances it is not. So I think that ejection fraction is the best parameter. I have to add something about atrial contribution; I think that this is an important problem, in fact we know that while some investigators have demonstrated that atrial contribution is proportionally higher in patients with left ventricular dysfunction than in patients with normal ventricular function, others have not confirmed this finding, and I would like to stress my conclusions, after restoration of atrio ventricular synchrony, we have a variable effect in cardiac output or ejection fraction and so on, it depends on the mechanical efficiency of the atrial contractions and the passive filling pressure and ventricular compliance. So when we have to face the problem of pacing patients we have to try to have information before implanting the pacemaker and to take into consideration all the problems.

DALLA VOLTA

I would like to stress one point about the exercise test in patients. I think that the study of the exercise test only through the study of electrocardiogram and the ST depression, either horizontal of downslope is a little reductive: the heart rate behavior, the drop of pressure and the recovery time should be considered. In our experience in the last 7 years, in patients the sensitivity of the test has been about 86% and specificity 86%. Moreover, in the first 30 cases of the Italian cooperative study, we have seen that if the exercise test does not show any change or improvement after 3 years of medical treatment, we have about 55% chance that the anatomical state of the coronary vessels has improved, while when there is a deterioration of the test the chance that there is a worsening of the anatomical lesions is about 80%. So probably a more comprehensive effort test should have a wider approach and leave more information and could be considered as a second instance test.

FORRESTER

I agree with you completely. In the United States we are far from saying that only about 25% of the information of a stress test is obtained from the ST segment.

CHAPTER III

DISEASES WITH ALTERED VENTRICULAR FUNCTION

CHAPTER III:

INTRODUCTORY REMARKS

J. J. Kellermann

The Chaim Shebe Medical Center
Tel-Harhomer, Israel

The use of non-invasive assessments of ventricular function has
caused an increased interest and greater attention for the evaluation
of ventricular dimensions in non-coronary heart disease.

In the past such an assessment has been obtained entirely by
catheterization, a method which in some places was and is still
reserved to exact clinical indications, and is rarely undertaken in
early stages of the disease. The assessment of ventricular function
in valvular aortic stenosis can greatly influence the decision for
early surgery. It has been found, that even in patients with
severely depressed function before operation the average E.F.
returned to normal. The observation has suggested that a mechanical
overload rather than irreversibly depressed myocardial contractility
is often responsible for the reduced function. On the other hand in
aortic regurgitation, especially in the symptomatic patients with
severe insufficiency it has been demonstrated that severely depressed
function often proved to be irreversible after operation. One of the
most important conditions requiring an early as possible assessment
of ventricular function is mitral regurgitation. It has been shown
that even in the relatively asymptomatic patients with severe mitral
regurgitation an irreversible myocardial dysfunction can develop.
Studies before and after mitral valve replacement indicate that in
contrast to aortic regurgitation, left ventricular function tends to
be reduced after operation, even if it was within a normal range
preoperation. It seems that early studies of ventricular function in
these patients and a timely initiation of surgery may have great
impact on prognosis. A special problem is the assessment of patients
with ischemic cardiomyopathy, in these patients too non-invasive
assessment by echoardiography and radionuclide imaging at rest and
during exercise is of great importance.

This assessment should be undertaken also in the so-called "silent ischemic patients", or in those with atypical chest pain or dyspnea at a constant A.T.H.R.

Needless to stress that the outcome of these assessments may have an important impact on therapy. Finally, one must acknowledge the prognostic importance of left ventricular function on prognosis. It is of special interest whether impaired left ventricular dysfunction after surgery or even after the other therapeutical procedure, is reversible. I am confident that we shall learn a great deal from the presentations in this session and that a fruitful discussion will follow some of the subjects mentioned in this short introduction.

RIGHT AND LEFT VENTRICULAR FUNCTION

IN PATIENTS WITH MITRAL STENOSIS

M. Mariani

Cattedra di Malattie dell'Apparato Cardiovascolare
Università di Pisa, Italy

The study of both left and right ventricular performance generally is interesting, not only from the physiopathological point of
view but also as it concerns the surgical prognosis.

In patients with mitral stenosis an impairment of left ventricular function has been pointed out in the last ten years. Referring
to the overall left ventricular function, many investigators have
observed a slight reduction of end diastolic volume, a moderate
increase of end systolic volume, and a reduction of ejection fraction[1-6]. In other words, there is a reduction of left ventricular
performance indices.

In addition, an abnormal behavior of left ventricular regional
contraction was observed[6-7]. However, from the surgical point of
view the real role of left ventricular performance indices is still
being discussed, and in recent years some investigators have emphasized the importance of right ventricular function particularly with
regard to the inteference of both ventricles[8-11].

The aim of this study is the evaluation of the right and left
ventricular function in patients with pure mitral stenosis in order
to verify their influence on the surgical prognosis.

The results obtained in sixty-three subjects affected by pure
mitral stenosis (45 females and 18 males; mean age 45 years) have
been reported. Before surgery all the patients underwent left and
right cardiac catheterization and left ventriculography was performed
at 45° right anterior oblique (RAO) projection. Left ventricular
function was evaluated by computerized quantitative cineangiocardio-
graphy, the ventriclograms being projected by a TV camera on to a

video monitor, and the left ventricular cavity shape was outlined using a "light pen". So, for the overall evaluation of ventricular function, end diastolic volume (EDV) and end systolic volume (ESV) were obtained by the "area length" method. The left ventricular regional function was also analysed with the same semiatomatic system in normal subjects. When a frame-by-frame analysis was made, volume variations of the whole cardiac cycle were noticed[12-14].

By plotting the single values, the relative curves and their first differentiations were obtained. In addition, it is possible to identify the various phases of contraction and relaxation by this method; we analysed volume variation velocity by considering the beginning of a new phase, the point where the volume velocity was significantly different from the velocity of previous frames.

In Figure 1 the volume curve obtained in normal subjects during relaxation and contraction is illustrated. During diastole it is possible to observe a slow relaxation period, a rapid relaxation period and then another slow relaxation period; during systole a slow contraction period, a rapid contraction period and another slow contraction period.

The percentage shortening of longitudinal axis and all semiaxis was obtained according to Hermann's technique (Figure 2). For regional performance, the single values of longitudinal axis and semiaxis were obtained and plotted against time. So the same parameters were derived to evaluate the regional function.

In our study for the overall and regional left ventricle evaluation we considered the following parameters: ejection fraction (EF), stroke volume/end diastolic volume relationship (SV/EDV), axes per cent shortening, maximum velocity of volume ejection rate or relaxation rate (dl/dt max/v), maximum velocity of axes shortening rate or relaxation rate (dl/dt max/l), mean left ventricular ejection rate (EF/ET).

As far as right ventricular function is concerned in particular for right EDV and right EF determinations, we used the radiocardiographic method[15] or the first pass obtained by a "gamma camera" after injection of ^{99m}Tc [16-17]. In both conditions the right ventricle wash-out was assumed to be an exponential curve and the RVEF was calculated with the following formula: $EF=1-e^{-k}$, where k is 1/0.37 sec.

Tricuspidal incompetence was also excluded by means of echocontrastography. Mono- and bi-dimensional echo were used to evaluate mitral valve apparatus.

The mean values of overall left ventricular performance indices in patients with mitral stenosis grouped on the basis of different

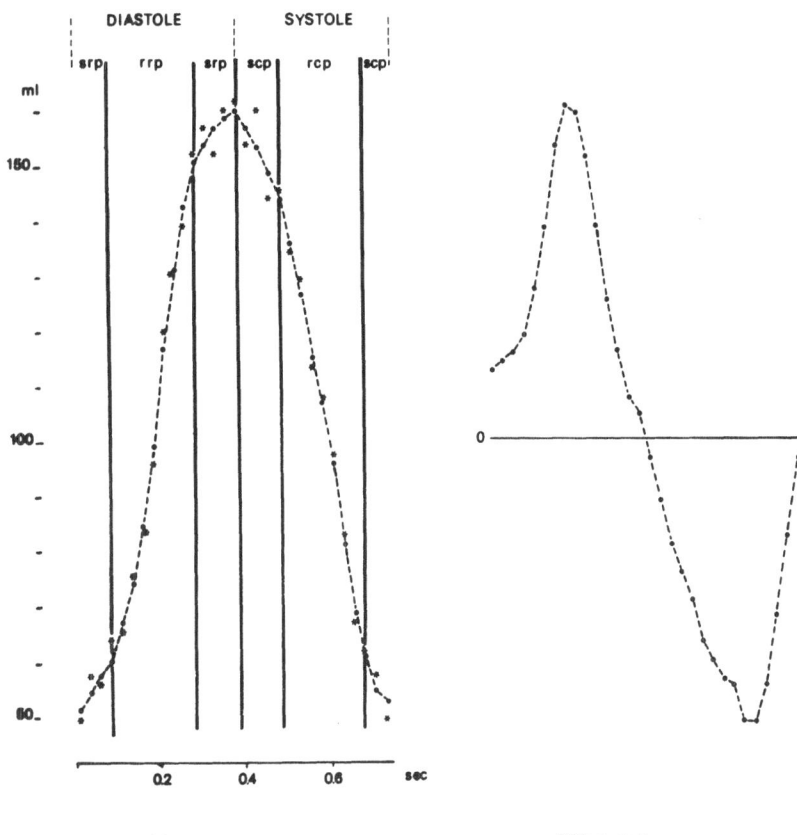

Volume curve   Differentiation

Fig. 1. Volume curve in normals during relaxation and contraction
         phases (on the left); first differentiation (on the right).
         srp = slow relaxation period; rrp = rapid relaxation period;
         scp = slow contraction period; rcp = rapid contraction
         period.

values of pulmonary vascular resistances (PRV) compared with normal
ones are reported Table 1.

     It is possible to observe a reduction of 18% of EDV, an increase
of 35% of ESV and a reduction of 26% of LVEF if compared with normal
subjects. This reduction becomes more important in patients with
high pulmonary vascular resistances. Mass and systolic stress appear
to be normal.

     In other words, an impairment of left ventricular performance is
displayed. This behavior is confirmed when maximum volume velocity
of contraction and relaxation are analysed. Figure 3 shows a reduction of the velocity not only during contraction but also during the
relaxation period. In addition, a different behavior of single

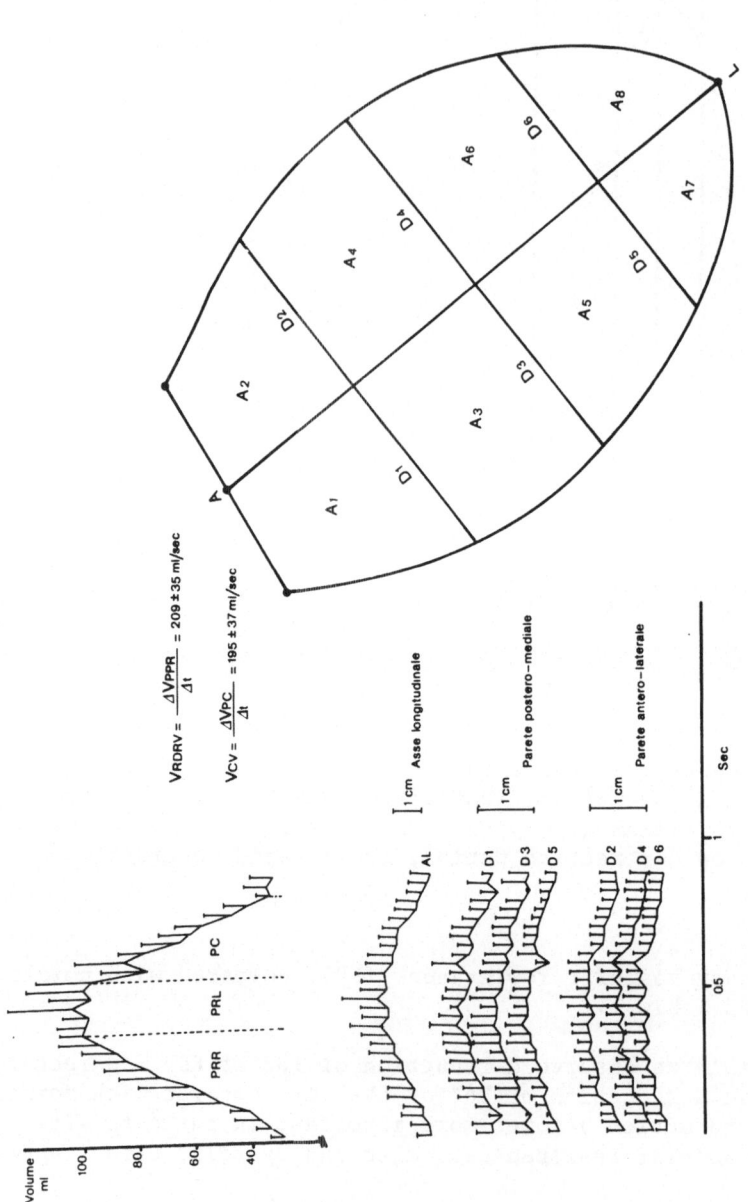

Fig. 2. Ventricular shape divided according to Herman's technique (on the right). Mean curves with standard deviations of volume, longitudinal axis and hemiaxes plotted against time in normal subjects (on the left).

Table 1. Mean Values of Overall Left Ventricular Performance
Indices in Patients with Mitral Stenosis Grouped on the
Basis of Different Values of PVR Compared with Normals

	MVA cm	EF %	EDV ml/m^2	ESV ml/m^2	PVR mmHg/l/min m^2	Mass g/m^2	EDS 10^3dyne cm^2	ESS 10^3dyne cm^2
I GROUP	1.38	53.8	77	35	1.74	93.6	18.2	91.5
n=21	±0.58	14.9	18	12	0.40	25.3	9.7	25.6
II GROUP	1.12	55.7	74	33	3.62	–	–	–
n=21	±0.32	11.4	23	13	0.70	–	–	–
III GROUP	0.66	48.4	74	38	13.20	89.6	11.7	108.3
n=21	±0.18	10.9	17	10	8.50	28.0	5.6	38.9
NORMALS	(4-6)	71.0	91	25	1.40	92.0	–	–
n=20		± 4.6	20	6	0.46	16.0	–	–

I GROUP : PVR < 2.5 mmHg; II GROUP : 2.6 < PVR < 5; III GROUP : PVR > 5.1
MVA = mitral valve area; EF = Ejection fraction; EDV = end-diastolic vo -
lume; ESV = end-systolic volume; PVR = pulmonary vascular resistences ;
Mass = left ventricular mass; EDS = end-diastolic stress; ESS = end-sy -
stolic stress.

phases of contraction and relaxation is observed: two phases compared with the three of normal subjects; namely, there is an impairment of both diastolic and systolic function.

Regarding regional analysis, the longitudinal axis and semiaxis
velocities, relative to the antero-lateral wall, show a significant
reduction. Therefore left ventricular performance, both in diastole
and in systole, is impaired; but it is still necessary to confirm
whether preload or afterload mismatch is responsible.

Regarding the correlation between preload and left ventricular
impairment in mitral stenosis, the EDV and SV relationship was
analysed (Figure 4). The continuous line with confidence limits
shows the relationship in normals, while the broken line refers to
mitral stenosis: it is possible to observe that EDV is lower than in
normals, and for EDV extremely low the relative stroke index is in
the confidence limits; but it is not possible to assert the same in
patients with high PVR. So in subjects with low PVR the preload
mismatch is not significant.

If the influence of afterload as assessed by end systolic
stress (ESS) on LVEF in normals is examined[18], it is possible to

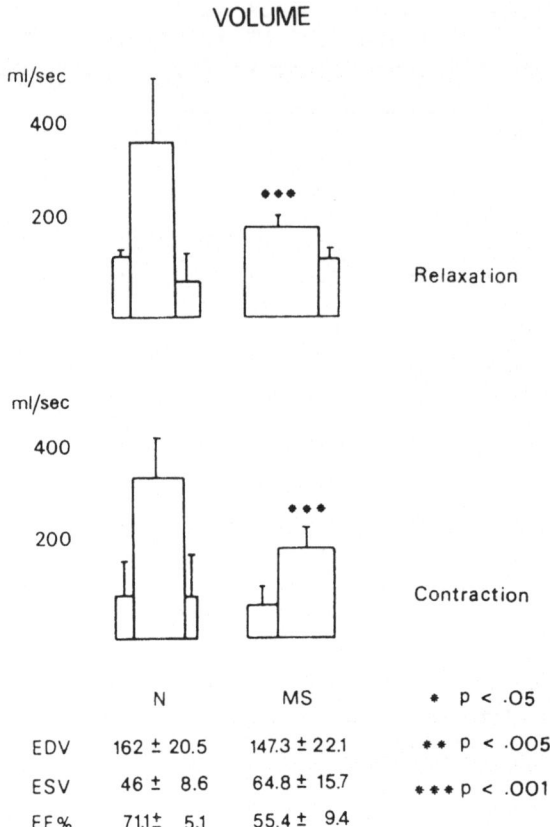

Fig. 3.   Maximum volume velocity of relaxation and contraction in
          normals, on the left, and in mitral stenosis, on the right.

assert that for an increases of ESS, even if it is in a normal
range, it corresponds to a decrease of EF as is shown by a negative
relationship between ESS and EF in patients with mitral stenosis.

    First of all, it is evident that ESS in mitral stenosis is in
the normal range, while EF may be reduced. However, a linear negative relationship between the parameters exists as in normals
(r=0.849, p<0.001).   In others words, with ESS being in basal conditions in the normal range, it is possible to think that the left
ventricle in mitral stenosis follows the general law of the heart.

    Finally, considering the mitral valve area influence on left
ventricular performance (EF), it is possible to observe the absence
of a correlation between the two parameters, while a good relationship was obtained between RVEF and mitral area (r=0.459, p<0.001).

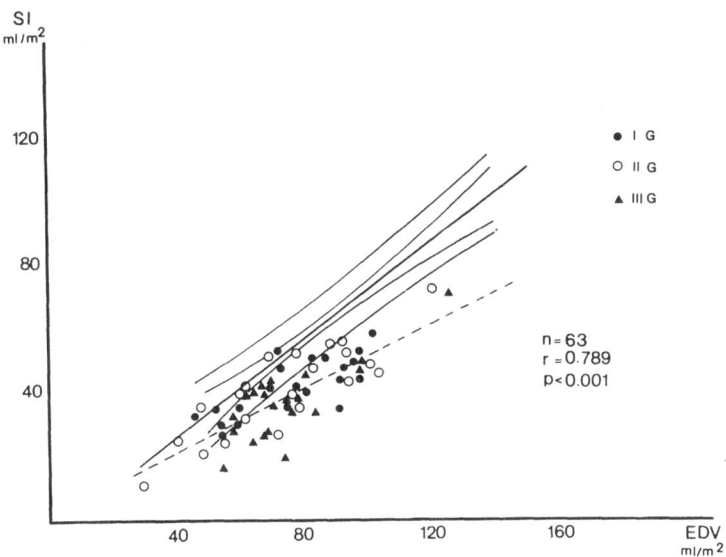

Fig. 4.   Left ventricular end-diastolic volume/.stroke index relation-
          ship (see text).

Table 2.   Mean Values of Overall Right Ventricular Performance
           Indices in Patients with Mitral Stenosis Grouped on the
           Basis of Different Values of RVP Compared with Normals

		CI 1/min/mq	SI ml/mq	EF %	EDV ml/mq	PVR mmHg/1/min/mq	MVA cmq
I	GROUP	3.09 ±0.7	40.2 ±9.8	32.1 ±10.5	135 ±39	1.74 ±0.40	1.38 ±0.58
II	GROUP	2.59 ±0.73	33.5 ±9.3	29.5 ±8.5	118 ±28	3.62 ±0.70	1.12 ±0.32
III	GROUP	2.09 ±0.41	24.4 ±6.8	20.3 ±6.8	122 ±27	13.20 ±8.5	0.66 ±0.18
	NORMALS	4.18 ±0.90	54.6 ±12.3	46.0 ±7.7	119 ±18	1.4 ±0.46	(4-6)

I Group: RVP<2.5 mmHg; II Group: 2.6<RVP<5.0; III Group RVP>5.1;
CI= cardiac index; SI= stroke index; EF= ejection fraction;
EDV=end-diastolic volume; ESV=end-systolic volume; PVR=pulmonary
vascular resistances; MVA=mitral valve area.

So we must consider the right ventricular function. The results
obtained from analysis of right ventricular function, performed in
order to discover if this is the main factor affecting clinical and
surgical prognosis in mitral stenosis, are reported in Table 2. The

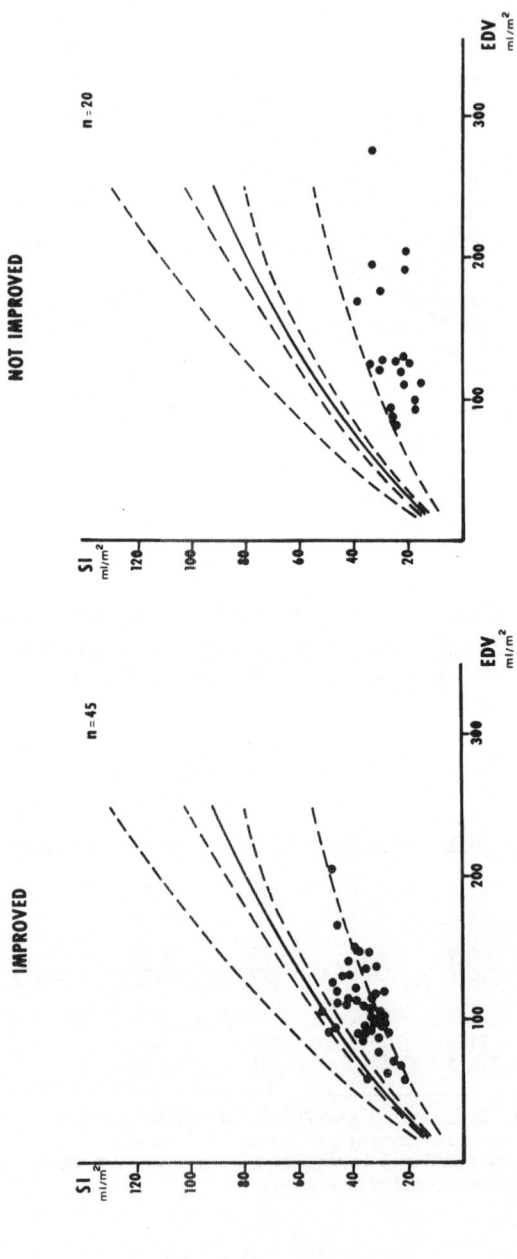

Fig. 5. Relationship between right ventricular volume and stroke index in respect to normals in patients with mitral stenosis improved (on the left) and not improved (on the right) after surgical correction.

REDV results were not significantly increased (5%) while EF appears
to be reduced by 19% with respect to normal subjects. In addition,
a progressive reduction of RVEF to a progressive increase of PVR is
seen.

The importance of right ventricular function on guiding the
cardiac output is shown by a very significant statistical relationship between the RVEF versus CI (r=0.78, p<0.01) and RVEF versus
SV (r=0.623, p<0.001), while a poor statistical relation is shown
between LVEF versus CI and SV. In other words, the right ventricular
function is responsible for the flow rate in mitral stenosis.

This interpretation may be supported by surgical data analysis
that shows REDV versus SI (Figure 5). Those patients who improved
are in the normal range, while those who did not improve after surgery, the same relationship is outside these limits. On the other
hand, the same relation referred to the left ventricle in the non-
improved patients is slightly out of the normal confidence limits.

In conclusion, the right ventricular function appears to be the
most important factor in conditioning the overall function in
patients with mitral stenosis and in predicting surgical results,
while the left ventricle seems to play a secondary role.

REFERENCES

1. S. J. Heller and R. A. Carleton, Abnormal left ventricular con-
    traction in patients with mitral stenosis, Circulation,
    42:1099 (1970).
2. G. C. Curry, L. P. Elliott, and H. W. Ramsey, Quantitative left
    ventricular angiocardiographic findings in mitral stenosis,
    Am.J.Cardiol., 29:621 (1972).
3. L. D. Horwitz, C. G. Mullins, R. M. Payne, and G. C. Curry, Left
    ventricular function in mitral stenosis, Chest, 64:609
    (1973).
4. J. A. Holzer, J. S. Karliner, R. A. O'Rourke, and K. L.
    Peterson, Quantitative angiographic analysis of the left
    ventricle in patients with isolated rheumatic mitral
    stenosis, Br.Heart J., 35:497 (1973).
5. J. P. Dubiel and J. S. Dubiel, Quantitative angiocardiography
    evaluation of left ventricular function in mitral stenosis,
    Pol.Med.Sci.Hist.Bull., 15:183 (1975).
6. A. Barsotti, A. Balbarnini, R. Mariotti, G. Gherarducci, and M.
    Mariani, Il volume ventricolare sinistro nella stenosi
    mitralica, Boll.Soc.It.Cardiol., Vo.XXI, 1983 (1976).
7. S. J. Heller and R. A. Carleton, Abnormal left ventricular con-
    traction in patients with mitral stenosis, Circulation,
    42:1099 (1970).

8. R. R. Taylor, J. W. Covell, E. H. Sonnenblick, and J. Ross, Jr.,
    Dependence of ventricular distensibility on filling of the
    opposite ventricle, Am.J.Physiol., 213:711 (1967).

9. C. E. Bemis, J. R. Serur, D. Borkenhagen, E. H. Sonnenblick, and
    C. W. Urshel, Influence of right ventricular filling pressure
    on left ventricular pressure and dimension, Circulation Res.,
    34:498 (1974).

10. G. Elzinga, R. Van Grondelle, N. Westerhof, and G. C. Van Den
    Bos, Ventricular interference, Am.J.Physiol., 226:941 (1974).

11. W. P. Santamore, P. R. Lynch, G. Meier, and J. Heckman, Myo-
    cardial interaction between the ventricles, J.Appl.Physiol.,
    4:362 (1976).

12. M. Mariani, G. Valli, A. Barsotti, R. Balocchi, A. Balbarini,
    and D. Pierotti, Cineangiografia quantitativa: elaborazione
    computerizzata delle immagini ventricolari sinistre, Boll.
    S.I.C. XXI, 6:1077 (1975).

13. A. Barsotti, R. Mariotti, A. Balbarini, and M. Mariani, Quanti-
    tative evaluation of the regional left ventricular function
    in normal subjects by means of cineangiocardiography,
    Cardiov.Res., 14:30 (1980).

14. M. Mariani, A. Barsotti, G. Valli, A. Dalle Luche, A. Balbarini,
    R. Mariotti, and G. Tartarini, Quantitative cineangiocardio-
    graphy: a computerized method for evaluating relaxation and
    contraction in health and disease. Abstract of VIII Europ.
    Congress of Cardiology, Paris, No.2620:215 (1980).

15. M. Mariani, C. Giuntini, A. Maseri, and L. Donato, Clinical of
    radiocardiograhpy (15 years experience). Symposium of
    Cardiology, Tokyo, Proceedings of the First World Congress of
    Nuclear Medicine, 1:241 (1974).

16. H. R. Schelbert, J. W. Verba, A. D. Johnson, G. W. Brock, N. P.
    Alazraki, F. J. Rose, and W. L. Ashburn, Non-traumatic deter-
    mination of left ventricular ejection fraction by radio-
    nuclide angiocardiography, Circulation, 51:902-909 (1975).

17. R. C. Marshall, H. J. Berger, J. C. Costin, G. S. Freedom, J.
    Wolberg, L. S. Cohen, A. Gottschalk, and B. L. Zaret, Assess-
    ment of cardiac performance with quantitative radionuclide
    angiocardiography, Circulation, 56:820-829 (1975).

18. J. L. Bolen, M. G. Lopes, D. C. Harrison, and E. L. Alderman,
    Analysis of left ventricular function in response to after-
    load changes in patients with mitral stenosis, Circulation,
    52:894 (1975).

EVALUATION OF LEFT VENTRICULAR FUNCTION

IN MITRAL REGURGITATION

K. L. Peterson

University of California
San Diego, USA

An understanding of the mechanics of left ventricular function
in mitral regurgitation necessitates consideration of an acute, as
opposed to a chronic, onset of the volume overload. In fact, it is
now recognized from both human and experimental animal observations
that acute and chronic regurgitation produce congestive heart failure
by significantly different pathogenetic mechanisms, Table 1.

ACUTE MITRAL REGURGITATION

In acute mitral regurgitation, due either to disruption of one
or more of the elements of leaflet support or rupture of a leaflet
itself, the relatively noncompliant left atrium is subject to an
immediate increase in systolic filling, resulting in an abrupt augmentation in pulmonary venous pressure[1]. Presumably the compliance
of the pulmonary veins and lungs are likewise relatively small and
contribute to the generation of a large, peaked "V" wave in the
pulmonary arterial wedge pressure tracing. Studies in the experimental animal suggest that during moderate degrees of acute mitral
regurgitation, the extent of shortening (passive and active) of the
left atrium is enhanced; however, with severe degrees of regurgitation, the atrium becomes over-stretched and appears to move into
a descending limb of its length-tension relation; the extent of
shortening then become significantly diminished[2].

Acute mitral regurgitation in the experimental animal, as well
as man, causes an immediate increase in left ventricular preload,
reflected by an increase in left ventricular end-diastolic volume and
pressure. However, the increment of increase in end-diastolic volume
is relatively modest and likely reflects an increase in sarcomere

191

Table 1. Pathophysiologic Characteristics of Acute and Chronic Mitral Regurgitation

	EDV	ESV	TOTAL SV	FORWARD SV	EJECTION FRACTION	LEFT ATRIAL PRESSURE	LEFT ATRIAL COMPLIANCE	LEFT ATRIAL VOLUME
ACUTE MR	↑	N or ↑	↑	N or ↓	↑	↑↑↑	N	N
CHRONIC MR	↑↑	↑↑	↑↑	N	N or ↑	N or ↑	↑	↑
CHRONIC MR WITH DEPRESSED MYOCARDIAL CONTRACTILITY	↑↑↑	↑↑↑	N or ↑	N or ↓	↓	↑↑	↑↑	↑↑

LEGEND: MR = MITRAL REGURGITATION; EDV = END-DIASTOLIC VOLUME; ESV = END-SYSTOLIC VOLUME; SV = STROKE VOLUME;
N = NORMAL; ↑ = MILDLY INCREASED; ↑↑ = MODERATELY INCREASED; ↑↑↑ = SEVERELY INCREASED; ↓ = DECREASED

stretch alone[3,4]; later on, other adaptations serve to increase the
end-diastolic volume to values which are 50-300% the control state
depending upon the duration of the valve leakage and the status of
myocardial contractility, Figure 1. Because of the low impedance
conditions in the left atrium and pulmonary veins, left ventricular
afterload is reduced and allows an increase in the extent of shortening; end-systolic volume remains approximately the same or increase
slightly. The extent of shortening (i.e. ejection fraction) increases initially, but then remains constant over a period of at
least four weeks, Figure 2.

Assessment of myocardial contractility in the presence of acute
mitral regurgitation is problematic. As noted above, the afterload
conditions during ejection are complex and thought to be responsible
for maintaining normal ejection phase indices (ejection fraction,
mean velocity of circumferential fiber shortening) even though myocardial contractility might be depressed. The effect of a volume
overload per se on left ventricular myocardial contractility has been
extensively studied; investigation of isolated papillary muscle
function in experimental animals exposed to various types of relatively acute volume overload (aorto-caval fistula, atrial septal
defect) have suggested that myocardial contractility remains
normal[5,6]. Similarly, when left ventricular function was analyzed
before and after production of a fistula between the aorta and
inferior vena cava, resulting in progressive left ventricular
dilatation and moderate myocardial hypertrophy without clinical
evidence of heart failure, the length-active tension relation remained essentially normal. Following chronic adjustment to the
shunt, the mean velocity of circumferential fiber shortening usually
remained normal, although in some dogs with very large shunts and
signs of congestive heart failure this index became depressed[7].
Left ventricular isovolumic indices which are dependent on an assumed
muscle model and a period of isometric contraction (e.g. Vmax, peak
Vce) are limited in the presence of mitral regurgitation by the
significant abbreviation or absence of the period of isovolumic
contraction. We have assessed dP/dt (a simpler index where there is
no assumption of a muscle model) in the same series of awake, instrumented experimental animals cited above and have found that this
index does not change appreciably as the left ventricle adapted over
four weeks to acutely imposed mitral regurgitation.

In recent years, Suga and Sagawa have published a number of
elegant isolated heart studies which have emphasized the utility of
the end-systolic pressure-volume and stress-volume relations as
unique descriptors of myocardial contractility independent of the
effect of preload and encompassing the effects of afterload[8,9].
They have shown that an increase in contractility is associated with
an increase in slope and a shift to the left of these relations;
depression of contractility moves the relations in the opposite
direction, i.e. to the right with decrease in slope. In our

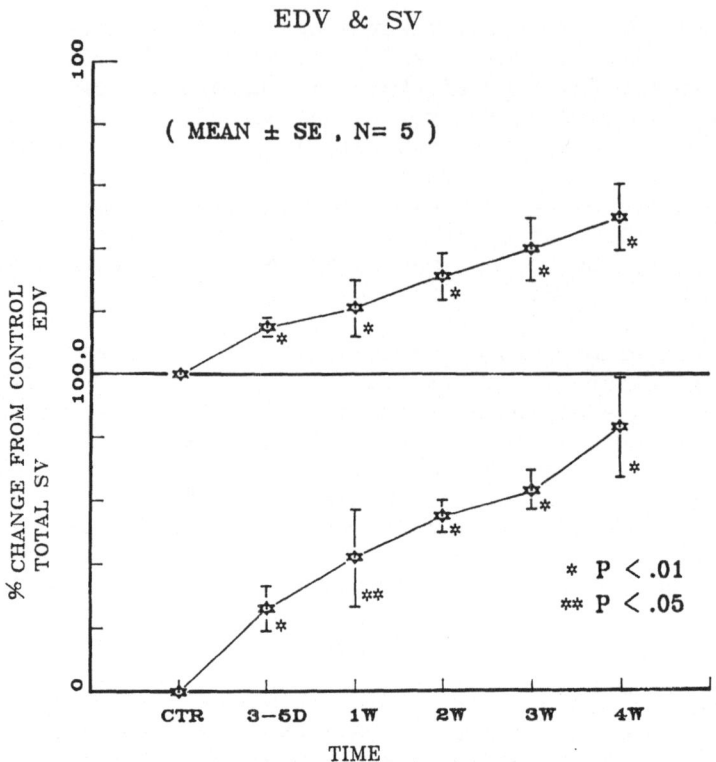

Fig. 1. Plot of serial changes over four weeks in end-diastolic
         volume (EDV) and stroke volume (SV) in five awake, instru-
         mented dogs with acute mitral regurgitation. Note the
         progressive increase in both parameters of left ventricular
         function, reflecting the myocardial adaptations resulting
         from the acute volume overload.

experimental model of mitral regurgitation over the first four weeks
after creation of the valve leakage and compared them with control
values. We have used angiotensin II to increase, and nitroprusside
to decrease, left ventricular afterload in order to determine the
slopes of the relations as well as the extrapolated volume at zero
force (Vo). Both the pressure-volume and stress-volume relations at
end-systole showed a progressive reduction in slope and an oper-
ational shift to the right as the left ventricle adapted to the
presence of mitral regurgitation, Figure 3. By 3-5 days after the
creation of the volume overload the slopes of the relations had
decreased by only approximately 10%; however, by one week, the slopes
for both of the relations were down to 50% of control values. Thereafter, the slopes remained relatively stable until the dogs expired
from pulmonary venous congestion, Figure 4. The volume intercept
at zero pressure (Vo) did not change significantly.

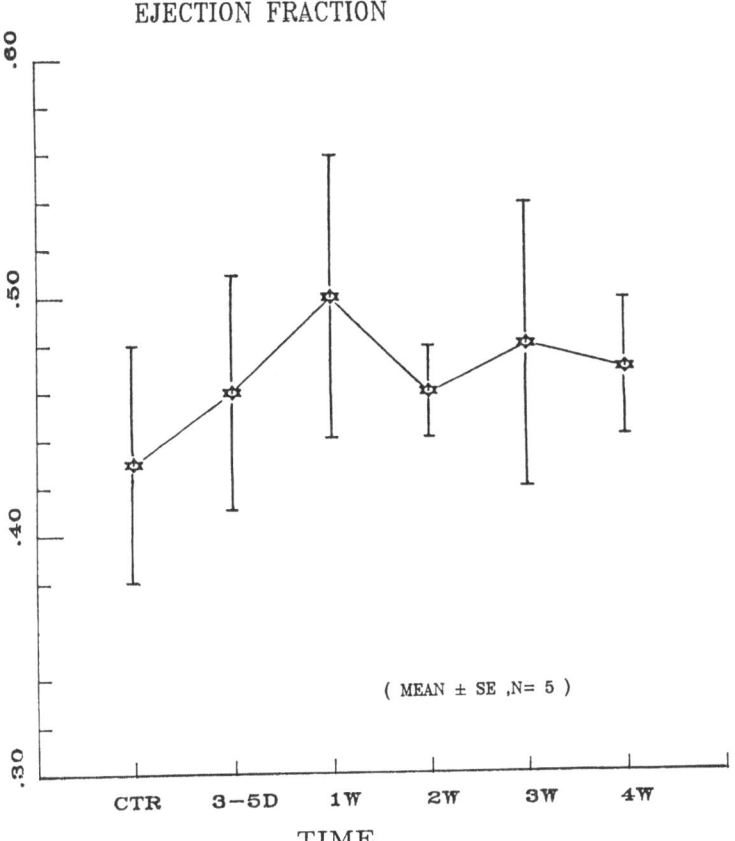

Fig. 2. Plot of serial changes over four weeks in the ejection
         fraction in the same five dogs with acute mitral
         regurgitation.

    The extent to which the rightward shift of the relation reflects
a reduction in basal contractility (as might be concluded on the
basis of isolated heart experiments), as opposed to a direct consequence of myocardial adaptations to the volume overload, remains
conjectural. Conceivably, serial replication of sarcomeres and fiber
slippage could lead to a larger end-systolic volume for any given
stress even though each sarcomere continued to exhibit the same
percent shortening. In the above cited animal experiment, if the
slopes of the end-systolic relations were standardized to an index of
the number of sarcomeres (e.g. end-diastolic volume), a shift to the
right in the relation was no longer detected. This suggests that
strain (extent of shortening normalized to the starting length) did
not change between the control state and four weeks after mitral
regurgitation and that the increase in the end-systolic volume is
attributable to an increase in the units of muscle, i.e. sarcomeres.
Presently, it is our conclusion that myocardial contractility remains

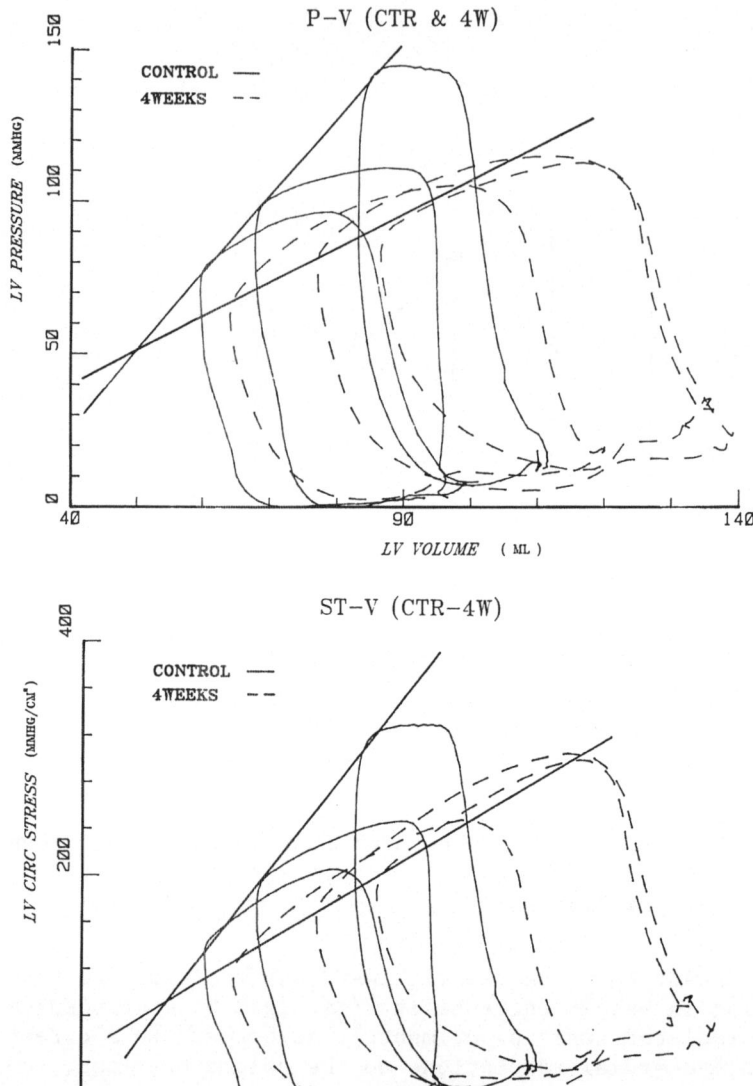

Fig. 3.   Plots of pressure-volume (P-V) loops (upper panel) and
          wall stress-volume (ST-V) loops (lower panel) at control
          (solid lines) and four weeks after the creation of mitral
          regurgitation (dashed lines). The loop with elevated
          systolic pressure was obtained during angiotensin II
          infusion; the loop with reduced systolic blood pressure
          was obtained during infusion of nitroprusside. The
          end-systolic coordinates for each loop (point of maximal
          elastance) are connected by a solid line for each state.

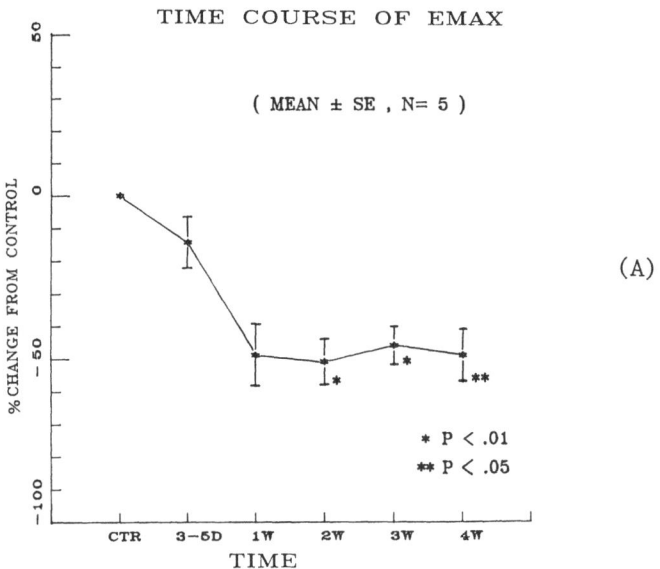

(A)

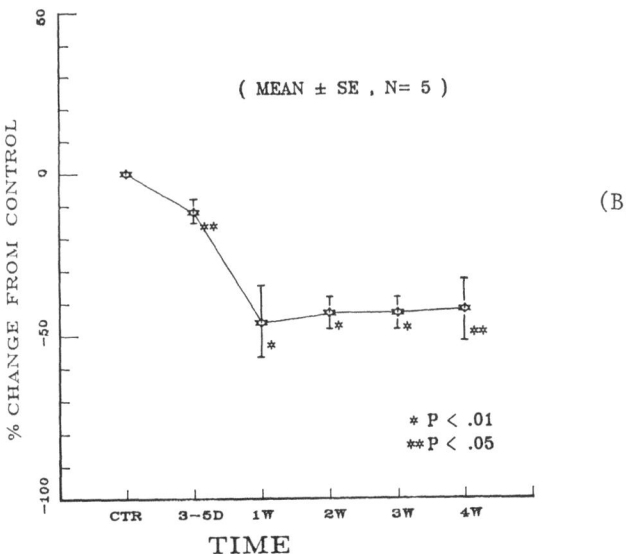

(B)

Fig. 4. Plots of time course over four weeks of the slope of the
         pressure-volume relation (Emax), shown in upper panel, and
         the slope of the stress-volume relation, shown in lower
         panel, in five dogs with acute mitral regurgitation. See
         text for discussion.

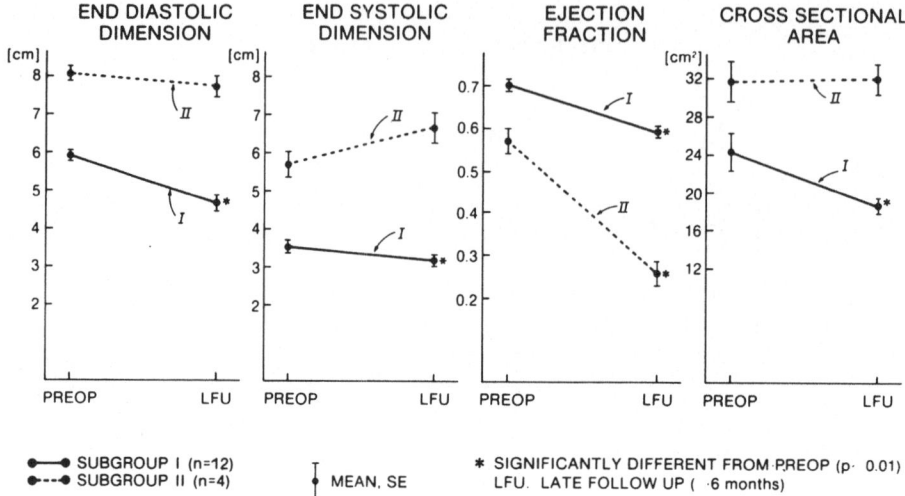

Fig. 5. Plots of indices of left ventricular function (end-diastolic
diameter, end-systolic diameter, ejection fraction, and
cross-sectional area of myocardium at minor axis) in subgroups of patients with chronic mitral regurgitation, before
and 6 months after mitral valve replacement[14]. Reproduced
with the permission of the authors and American Heart Association. See text for discussion.

normal as the ventricle adapts to the development of acute mitral
regurgitation and that the shift to the right and reduction in slope
of the end-systolic pressure–volume and stress–volume relations are
consequences of eccentric hypertrophy. This conclusion is also
consistent with previous experiments which demonstrate that a volume
overload _per se_ does not affect adversely myocardial contractility.

CHRONIC MITRAL REGURGITATION

Insight into the mechanics of contraction of chronic mitral
valve leakage is provided solely from human studies since an animal
model of chronic mitral regurgitation has never been successfully
produced. The functional and morphological characteristics of
chronic, as opposed to acute, mitral valve leakage are shown in
Table 1. Once the heart adapts to the leakage of the mitral valve
over an extended period of time, four compensatory mechanisms are
noted: 1) dilatation of the left atrium with coincident reduction in
its compliance, 2) dilatation of the left ventricle with a shift to
the right in the diastolic pressure–volume relation, 3) maintenance
of a normal forward stroke volume by preservation of normal or

increased percentage shortening from a significantly larger end-diastolic volume, and 4) increased sphericity of the left ventricular chamber so as to minimize increases in wall stress.

The dilatation of the left atrium and resultant increase in its compliance and capacitance allows accommodation of the regurgitant volume at relatively normal pulmonary venous pressure. For example, ten patients with long-standing mitral regurgitation have been described in whom left atrial and pulmonary arterial wedge pressures despite the presence of atrial fibrillation[10]. This mechanism serves to maintain pulmonary venous pressure within normal limits, and the symptoms of pulmonary congestion are diminished or absent. Generally, if normal sinus rhythm persists and the volume overload on the chamber is severe, the percentage active shortening of the atrium diminishes. By contrast, with moderate degrees of regurgitation, the overall extent of shortening of the left atrium can actually increase, showing evidence of preload reserve. If atrial fibrillation develops, the left atrium continues to demonstrate a modest amount of elastic recoil in early diastole but mainly serves as a conduit between the pulmonary veins and the left ventricle.

The progressive increase in left ventricular end-diastolic volume results from the development of eccentric hypertrophy[11]. This form of hypertrophy is characterized by preservation of the ratio of wall thickness to radius (or mass to volume) despite the increase in chamber size. The stimulus (i) which initiates and perpetuates the increase in muscle mass and myofiber size is unknown, but there is strong circumstantial evidence that increase in diastolic wall forces play an important role. For many years it has been noted that the mean pulmonary arterial wedge pressure and the left ventricular end-diastolic pressure often remain within the normal range (<12 mm Hg) despite a two- to three-fold increase in end-diastolic volume. It has been inferred from these measurements that the ventricle undergoes "muscle creep" to explain this rightward shift in the passive pressure-volume curve. However, the pattern of hypertrophy seen in chronic mitral regurgitation, as well as the rightward shift in the diastolic pressure-volume coordinates, can both be explained on the basis of more sarcomeres being laid down in series where each sarcomere unit maintains a normal stiffness (or compliance) constant. Whatever the mechanism, it is clear that the diastolic adaptations of the left ventricle are of fundamental importance in minimizing symptoms for the patient with significant mitral valve leakage.

As with acute mitral regurgitation, the low impedance pathway, coupled with a normal aortic impedance, serves to maintain left ventricular afterload within a favorable range. Also, the development of eccentric hypertrophy, where more sarcomeres are laid down in both series and in parallel, serves to keep wall stress within normal range while allowing delivery of an increased stroke volume (forward plus regurgitation volume). Moreover, the left ventricular chamber becomes more spherical as the myocardium adapts to the volume over-load of mitral regurgitation. This geometric change has the bene-ficial effect of minimizing the increase in wall forces along the minor axes where most of the fiber shortening is occurring and con-tributes to preservation of normal muscle shortening. Thus, the patient with long-standing mitral regurgitation can compensate over many years for the increased work imposed by the mitral valve leakage and can maintain a normal forward stroke volume without incurring any adverse effects on myocardial contractility.

If the volume overload is excessively prolonged, however, left ventricular myocardial function begins to diminish, left ventricular empyting becomes less complete, and the end-systolic volume in-creases. In turn, the ventricle dilates further and moves onto a steeper portion of the pressure-volume relation. The patient then becomes symptomatic in concert with these pathophysiologic develop-ments. For the cardiologist, recognition of this depression of myocardial contractility may be a difficult clinical task since continued favorable unloading into the left atrium serves to maintain the ejection fraction within a normal range. Depression of myo-cardial contractility may only be uncovered once the mitral valve is replaced and there is an acute increase in the afterload on the left ventricle[12]. The task for the cardiologist, therefore, is to find parameter(s) of left ventricular size and function which allow detection of progressive dilatation and "silent" depression of myo-cardial contractility, preferably by techniques which can be per-formed serially and without risk.

At our institution, we have applied serially M-mode and 2-D echocardiography, as well as radionuclide angiography, to a cohort of patients with chronic mitral regurgitation and have found that end-diastolic and end-systolic left ventricular size, fractional shorten-ing, and muscle cross-section area are predictive of persistence of symptoms postoperatively, and in some cases, progressive worsening of left ventricular function. Patients with an end-diastolic diameter of 8 cm or greater, as well as an end-systolic diameter of 5.8 cm or greater, manifested a striking reduction in their ejection fraction over the first six months after surgery. Concomitantly, there was no detectable regression in myocardial hypertrophy. By contrast, patients with preoperative end-diastolic and end-systolic diameters of approximately 6 cm and 3.5 cm, respectively, demonstrated a slight reduction in their calculated ejection fraction at 6 months or more postoperatively and exhibited significant regression of hypertrophy, Figure 6. We have interpreted the fall in the ejection fraction and absence of regression of hypertrophy in the former group as indications that left ventricular afterload either increased or at least remained mismatched to the contractile state despite the cor-rection of the mitral valve leakage[12]. In the latter group, any increase in systolic afterload brought about by the closure of the low impedance leak was offset by the concomitant reduction of the left ventricular volume.

Recently, others have reported on the retrospective application of similar criteria to a group of patients with chronic mitral regurgitation and found that the combination of end-systolic stress and dimension measurements segregated patients with poor clinical outcomes and lack of regression of hypertrophy[13]. The ratio of end-systolic stress to end-systolic volume index has also been found to be more useful than ejection fraction and other hemodynamic measurements in predicting the outcome of mitral valve replacement[14]. It should be appreciated, however, that this index is only on coordinate on the end-systolic stress-volume relation and does not give the true slope. Nevertheless, from an empirical standpoint, the index appears to have the potential for identifying those patients who might have an unfavorable outcome once the low impedance leak is closed by corrective mitral surgery.

Presently, we utilize a constellation of criteria for recommendation of valvular surgery in the patient with long-standing mitral regurgitation, including: 1) symptomatic status (asymptomatic patients are not operated upon), 2) size of the left ventricle by echocardiography, 3) the end-systolic stress-volume ratio, 4) exercise performance on the treadmill, and 5) abnormal hemodynamic performance during cardiac catheterization, including an elevated pulmonary arterial wedge and pulmonary artery pressures at rest or exercise, and a subnormal cardiac output either at rest or exercise. Further careful pre- and postoperative analyses of patients undergoing mitral valve surgery, as well as improved constructs for assessment of myocardial contractility in the presence of mitral regurgitation, will be needed before more refined criteria can be developed for optimal timing of surgical intervention.

REFERENCES

1. W. C. Roberts, E. Braunwald, and A. G. Morrow, Acute severe mitral regurgitation secondary to ruptured chordae tendineae, Circulation, 33:58 (1966).
2. S. Sasayama, M. Takahashi, G. Osakada, K. Hirose, H. Hamashima, E. Nishimura, and C. Kawai, Dynamic geometry of the left atrium and left ventricle in acute mitral regurgitation, Circulation, 60:177 (1979).
3. E. H. Sonnenblick, J. Ross Jr., J. W. Covell, H. M. Spotnitz, and D. Spiro, Ultrastructure of the heart in systole and diastole, Circ.Res., 21:423 (1967).
4. C. Yoran, J. W. Covell, and J. Ross Jr., Structural basis for the ascending limb of left ventricular function, Circ.Res., 32:297 (1973).
5. M. Turina, W. D. Bussman, and H. P. Krayenbuehl, Contractility

of hypertrophied canine heart in chronic volume overload, Cardiovasc.Res., 3:486 (1969).

6. G. Cooper, F. Puga, K. J. Zujko, C. E. Harrison, and H. N. Coleman, Normal myocardial function and energetics in volume-overload hypertrophy in the cat, Circ.Res., 32:140 (1973).

7. J. Ross Jr., and W. H. McCullagh, The nature of enhanced performance of the dilated left ventricle during chronic volume overload, Circ.Res., 30:549 (1972).

8. H. Suga, K. Sagawa, and A. A. Shoukas, Load independence of the instantaneous pressure-volume ratio of the canine left ventricle and effects of epinephrine and heart rate on the ratio, Circ.Res., 32:314 (1973).

9. K. Sagawa, The end-systolic pressure-volume relation of the ventricle: definition, modifications and clinical use, Circulation, 63:1223 (1981).

10. E. Braunwald and W. C. Awe, The syndrome of severe mitral regurgitation with normal left atrial pressure, Circulation, 27:29 (1963).

11. W. Grossman, D. Jones, P. McLaurin, Wall stress and patterns of hypertrophy in the human left ventricle, J.Clin.Invest., 56:56 (1975).

12. G. Schuler, K. L. Peterson, A. Johnson, G. Francis, G. Dennish, J. Utley, P. O. Daily, W. Ashburn, and J. Ross Jr., Temporal response of left ventricular performance to mitral valve surgery, Circulation, 59:1218 (1979).

13. M. R. Ziles, W. H. Gaasch, J. D. Carroll, and H. J. Levine, Chronic mitral regurgitation: predictive value of preoperative echocardiographic indexes of left ventricular function and wall stress, J.Amer.Coll.Cardiol., 3:235 (1984).

14. B. A. Carabello, S. P. Nolan, and B. M. Lockhart, Assessment of preoperative left ventricular function in patients with mitral regurgitation: value of the end-systolic wall stress-end-systolic volume ratio, Circulation, 64:1212 (1981).

LEFT VENTRICULAR FUNCTION AND SURGICAL

TREATMENT OF MITRAL REGURGITATION

M. E. Rothlin

Departments of Medicine and Surgery
University Hospital Zürich
Switzerland

J. N. Corvisart in 1808 was the first to recognize mitral regurgitation in a 39 year-old man with an apical systolic murmur, who died subsequently of cardiac failure. An autopsy rupture of the chordae to the mitral valve leaflet was found. At the beginning of this century Sir Thomas Lewis tended to attribute cardiac failure more to myocardial disease and dyfunction than the mechanical effects of mitral regurgitation. Also the latter view is correct in a large number of patients presenting with an apical murmur in ischemic or hypertensive heart disease. The present discussion deals with the dysfunction of the mitral valve causing left ventricular failure.

Proper mitral valve opening and effective closure is influenced by:

1. Size, position and integrity of the anterior and posterior valve leaflets.
2. Length, integrity and insertion of the chordae tendineae.
3. Functional integrity of the papillary muscles.
4. Size, shape and functional integrity of the left ventricle.
5. Size and integrity of the mitral valve annulus.

According to Selzer (1972) chronic rheumatic heart disease remains the most common cause of chronic mitral regurgitation. The etiology of mitral regurgitation in 315 patients operated at the University Hospital in Zürich between 1972 and 1982 is depicted on Table 1. Mitral valve prolapse is a common condition occurring in a few percent of the general population, About 5 out of 100 subjects with mitral prolapse develop progressive mitral regurgitation. The primary pathologic process appears to be myxomatous degeneration of the fibrous skeleton of the valve leaflets and chordae tendineae.

Table 1.   Etiology of Mitral Regurgitation in 315 Patients
           Operated Between 1972 and 1983 at the University
           Hospital, Zürich

MYXOMATOUS DEGENERATION	35 %
RHEUMATIC	17 %
ISCHEMIC	5 %
INFECTIVE ENDOCARDITIS	12 %
COMBINED WITH CONGENITAL DEFECT	12 %
CARDIOMYOPATHY	5 %
REOPERATION	14 %

This etiology accounts for 35% of our surgical series of mitral
regurgitation.   Rheumatic heart disease was present in 17%.   Almost
half of the operated patients had mitral regurgitation caused by
infective endocarditis, ischemic heart disease, congenital defects,
cardiomyopathy or following a previous operation on the mitral valve.
In a recent series of patients with isolated, severe, chronic, pure
mitral regurgitation without any degree of stenosis, mitral valve
prolapse accounted for 62%, while rheumatic origin was shown in 3%
only (Waller and coll., 1982).

The surgical result of mitral valve replacement depends to
some extent on the etiology of mitral regurgitation as it will be
seen from reviewing our experience with mitral valve surgery in
315 patients operated between 1972 and 1982, who are listed on
Table 2.

Mitral Regurgitation as Part of Congenital Heart Disease

There were 7 patients with A-V-canal, 4 patients with coarctation and 3 patients with uncommon malformations like common
ventricle, requiring mitral valve surgery for mitral regurgitation.
The age of these patients varied between 1 and 48 years with a mean
of 18 years.   In a about half of these patients the mitral valve
could be reconstructed and in the other half it had to be replaced.
The former method required 2 reoperations.   Surgical mortality was
very high due to the complexity of the congenital lesion in 2
patients and to bad left ventricular function in 2 other patients
with concomitant coarctation of the aorta.   Late results were
satisfactory.

On the other hand there was 20 patients operated for mitral regurgitation due to mitral valve prolapse with a concomitant or previously operated atrial septal defect. Mitral valve prolapse is a common condition in atrial septal defect. These patients were much older, on the average 49 years with a range from 14 to 64. There was no early mortality and late results were good. Valve reconstruction or replacement were used in the same proportion. Two patients with plastic reconstruction required reoperation.

Mitral Regurgitation in Cardiomyopathy

In hypertrophic obstructive cardiomyopathy mild systolic reflux over the mitral valve is the rule. In 5 out of 67 patients operated for this condition alterations of the mitral valve by endocarditis or by other mechanisms caused considerable mitral regurgitation requiring surgical repair. Plastic construction of the mitral valve resulted in two reoperations out of four reconstructions. Therefore in this situation mitral valve replacement by a low profile prosthesis is now preferred (Senning and Rothlin, 1982).

The poor results of mitral valve surgery in 4 patients with dilatative cardiomyopathy do not justify this indication, there were 1 early and 2 late deaths.

The surgical removal of endomyocardial fibrosis necessitated replacement of the mitral valve in 5 patients to prevent severe postoperative mitral regurgitation (Hess and coll., 1979).

As can be expected, in primary myocardial disease the overall results of mitral valve surgery are less satisfactory.

Mitral Regurgitation Due to Ischemic Heart Disease

In the present series only 15 mitral regurgitations were caused by ischemic heart disease. The presented results appear to be more favorable than it has been reported previously; this may be due in part to a more selective indication for this operation. Salomon and coll. (1977) reported an over 50% mortality two years after mitral valve replacement for ischemic mitral regurgitation. In a more recent study on combined mitral valve surgery with aorto-coronary bypass grafting Di Sesa and coll. (1982) found a 10% operative mortality in elective cases and a 60% mortality in emergency operations. Their patients with combined coronary artery bypass grafting and mitral valve replacement had a similarly elevated surgical risk regardless of the etiology of mitral regurgitation.

Mitral Regurgitation Due to Infective Endocarditis

A considerable proportion of patients with infective endocarditis present with severe acute mitral regurgitation due to
ruptured chordae tendineae. Urgent or even emergency surgery may
therefore be necessary. Nine out of 37 patients reported were
operated urgently before antibiotic cure of the infection, mainly
because of shock or rapid hemodynamic deterioration. On the other
hand, once the risk of infection has been overcome, patients with
mitral regurgitation due to infective endocarditis, who did not have
long standing severe mitral disease before surgery, have most grati-
fying functional results.

Mitral Regurgitation After Previous Valve Surgery

The majority of patients with a reoperation on the mitral valve
had previously been operated by mitral commissurotomy or plastic
reconstruction of the mitral valve, some had severe paravalvular leak
or prosthetic valve dysfunction. The results in these 46 patients
show a higher operative and late mortality and a higher rate of
reoperations than in other etiology patient groups. The main cause
of this increased risk is long-standing and more advanced heart
disease apart from technical problems with reoperations (Rothlin et
al., 1980).

Mitral Valve Surgery for Mitral Regurgitation Due to Rheumatic
Disease and Myxomatous Degeneration

Fifty-four patients with rheumatic mitral regurgitation and
112 patients with mitral valve prolapse - mitral regurgitation form
the largest groups in this surgical experience. Dysfunction of the
mitral valve was the sole cause of left ventricular volume overload
and left ventricular failure at rest or on exercise in these
patients. The indication for surgery was severe functional limitation (NYHA III) or objective signs of left heart failure. The vast
majority of the operations were performed electively. For both
subgroups the early (one month) mortality was 2% and late mortality
during a mean observation time of 3 years was 8%.

Only the patients with atrial septal defect and mitral prolapse
had a similar success whereas all other subgroups had an average
early mortality of 8% and a late mortality of 16%, within a comparable follow-up of 3 years. Additional risk factors such as complex
congenital defect, primary or ischemic myocardial disease, infection
and technical problems in reoperations were responsible for the less
favorable results of these operations.

The cause of late postoperative mortality on patients with
rheumatic or mitral valve prolapse - mitral regurgitation was sudden

death in 4 and heart failure in 3. Valve related death, like prosthetic dysfunction (1) cerebral hemorrhage (2) and endocarditis (1)
were responsible for more than one third of the late mortality.

For the operative indication for mitral valve repair or mitral
valve replacement in mitral regurgitation we have to take into account the following factors:

1. As we have seen in a considerable number of patients additional
    congenitial defects, primary or ischemic myocardial disease,
    infection or previous surgery have to be considered.
2. None of the surgical methods for repair of mitral replacement is
    ideal yet. Mitral valve repair and bioprostheses have a low
    risk of thromboembolism, varying between 1% and 2% per annum in
    our own experience. In case of sinusrhythm patients do not
    require anticoagulants after these operations. The durability
    of bioprostheses and of plastic mitral valve repair, however, is
    limited; 5 years after implantation 11% of porcine bioprostheses
    were replaced in our experience and in another 22% of operated
    patients signs of late deterioration could be detected by echo-
    cardiography; also there were no corresponding manifestations.
    In case of mitral valve repair we observed 2% of thromboembolism
    per annum, but after 10 years 20% of the patients were re-
    operated. In contrast the mechanic prostheses require few
    reoperations but some patients die of failure of the prostheses.
    Thromboembolism occurs in about 8% per annum in mechanical
    mitral protheses and all patients need anticoagulation with its
    own risks. Therefore a considerable prosthesis related mor-
    tality must be weighed against the benefits that can be expected
    from the operation.
3. The operative and late mortality increase with severity and the
    advancing stage of disease. It is for this latter problem, that
    better understanding of left ventricular function in mitral
    regurgitation can help in finding the best timing for mitral
    valve surgery.

Experience from Long-term Follow-up After Mitral Valve Replacement

Searching for criteria indicating proper timing of mitral valve
replacement in mitral regurgitation several centers have analysed
their long-term surgical results in relations to numerous preoperative variables.

Dalby et al. (1981) found severe cardiomegaly., i.e. a cardiothoracic ratio of 0,75 or more and a left atrial diameter of 12 cm or
more to be a predictor of increased early and late mortality. Hemodynamic variables such as a low cardiac index, an elevated LVEDP or
an elevated pulmonary vascular resistance affected the operative
mortality only, but had no effect on late survival.

Hammermeister et al. (1978), using univariate life table analysis, reported on three variables that influence survival; namely age EDV and EF.

In a study by Salomon et al. (1977), age below 60, functional class III or less, cardiac index of 2.00 l/min/m2 or more and LVEDP of 12 mm or less were significant correlates of improved early and long-term survival.

In all of these retrospective studies on large surgical series an elevated LVEDV or LVEDP and a decrease of ejection fraction and cardiac index, reflecting impairment of left ventricular pump function, were found to influence perioperative and late results.

Effect of Mitral Valve Surgery for Mitral Regurgitation on Left Ventricular Function

The effect of surgical relief of mitral regurgitation on left ventricular performance as assessed by left heart catheteritation and quantitative angiography was documented by Kennedy et al. (1979) One year after operation cardiac index remained unchanged, but LVEDP fell markedly from 17 (preoperative) to 11 mm Hg postoperative. Mitral reflux was abolished, EDVI had slightly decreased but EDVI remained unchanged, reflecting a significant decrease of ejection fraction from 55% to 43%. Left ventricular mass remained constant, thus left ventricular hypertrophy had failed to regress. The author concluded that the decrease of ejection fraction was caused by an increase in afterload, which results from removal of the low resistance outflow from left ventricle to left atrium by mitral valve replacement.

Using angiographic and echocardiographic techniques Schuler and coll. (1979) confirmed these observations to some extent. However, in the latter study severely elevated left ventricular dimensions of more than 7 cm in diastole and more than 5 cm in systole indentified a subgroup of patients presenting with severely deteriorated post-operative pump function illustrated by marked decrease in ejection fraction and unchanged cross section area of the left ventricular wall representing absent regression of left ventriculars hypertrophy On the other hand patients with mitral regurgitation and moderately elevated end-diastolic and end-systolic dimensions and a normal or supranormal shortening fraction manifest postoperatively persisting normal left ventricular shortening and reduced left ventricular size Moreover they showed regression of left ventricular hypertrophy as assessed in the echocardiogram by reduced cross section area. It wa concluded, that mitral valve replacement or mitral regurgitation should be performed before irreversible dependence of left ventricular function on afterload reduction develops.

A decrease of ejection fraction after mitral valve replacement for mitral regurgitation was also reported by Phillips and coll. (1981) in the majority of a larger series of patients. A preoperative ejection fraction of 0.40 or less was correlated with reduced long-term survival in this report. Postoperatively ejection fraction decreased or failed to increase with exercise in about two-thirds of patients illustrating an abnormal response of left ventricular performance. It is of interest, that subjective results in these patients were favorable in spite of left ventricular deterioration.

In contrast to the before-mentioned studies Peter and coll. (1981) observed a significant increase of ejection fraction with exercise in their patients with mitral regurgitation before and after mitral valve replacement. Moreover these authors did not find a significant decrease of ejection fraction after mitral valve replacement, an observation, which is in line with findings reported by Morton (1980) and Kirschbaum (1981). Patient selection for surgery, timing of the postoperative studies, methods of assessment and operative techniques are some factors, that can be discussed to explain these discrepancies.

A major problem in assessing the effect of mitral valve replacement on left ventricular pump function is the difficulty experienced in separating the influence of removal of abnormal loading conditions from those of a primary change of myocardial function. Huikuri (1983) has published most recently, data on endsystolic wall stress/ endsystolic volume-ratio obtained before, and 12 months after, mitral valve replacement for mitral regurgitation. This hemodynamic index is much less dependant on loading conditions of the ventricle. After mitral valve replacement stress volume-ratio increased significantly in spite of a decrease of ejection fraction caused by changing loading conditions. This was interpreted as indication of improvement of left ventricular function. Moreover preoperative isometric exercise caused a significant fall of ejection fraction with no change of stress/volume-ratio. One year after mitral valve replacement, however, ejection fraction remained constant and stress/volume-ratio increased with exercise, suggesting an improving response of the left ventricle to exercise after surgical removal of mitral regurgitation.

Carabello et al. (1981) had previously reported in a small group of patients, that left ventricular stress/volume-ratio determined preoperatively was separating patients with improvement after mitral valve replacement for mitral regurgitation if the value was above 2,5 from those who died or did not improve, having a value below 2,5. Comparison of Carabellos observations with Huikuris data do not confirm, however, the selecting property of stress/volume-ratio for good or bad operative results, because one-half of Huikuris patients, who all survived with improvement, would meet with Carabellos criteria for an unfavorable result.

Importance of Mitral Apparatus for Left Ventricular Peformance

Finally we come to the importance of mitral apparatus in left ventricular function after surgical correction of mitral regurgitation.

Wiggers and Katz (1922) and Rushmer (1956) pointed out the role of continuity between mitral annulus and left ventricular wall through chordae tendineae and papillary muscles for normal ventricular function. Lellehei et al. (1964) applied this concept to surgery by preserving papillary muscles and chordae during mitral valve replacement. They observed a decrease in mortality and in postoperative low output syndrome.

In contrast Bjork et al. (1964) presented clinical observations and Rastelli et al. (1967) experimental findings indicating that preservation of papillary muscles and chordae does not affect left ventricular performance after mitral valve replacement.

Numerous reports and also our experience in Zurich during the 1960's revealed a lower operative mortality for mitral valve reconstruction than for mitral valve replacement. Preservation of the mitral apparatus with the former method may be important for this difference.

David et al. (1981) took up this concept again in 1981 performing mitral valve replacement in dogs with and without preservation of papillary muscle-chordae tendineae continuity, and found that it is important for left ventricular function. In their most recent study on patients with porcine prosthesis implantation preserving the posterior mitral leaflet with its suspension, they made most interesting findings as compared to conventional mitral valve replacement. With the two techniques preserving the mitral apparatus left ventricular end-diastolic volume and left ventricular end-systolic volume decreased significantly in contrast to an increase of left ventricular endsystolic volume after conventional mitral valve replacement. Ejection fraction fell only in the patients with resection of the mitral apparatus. It shall be most interesting to see, whether these exciting observations are going to be confirmed in a larger number of patients.

SUMMARY

Analysing the results of mitral valve surgery for mitral regurgitation in 315 patients revealed a number of factors depending on the etiology of mitral regurgitation, which influence the early and late results. Such determinators are complex congenital lesions, primary or ischemic myocardial disease, infection or previous surgery.

Table 2. Surgical Results Including Mortality, Postoperative Functional Status According to NYHA and Reoperations in 315 Patients Operated Between 1972 and 1982 as Related to the Etiology of Mitral Regurgitation.

Etiology of mitral regurgitation	n	operative death (< 30 days)	late death	functional class I	II	III	reoperated	no follow up
ASD secundum	22	0	2	3	11		2	4
other congenital heart disease	15	4	1	4	2		2	2
cardiomyopathy	14	1	4	0	4	1	0	4
due to coronary heart disease	15	0	2	3	5	1	0	4
infective endocarditis	37	1	5	11	10	2	2	6
reoperation	46	4	8	5	17	2	5	5
rheumatic	54	1	4	14	21	4	7	3
myxomatous de- generation	112	2	8	29	49	4	6	14
total	315 100 %	13 4.1 %	34 10.8 %	69 21.9%	119 37.6%	14 4.6%	24 7.6 %	42 13.4 %

For the proper timing of operation in mitral regurgitation
understanding of left ventricular pump function and of its response
to mitral valve replacement is important. Reviewing the literature
we found a relatively small number of studies dealing with this
subject.

A single hemodynamic parameter indicating the right timing of
surgery has not been established, and it is unlikely that it will be.
However, it becomes clear from the various reported findings, that
mitral valve replacement must be performed in chronic mitral regurgitation before irreversible damage of the left ventricle develops.

There is a renaissance of the old concept of the importance of
the mitral apparatus for normal ventricular function and it is possible, that it will influence surgical techniques in the future.

REFERENCES

Björk, V. O., Björk, L., and Malers, E., Left ventricular function
     after resection of the papillary muscles in patients with
     total mitral valve replacement, J.Thorac.Cardiovasc. Surg.,
     48:635 (1964).
Carabello, B. A., Nolan, S. P., and McGuire, L. B., Assessment of
     preoperative left ventricular function in patients with
     mitral regurgitation: Value of the end-systolic wall stress-
     end-systolic volume ratio, Circulation, 64:1212 (1981).
Dalby, A. J., Firth, B. G., and Forman, R., Preoperative factors
     affecting the outcome of isolated mitral valve replacement: a
     10 year review, Am.J.Cardiology, 47:826 (1981).
David, T. E., Uden, D. E., and Strauss, H. D., The importance of the
     mitral apparatus in left ventricular function after
     correction of mitral regurgitation, Circulation, 68:II-76
     (1983).
Di Sesa, V. J., Cohn, L. H., Collins, J. J., Koster, J. K., and Van
     Devanter, S., Determinants of operative survival following
     combined mitral valve replacement and coronary revascular-
     ization, Ann.Thorac.Surg., 34:482 (1982).
Hammermeister, K. E., Fisher, L., Kennedy, J. W., Samuels, S., and
     Dodge, H. T., Prediction of late survival in patients with
     mitral valve disease from clinical, hemodynamic, and quanti-
     tative angiographic variables, Circulation, 57:341 (1978).
Hess, O. M., Tunina, M., Senning, Å., Goebel, N. H., Schóler, Y.,
     and Krayenbühl, H. P., Pre- and post-operative findings in
     patients with endomyocardial fibrosis, Brit.Heart J., 40:406
     (1978).
Kennedy, J. W., Doces, J. G., and Stewart, D. K., Left ventricular
     function before and following surgical treatment of mitral
     valve disease, A.Heart J., 97:592 (1979).
Kirschbaum, M., Luncia, F., Germon, P., Maranhao, V., Cha, S. D., and
     Lemole, G., 1981, Ventricular function before and after

mitral valve replacement, J.Thorac.Cardiovasc.Surg., 82:752 (1981).

Lillehei, C. W., Levy, M. J., and Bonnabeau, R. C., Mitral valve replacement with preservation of papillary muscles and chordae tendineae, J.Thorac.Surg., 47:532 (1964).

Morton, M. J., Bohnsted, S. W., Pantely, G. A., and Rahimtoola, S., Effect of successful mitral valve replacement on left ventricular function, Circulation, 62:III-208 (1980).

Peter, C. A., Austin, E. H., and Jones, R. H., Effect of valve replacement for chronic mitral insufficiency on left ventricular function during rest and exercise, J.Thorac.Cardiovasc.Surg., 82:127 (1981).

Phillips, H. R., Levine, F. H., Carter, J. E., Boucher, C. A., Mitral valve replacement for isolated mitral regurgitation: Analysis of clinical course and late postoperative left ventricular ejection fraction, Am.J.Cardiology, 48:647 (1981).

Rastelli, G. C., Tsakiris, A. G., Frye, R. L., and Kirklin, J. W., Exercise tolerance and hemodynamic studies after replacement of canine mitral valve with and without preservation of chordae tendineae, Circulation, 35:I-34 (1967).

Rothlin, M. E., Egloff, L., Kugelmeier, J., Turina, M., and Senning, Å., Reoperations after valvular heart surgery, indications and late results, Thorac.Cardiovasc.Surg., 28:71 (1980).

Rushmer, R. F., Initial phase of ventricular systole: Asynchonour contraction, Am.J.Physiol., 184:188 (1956).

Salomon, N. W., Stinson, E. B., Griepp, R. B., and Shumway, N. E., Patient-related risk: Factors as predictors of results, following isolated mitral valve replacement, Ann.Thorac. Surg., 24:519 (1977).

Schuler, G., Peterson, K. L., Johnson, A., Francis, G., Temporal response of left ventricular performance to mitral valve surgery, Circulation, 59:1218 (1979).

Selzer, A., Nonrheumatic mitral regurgitation, Med.Concepts Cardiovasc.Dis., 48:25 (1979).

Senning, Å., and Rothlin, M. E., Chirurgie bei hypertropher obstruktiver kardiomyopathie, Z.Kardiologie, 71:806 (1982).

Waller, B. F., Morrow, A. G., Maron, B. J., and Del Negro, A. A., Etiology of clinically isolated, severe, chronic, pure mitral regurgitation: Analysis of 97 patients over 30 years of age having mitral valve replacement, Am.Heart J., 104:276 (1982).

Wiggers, C. S., and Katz, L. M., Contour of the ventricular volume curves under different conditions, Am.J.Physiol., 58:439 (1922).

VENTRICULAR DYSFUNCTION IN HYPERTROPHIC CARDIOMYOPATHY

R. D. Leachman

Texas Heart Institute
Houston, USA

More than 25 years have passed since Sir Russell Brock[1] proposed the idea of functional obstruction to left ventricular outlet.
Within another year from that date, Donald Teare[2] published the
pathological finding encountered in eight hearts with severe but
asymmetric hypertrophy affecting the interventricular septum and
because of cellular disarray likened it to a "muscular hamartoma".
Spodick and Littman[3] emphasized the malignant nature of idiopathic
hypertrophy when symptoms had appeared in a publication about the
same time. The physiologic and clinical features of obstructive
cardiomyopathy were detailed by Braunwald with the conclusion that
there was a dynamic obstruction to left ventricular outlet during
systole.

Although twenty years have passed since Braunwald's presentation, there is still some disagreement as to the presence or
absence of obstruction[4,5] whether the hypertrophy is primary or
secondary, and the role of coronary blood flow in causing pain and
ventricular dysfunction. There is general agreement that there is
resistance to left ventricular filling with a decrease in ventricular
compliance and in ventricular distensibility, a point emphasized by
Goodwin[6].

The disease is clinically characterized by the presence of:
1) asymmetric septal hypertrophy greater than the posterior wall and
2) systolic anterior motion of the mitral valve. These findings are
not specific but are almost always found in those patients with
obstruction. Similarly, the histologic appearance of myocardial
fiber disarray occurs in a high percentage of patients with HOCM,
but also occurs in other forms of secondary ventricular hypertrophy.
Whether this histologic appearance is related to the degree of

severity of hypertrophy rather than to a specific disease is an open
question. There seems little doubt that the anterior leaflet of the
mitral valve and hypertrophy of the interventricular septum are
somehow related in this disease. The question is whether one causes
the other.

The combinations of abnormalities that can occur in this disease
make it difficult to find a homogeneous group of patients for study.
The degree of left ventricular outflow obstruction, presence or
absence of mitral regurgitation, and the presence or absence of
coronary artery disease complicate the analysis of ventricular function.

The abnormality of cardiac function can be related to: 1) left
ventricular outlet obstruction, 2) resistance to left ventricular
filling, 3) mitral valve regurgitation, 4) associated coronary artery
disease, and possibly 5) functional myocardial ischemia and 6) wall
stress factors produced by regional contraction differences (mid-
cavity obliteration).

There are those who continue to believe there is no obstruction[4,5]. The argument against the existence of left ventricular
outflow impedance has been based principally on the observation that
the ventricle empties more completely and at least at the beginning
of systole more rapidly than normal. They have concluded that all
the blood is ejected in early systole. Left ventricular emptying is
undoubtedly greater than normal, and the ventricle does appear to be
hypercontractile. Ejection fractions at rest of 75-80% are usual.
There is, however, a prolonged ventricular ejection time. There is a
pressure gradient between left ventricle and aorta. The presence of
"jet lesions" on the septal endocardium[7] the ventricular aspect of
the anterior mitral leaflet and the ventricular surface of aortic-
valve leaflets, I believe, is anatomic evidence of obstructed flow.
Abnormal position and attachment of the mitral valve has also been
described in some patients[8,9,10,11]. Two of six hypertrophic
cardiomyopathic hearts that we reviewed had abnormal chordae
tendinae, passing directly from the valve leaflet to the interventricular septum, an anomaly resulting in an abnormal position of
the anterior leaflet. The good correlation of systolic anterior
motion of the anterior leaflet with left ventricular outlet
gradient[12] is further support of obstruction.

The fact that mitral valve replacement alone in patients with
HOCM results in disappearance of resting and provocable outflow
gradients is also a strong argument that there is hemodynamic obstruction and that it is indeed produced by the anterior leaflet of
the mitral valve. In these surgically treated cases[13-15], the
disappearance of early systolic closure of the aortic valve is evidence that the obstruction has been eliminated[16].

Angina, dyspnea, and syncope or near syncope can all be the consequence of left ventricular outlet obstruction. They can be caused by other abnormalities that occur in this disease.

If the obstruction to systolic emptying is an important part of the pathophysiology of HOCM, then relief of the obstruction should result in clinical improvement. This has indeed proved to be true with most patients improved with regard to angina, dyspnea, and syncope[17,18] as well as an apparent improved survival.

Additionally, we[19] have studied by echocardiogram five patients who had only mitral valve replacement as treatment of HOCM an average of 48 months previously. No muscle was resected. After relief of the obstruction, the ventricular septal thickness, measured at outflow tract level, decreased from 2.2 cm to 1.6 cm. This leads us to conclude that there is a causal relation between the obstruction and the hypertrophy. If the obstruction is unimportant, it seems unlikely that the hypertrophy would disappear by only replacing the mitral valve.

Reduction of left ventricular volume in patients with HOCM results in deterioration of left ventricular function with decreased cardiac output and increased obstruction. Afterload reduction such as occurs with amyl nitrate or decrease in pre-load such as with blood loss or valsalva maneuver produce an increase in outflow gradient and often exaggerate symptoms.

Mitral regurgitation results in such change by decreasing resistance to ventricular emptying. In our experience 30% of patients with HOCM have moderate to severe mitral regurgitation, a finding similar to other reports[20]. It is our belief that the development and progression of mitral regurgitation correlates with clinical worsening in some patients. When there is severe regurgitation, medical treatment results in little improvement in our experience, and in these patients, the onset of atrial fibrillation with rapid ventricular rate results in deterioration.

Asymmetric hypertrophy and systolic anterior motion of the mitral valve are the echographic characteristics of HOCM, resistance to left ventricular filling is the functional marker. The ventricular distensibility is reduced and resistance to diastolic filling is increased. These diastolic abnormalities may be secondary to the severe ventricular hypertrophy or there may be associated dynamic changes in wall relaxation that resist normal ventricular filling. The result is an elevated left ventricular end-diastolic pressure with an associated prominent A wave without an increased diastolic volume. There may be left atrial and pulmonary venous hypertension as a consequence. Ventricular filling seems dependent on the atrial contraction and atrial fibrillation results in a great drop in cardiac output. Gotmans and Lewis[21] studied 14 patients with HOCM

and measured compliance by five methods. Diastolic pressure in early
and late diastole was increased over normal at corresponding left
ventricular volumes. They had no data to study the temporal changes
in ventricular filling.

Sanderson[22] and others studied 20 patients with HOCM and
measured changes in left ventricular volume and shape during one
cardiac cycle. By relating volume changes to mitral valve motion,
they calculated the isovolumic relaxation period, peak filling rates
and duration of rapid filling. They concluded that there was a
disorder of ventricular relaxation with prolonged isovolumic relaxation time and prolonged rapid filling phase with normal peak
rates of LV volume increase. Only in late diastole did the elastic
behavior of the ventricle seem to determine filling. They speculated
that the prolonged isovolumic relaxation might reduce subendocardial
blood flow. The resemblances of delayed relaxation to ischemic
disease was noted and their patients with chest pain had a longer
isovolumic relaxation period than those without. Coronary arterio-
grams were not described in this group.

Hanrath[23] and associates came to similar conclusions from an
analysis of echocardiograms in patients with HOCM and chronic pressure overload from aortic stenosis or severe systemic arterial hypertension as compared to normals. The status of the coronary arteries
was not mentioned. They found that end-systolic and end-diastolic
volumes were smaller than normal in patients with HOCM.

The isovolumic relaxation period (minimal LV dimension to beginning mitral valve opening) was prolonged. The rapid filling period
was noted to be shorter than normal. Similar changes were noted in
those patients with chronic left ventricular overload but were quan-
titatively less pronounced.

It seems likely therefore that these changes in diastolic filling are a manifestation of the degree of hypertrophy and not necessarily a specific feature of HOCM. An imbalance between oxygen
demand and delivery in severe hypertrophy may contribute to some of
these functional changes. It is of interest, however, that
Hanrath[25] was able to improve (shorten) isovolumic relaxation time
in patients with HOCM by giving intravenous verapamil. A similar
effect has been shown to occur with nifedipine[26]. Whether this
will have a specific effect on angina remains to be seen.

The changes in diastolic function after surgery are inconsis-
tent[17,24]. End-diastolic pressure has been improved in some but
not all. In a group of patients with only mitral valve replacement,
it was noted that end-diastolic pressure had normalized within two
weeks after surgery. The left ventricular volumes calculated from
single projection angiograms were larger than pre-operatively. This
observation led us to speculate that the mitral valve, chordae

tendinae, and papillary muscles contribute to the dynamic resistance of left ventricular filling, a part of the mechanism regulating total ventricular compliance. Removal of these structures could eliminate a check-rein effect and permit greater filling.

Ischemic heart disease results in slow ventricular relaxation and can change ejection fraction and patterns of left ventricular contraction. Since many patients with HOCM do have angina, we reviewed our own experience with regard to the frequency of significant coronary artery disease in a group of patients with HOCM who had coronary arteriographic study. It has been said that the epicardial coronary arteries are most often normal[27]. We found a rather striking 16 of 75 patients with significant coronary disease. There were even 12 of the 16 patients who had saphenous vein bypass grafting of coronary arteries with or without additional surgical treatment of HOCM. This finding, particularly in those patients with HOCM over 40 years of age, in our opinion, emphasizes the need for coronary arterial study. To assume that angina and other physiologic disturbances result from HOCM in the absence of this information, seem illogical.

Scattered myocardial necrosis and actual infarct in patients with normal epicardial arteries certainly occurs. Small vessel compression, the well-known "milking effect" seem angiographically, or spasm, may produce a more global ischemia. The contribution of such ischemia to the abnormal ventricular function is known in any patient, but is one of the factors that make analysis difficult.

Exercise in the patient with HOCM results in an increase in heart rate, increased cardiac output, an increase in left ventricular end-diastolic pressure and an increase in pulmonary capillary wedge pressure. Stroke volume at rest and exercise in three patients that we have studied (cardiogreen output ÷ heart rate) did not change.

The ability of the patient with HOCM to increase cardiac output appears to be principally dependent on his ability to increase heart rate and not end-diastolic volume. This likely explains why many such patients refuse to take large dose propanalol, since the heart rate slowing effect of that drug interferes with the ability to increase cardiac output during exercise.

We observed that phenomenon in the three above mentioned patients who, about 1/2 hour after a total of 5 mg of intravenous propanolol, were unable to increase their heart rates with exercise as before, and could not increase cardiac outputs since stroke volume remained constant.

Recent calculations of ejection fraction at rest and exercise[28] by gated nuclear angiography confirm that at rest there is

hypercontractility with resting ejection fraction of 75%-80% with
little further change with exercise.

We have selected a subset of patients with hypertrophic cardiomyopathy with angiographic features of mid-cavity obliteration.
There appear to be different pathophysiologic consequences[29]. The
measured pressure gradient is well into the cavity of the left ventricle (raising the question of catheter entrapment) with an apical
aneurysm being present and with no blood ejected from the apex in
systole. The diastolic hemodynamics are similar if not identical to
HOCM. Histologic examination of these aneurysms resected at surgery
confirm the presence of scarring consistent with ischemia, but in the
presence of normal coronary arteries.

It is tempting to speculate that in the apical cavity because of
the midcavity obliteration, the very high systolic pressure and
perhaps intramyocardial arterial compression may produce ischemic
necrosis, but principally, at the apex, where the pressures are so
high. Two of four patients observed under medical treatment for an
average of 10 years developed apical aneurysms and the other two has
severe global hypokinesia.

SUMMARY

Hypertrophic cardiomyopathy is not a rare disease that occurs of
unsure cause and has a strong familial tendency. The characteristics
of the disease are: 1) left ventricular outlet obstruction produced
by the anterior leaflet of the mitral valve, 2) severe ventricular
hypertrophy with the interventricular septum usually thicker than the
left posterior wall.

The above abnormalities result in abnormal systolic and diastolic function of the left ventricle. There is prolonged left ventricular ejection, hypercontractility with almost complete ventricular emptying in systole, emptying of the intramyocardial
arteries during systolic compression, resistance to diastolic filling
with left atrial hypertension both in late diastole (elastic compliance) and in early diastole (dynamic relaxation). These physiologic variables are complicated by the frequent association of
coronary arteriosclerosis, varying degrees of myocardial fibrosis,
and the presence of variable degrees of mitral valve regurgitation.

There is little argument that there is abnormal anatomic
relation between the anterior leaflet of the mitral valve and the
interventricular septum. It can be assumed that the "hamartoma" of
the septum distorts the normal relation. I believe that the mitral
valve, hereditarily, is abnormal in its position and that the hypertrophy is secondary. This idea is supported by the number of
patients with abnormal attachments of the mitral valve to the upper

septum. It is further supported by the observed regression of hypertrophy after isolated mitral valve replacement, a change that should not occur unless the hypertrophy is secondary.

It is, lastly, tempting to speculate that the greater hypertrophy of the ventricular septum and some of the fiber disarray is indeed a result of selectric hypertrophy of the portion of the septum related to the papillary muscle, a consequence of the unusual stress on these structures.

REFERENCES

1. R. Brock, Functional obstruction of the left ventricle (acquired aortic subvalvar stenosis), Guy's Hospital Report, 221-238 (1957).

2. D. Teare, Asymmetrical hypertrophy of the heart in young adults, Brit.Heart J., 20:1-8 (1958).

3. D. H. Spodick and D. Littman, Idiopathic myocardial hypertrophy, Am.J.Cardiol., 1:610-623 May (1958).

4. R. Shabetai, Cardiomyopathy, J.Am.Coll.Cardiol., 1:252-263 (1983).

5. J. P. Murgo, B. R. Alter, J. F. Dorethy, and S. A. Altobelli, Dynamics of left ventricular ejection in obstructive and non-obstructive cardiomyopathy, J.Clin.Invest., 66:1369-1382 (1980).

6. J. F. Goodwin, Prospects and predictions for the cardiomyopathies, Circulation, 50:210-219 (1974).

7. P M. Shah, Newer concepts in hypertrophic obstructive cardiomyopathy, JAMA, 242:1663-1665 (1970).

8. V. O. Bjork, G. Hultquist, and H. Lodin, Subaortic stenosis produced by an abnormally placed anterior mitral leaflet, J.Thorac.Cardiovasc.Surg., 41:5, 659-669 (1961).

9. P. F. Moberg and H. Soderberg, On the pathogenesis of idiopathic hypertrophic subaortic stenosis, J.Cardiovasc.Surg., 4:602 (1963).

10. J. E. Edwards, Pathology of left ventricular outflow tract obstruction, Circulation, XXXI:586-599 (1965).

11. R. D. Sellers, C. W. Lillehei, and J. E. Edwards, Subaortic stenosis caused by anomalies of the atrioventricular valves, J.Thorac.Cardiovasc.Surg., 48:2, 289-302 (1964).

12. W. L. Henry, C. E. Clark, J. M. Griffith, and S. E. Epstein, Mechanism of left ventricular outflow obstruction in patients with obstructive asymmetric septal hypertrophy (Idiopathic hypertrophic subaortic stenosis), Am.J.Cardiol., 35:337-345 (1975).

13. J. G. Eillis, O. J. Terneny, W. L. Winters, and R. D. Leachman, Critical role of the mitral valve leaflet in hypertrophic subaortic stenosis and amelioration of the disease by mitral valve replacement, Chest, 59:4.378-382 (1971).

14. H. B. Schumacker and H. King, New operative approach in the management of hypertrophic subaortic stenosis, J.Thorac. Cardiovasc.Surg., 49:3,497-503, March (1965).

15. D. A. Cooley, R. R. Grace, D. C. Wukasch, and R. D. Leachman, Replacement and/or repair of the mitral valve as treatment of idiopathic hypertrophic subaortic stenosis, Cardiovasc.Dis., 3:4,381-393 (1976).

16. Z. Krajcer, F. Orzan, L. W. Pechacek, E. Garcia, and R. D. Leachman, Early systolic closure of the aortic valve in patients with hypertrophic subaortic stenosis and discrete subaortic stenosis: Correlation with preoperative and post-operative hemodynamics, Am.J.Cardiol., 41:823 (1978).

17. B. J. Maron, W. H. Merrill, P. A. Freier, K. M. Kent, S. E. Epstein, and A. G. Morrow, Long-term clinical course and symptomatic status of patients after operation for hyper-trophic subaortic stenosis, Circulation, 57:6,1205-1213 (1978).

18. R. L. Reis, M. R. Bolton, J. F. King, D. M. Pugh, M. I. Dunn, and D. T. Mason, Anterior-superior displacement of papillary muscles producing obstruction and mitral regurgitation in idiopathic hypertrophic sub-aortic stenosis, Circulation, 49 and 50:Suppl.II, 181-188 (1974).

19. R. D. Leachman and Z. Krajcer, Significant reduction of asym-metrical septal hypertrophy following mitral valve replace-ment in patients with idiopathic hypertrophic subaortic stenosis, Abs.Circulation, 64:Suppl.IV:314 (1981).

20. A. G. Adelman, M. J. McLoughlin, and Y. Marquis, et al., Left ventricular cineangiographic observations in muscular sub-aortic stenosis, Am.J.Cardiol., 24:689 (1969).

21. M. S. Gotsman and B. S. Lewis, Left ventricular volumes and compliance in hypertrophic cardiomyopathy, Chest, 66:498-505 (1974).

22. J. E. Sanderson, D. G. Gibson, D. J. Brown, and J. F. Goodwin, Left ventricular filling in hypertrophic cardiomyopathy: An angiographic study, Brit.Heart J., 39:661-670 (1977).

23. P. Hanrath, D. G. Mathey, R. Siegart, and W. Bleifeld, Left ventricular relaxation and filling pattern in different forms of left ventricular hypertrophy: An echocardiographic study, Am.J.Cardiol., 45:15-23 (1980).

24. A. G. Adelman, E. D. Wigle, N. Ranganathan, G. D. Webb, B. S. L. Kidd, W. G. Bigelow, and M. D. Silber, The clinical course in muscular subaortic stenosis: A retrospective and prospective study of 60 hemodynamically proved cases, Ann.Int.Med., 77:515-525 (1972).

25. P. Hanrath, D. G. Mathey, P. Kremer, F. Sonntag, and W. Bleifeld, Effect of verapamil on left ventricular isovolumic relaxation time and regional left ventricular filling in hypertrophic cardiomyopathy, Am.J.Cardiol., 45:1258 (1980).

26. B. Lorell, W. Paulus, and W. Grossman et al., Improved diastolic compliance on hypertrophic cardiomyopathy treated with

nifedipine, Abs. 53rd Scientific Session, <u>Am.Heart Assoc.</u>, III:317 (1980).

27. S. E. Epstein, W. L. Henry, and C. E. Clark et al., Asymmetric septal hypertrophy, <u>Ann.Int.Med.</u>, 81:650-680 (1974).

28. J. S. Borer, S. L. Bacharach, and M. V. Green, et al., Effect of septal myotomy and myomectomy on left ventricular function at rest and during exercise in patients with IHSS, <u>Circulation</u>, Suppl:60, I-82-87 (1979).

29. R. D. Leachman and Z. Krajcer, Surgical versus medical therapy in patients with hypertrophic cardiomyopathy and mid-cavity obliteration, Abstract submitted to 56th Scientific Sessions Meeting, <u>Am.Heart Assoc.</u>, November (1983).

RIGHT VENTRICULAR FAILURE IN CONGENITAL HEART DISEASE

S. Dalla Volta, R. Razzolini, L. Daliento, R. Chioin
N. John and G. Cuman

Cattedra di Cardiologia Università
Padova, Italy

INTRODUCTION

Angiographic estimation of right ventricular volumes in congenital heart disease is now beginning as an interesting procedure in many catheterization laboratories[1.2]. Associated to the anatomical analysis of the different lesions and to the study of the hemodynamic parameters (pressures, flows and shunts), it has proved its role in the prediction of the outcome of many congenital lesions, with and without corrective surgery. Moreover, in our experience in some inborn defect with a spectrum of functional anomalies, as pulmonic atresia with intact ventricular septum, the analysis of the morphology of the ventricles and of their volumes has proved the possibility of suggesting the most adequate surgical correction in single cases.

In spite of some well known drawbacks, the Simpson's rule[3] has shown to be an accurate method of assessment of the right ventricular volumes, even in cases of complex internal geometry of the right ventricle in comparison to the left.

The results so far obtained with this approach to the complete understanding of many congenital heart defects have prompted the revision of our data, as obtained during the usual heart catheterization.

PATIENTS AND METHODS

The study has been performed on 24 patients with Tetralogy of Fallot, 10 patients with Transposition of the great arteries

(4 simple, 6 with VSD or VSD and PS), 1 patient with tricuspid insufficiency of the newborn, 15 patients with pulmonary atresia with intact ventricular septum, 24 patients with pulmonary stenosis, and 36 patients with ventricular septum defect.

In all cases a right ventricular cineangiogram was obtained with biplane technique, and submitted to quantitative analysis.

The right ventricular volume was calculated with Graham's modification of Simpson's rule[2]. In cases of pulmonary atresia, the tricuspid valve diameter was measured in end diastole using antero-posterior view of the right ventricular cineangiogram.

Statistical analysis was performed: linear correlation was calculated with least square technique, comparison between groups was obtained with t-test.

RESULTS

Tetralogy of Fallot

Before total repair, right ventricular volumes are essentially normal (Table 1), with a slightly reduced ejection fraction, probably in relation to the myocardial changes from the anoxia. This is in agreement with the rare occurrence of heart failure in TOF. After surgery, volumes are increased, while ejection fraction remains constant. This behavior is due to multiple factors: 1) relief of pulmonary stenosis, which lowers right ventricular afterload; 2) site and extension of ventriculotomy and influence of the inert material of the patch; 3) right ventricular volume overload due pulmonary or tricuspid valve incompetence or residual left to right shunts[4,5]. In Figure 1 a semiquantitative relationship is shown between right ventricular volume and severity of right ventricular volume overload.

Transposition of the Great Arteries

Right ventricular end-diastolic and end-systolic volumes are significantly larger in simple TGA than in complicated forms (PS,

Table 1. Tetralogy of Fallot

	PREOP	POSTOP	
RVEDV	56.4 + 20.4	105 + 45.9	.001
RVESV	29 + 13.7	53.6+ 33	.001
RVEF	48.6 + 12.7	51.5+ 11	NS
P/RVESV	3.86 + 1.5	1.08+ .6	.001

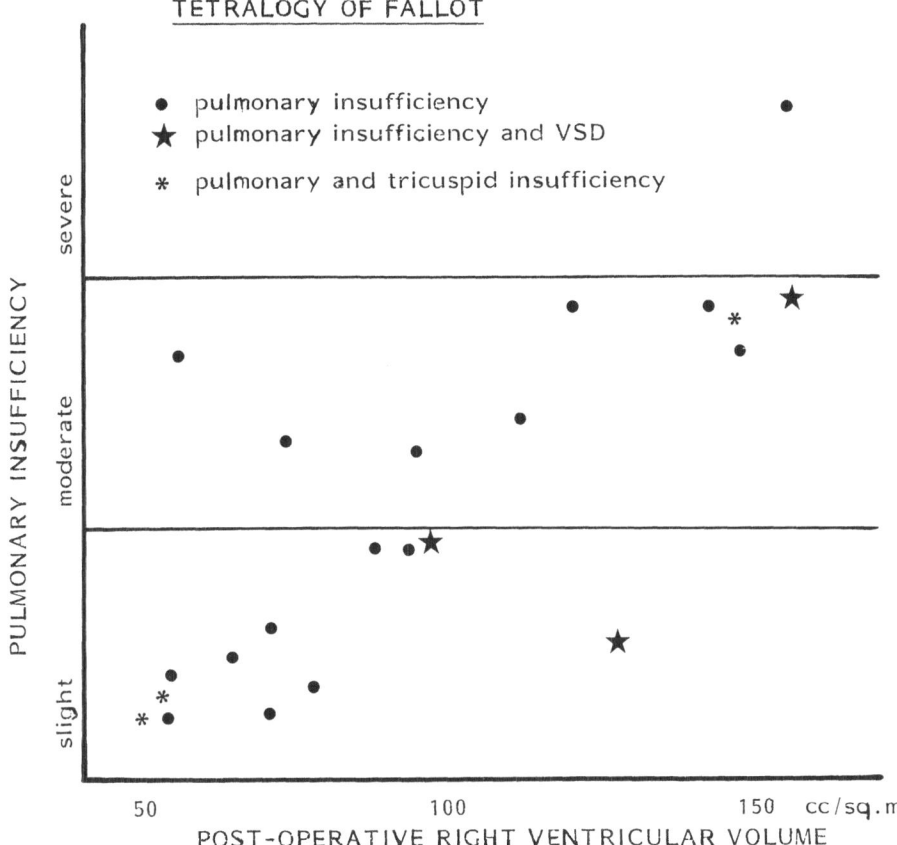

Fig. 1. Semiquantitative relationship between right ventricular
         volume and severity of right ventricular volume overload in
         TOF after surgery.

VSD + PS, VSD + aortic coarctation). Ejection fraction is cor-
respondingly reduced (Table 2). Arterial oxygen saturation seems to
be inversely related to volumes (Figure 2). Therefore, simple TGA
appears as the most severe form of transposition in terms of right
ventricular function and oxygen saturation.

Transient Tricuspid Insufficiency

In our case (Table 3) RVEDV was large at birth and diminished
toward normal size at 4 months. However, ejection fraction did not
increase at all. This could mean that some myocardial depression is
present together with diastolic overload. Anoxia in perinatal period
due to various causes, induces vasoconstriction of the exceedingly
sensitive pulmonary vasculature. This could explain both tricuspid
incompetence and right ventricular failure[6].

Table 2.   Transposition of the Great Arteries

	Simple	TGA + PS	TGA + CIV	TGA+CIV+PS	TGA+CIV+CoA	TGA+CIV+PDA
RVEDV	88.3 +/− 29.3	64.8 +/− 17.7	45.4 +/− 15.5	66.9 +/− 21.6	63.1 +/− 21.9	51
RVESV	41.8 +/− 12.9	27.3 +/− 7.0	16.5 +/− 7.2	25.9 +/− 13.8	23.7 +/− 16.3	21.5
RVEF	.51 +/− .06	.56 +/− .04	.63 +/− .09	.61 +/− .09	.63 +/− .14	.58

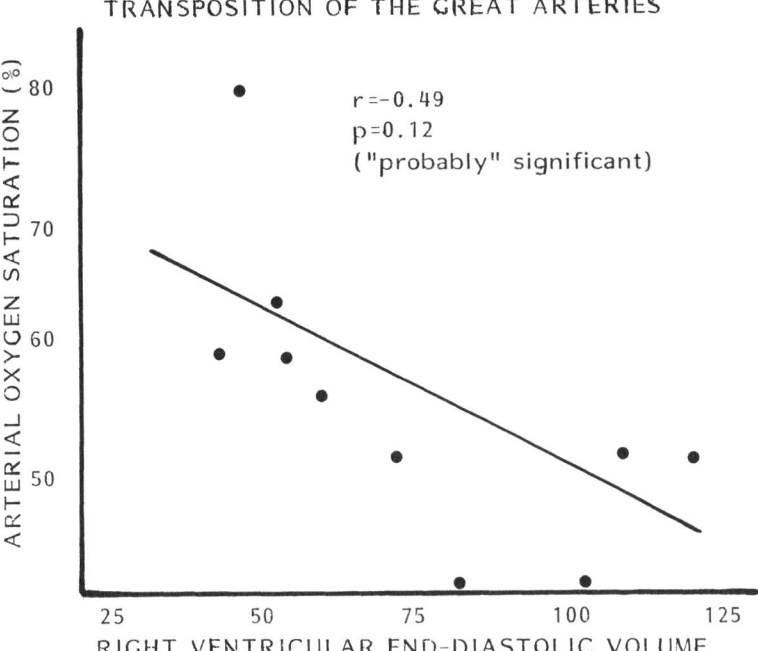

Fig. 2.   Inverse relationship between arterial oxygen saturation and
          right ventricular end-diastolic volume in transposition of
          the great arteries.

Pulmonary Atresia with Intact Ventricular Septum

In this disease right ventricular volumes do not follow a unique
pattern. In some cases volumes are strikingly reduced, whereas in
others are normal or even increased. This is related to the morphology of the tricuspid valve, whether restrictive or incompetent. We
found a linear relationship between diameter of the tricuspid orifice
and right ventricular end-diastolic volume (r=.857, p<.001). Ejection fraction is normal in normal sized ventricles and depressed both
in undeveloped or excessively large ventricles (Figure 3). Right
ventricular mass evaluation is not possible with angiography: anatomical studies, however, demonstrated that in every case the reduced
function is related to a deficient mass. The right ventricle appears
hypoplastic, with thin wall rich in sinusoids, and more or less
similar to Uhl's disease.

Pulmonary Stenosis

Right atrial pressure increases rapidly according to gradient
(Figure 4), while end-diastolic volume does not. This denotes a
substantial increase in stiffness, which may cause failure. Pumping
capability assessed as pressure to volume ratio shows a non-linear

Table 3.   Transient Tricuspid
           Insufficiency of the
           Newborn

	BIRTH	4 MONTHS
C/T ratio	0.8	0.57
Art. ox. sat.	68%	95%
RVEDV	109	84
RVESV	46	34
RVSV	63	50
RVEF	58%	59%

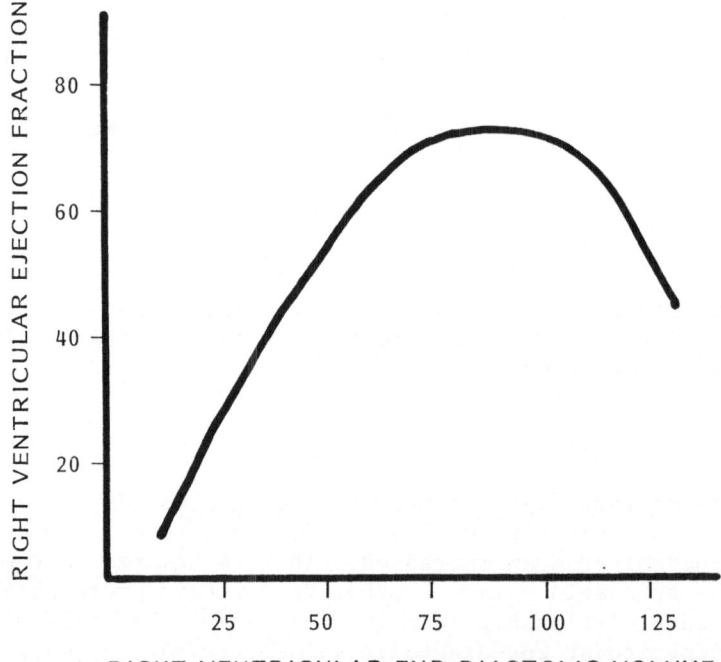

Fig. 3.   Depression of right ventricular ejection fraction in
          undeveloped or excessively large ventricles in pulmonary
          atresia with intact ventricular septum.

behavior, as its peak value is in moderate gradient range, and is
depressed both for lowest and for highest pressures[7].

Ventricular Septum Defect

Cardiac failure is usually present, together with pulmonary
hypertension, usually from the third month of life. The Qp/Qs ratio

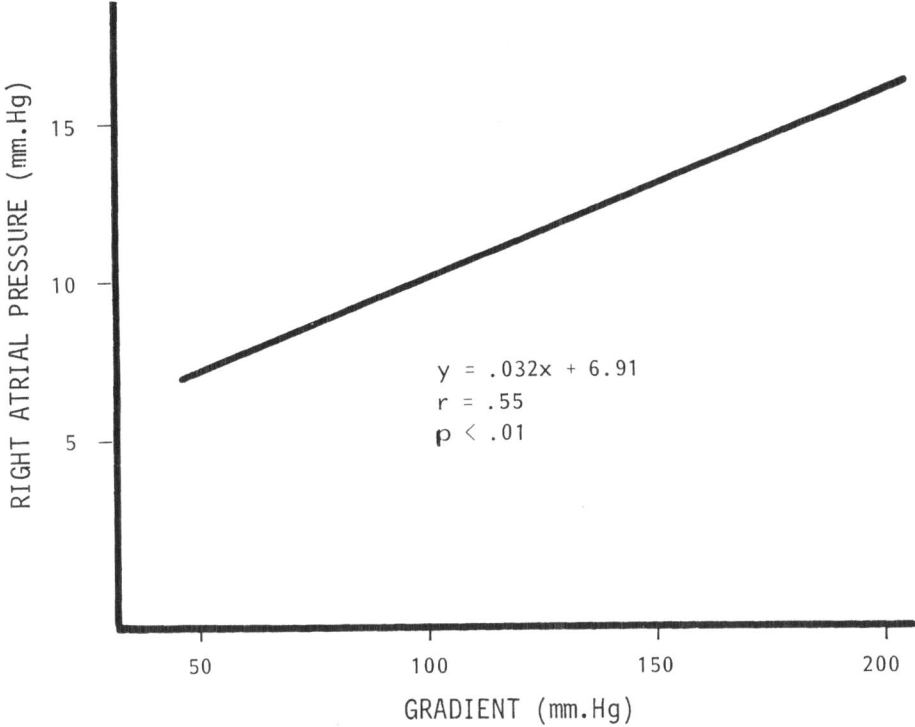

Fig. 4.   Positive linear correlation between gradient and right
          atrial pressure in pulmonary atenosis.

is usually high, above 4:1, the pulmonary vascular resistances are
below 5 units. The size of the defect and the ratio of left to right
shunt are the most reliable indices of the importance of RVF for a
same level of pulmonary artery pressure, both in isolated large VSD
and in complete common A/V canal defect[8].

DISCUSSION

     This study has proved at the same time the technical feasibility
and the prognostic-pathophysiological importance of the study of the
right ventricular volumes in a non selected group of congenital
cardiac defect, as integrant part of the intracavitary study.

     With particular reference to the topic "congenital defect and
right ventricular failure", some points deserve mention, in comparison with the conclusions drawn solely on the basis of hemodynamic
study and anatomical description of the different lesions.

     Congenital heart disease are the usual cause of the right ven-
tricular failure in babies, children and adolescents, rarely in

adults. In spite of the rather uniform clinical presentation, the
mechanisms of the failure are multiple and related to the basic
disfunction and to the compensative adaption. The inadequacy of the
right ventricle to sustain the burden of increased volume and/or
pressure load or to compensate the reduced cardiac contractility
results in the clinical picture of the congestive heart failure.

The various and not constant interplay between these mechanisms
may be summarized in few fundamental causes:

a) the altered model of the prenatal circulation;
b) the combination of the systolic and diastolic overloads
c) a rare primitive impairment of the inotropic state of the
   ventricle(s);
d) the sudden change in the pattern of the circulation, paradox-
   ically, improved by the intracardiac anatomic repair of some
   complex inborn lesions.

While the ages does not seem to play an important role in the
aspects of the same disease, the characteristics of the altered
contraction and of the pump mechanism of the ventricles in the
different disease are relevant.

Before the birth the right ventricle sustains 60% of the combined cardiac output at the same level of pressure and resistances of
the left ventricle: the large shunts between the two circulations
help to maintain the flows. After birth the appearance of RVF is
more or less rapid, due to different combination of the above listed
mechanisms; in some cases the cardiac failure is delayed and appears
only at an advanced age. The main causes and most relevant mechanisms leading to RVF can be summarized as follows:

a) in the tricuspid insufficiency of the newborn the right ventricular failure is usually transient, and associated to the delayed
regression of the prenatal pattern of pulmonary circulation. The
right end diastolic volumes are high (about 110 ml), the ejection
fraction is only slightly depressed (55-57%), the cardiac index
mildly reduced. These parameters show a tendency toward normalization (but do not reach totally normal values) six to eight months
after clinical recovery.

b) Pulmonary atresia with intact ventricular septum invariably
leads RVF and shows three different patterns of insufficiency: small
right ventricular end diastolic volumes in 55% of cases; normal end
diastolic volumes in 25% and slightly increased volume in 20%. In
the first two subsets the ejection fraction is reduced. The volumes
are well correlated to the size of the tricuspid valve. Surgical
risk can now be accurately predicted on the basis of these patterns.

c) In the Ebstein disease the cardiac failure can be associated either to a large and to a small right "ventricular cavity": the anomaly of the insertion of the tricuspid valve and the degree of thinness of the true right ventricular cavity show the best relationship with the degree of the RVF.

d) Pulmonary stenosis with intact ventricular septum is a rather rare cause of RVF, due to the adequate hypertrophy of the right ventricle. However, in presence of cardiac insufficiency, the preload reserve is insufficient, the contractility of the ventricle mildly depressed, the compliance of the ventricle greatly reduced. The difficulty to perform adequate quantitative measurements of the right ventricle prevents calculation of afterload, which is probably increased.

e) In the defect of the ventricular septum with pulmonary hypertension, cardiac failure is usually present from the third month of life. Usually the Qp/Qs ratio is high, above 4:1, the pulmonary vascular resistances below 5 units. The size of the defect and the amount of the left to right shunt are the most reliable indices of the importance of RVF for a same level of pulmonary artery pressure, both in isolated large VSD and in the complete common A/V canal defect.

f) Very infrequently, except in patients older than 55 years of age, an isolated atrial septum defect is a cause of the RVF. A very large shunt with increased pulmonary artery pressures and atrial fibrillation are accompanied by very large end diastolic volumes, reduced ejection fraction, usually normal coronary arteries. The long duration of the disease is an important mechanism favoring the appearance of the cardiac insufficiency, usually correctable through the surgery.

g) Transposition of the great arteries exhibit the clinical picture of the cardiac failure after the first week of age, in coincidence with the rapid increase of the heart. In simple TGV, the RV end diastolic volumes are the highest and the ejection fraction the most reduced in comparison with the other forms of transposition. with VSD or VSD and subpulmonary stenosis (where the left ventricular volumes too are increased). The reduction of arterial pO2 shows the better correlation with the degree of impairment of the RV, suggesting the predominant role of myocardial anoxia.

h) Tetralogy of Fallot, unless associated to pulmonic insufficiency, is a rare cause of RVF. The end diastolic volumes of the right ventricle are normal, reduced in the left ventricle. The right ventricular ejection fraction is moderately reduced, probably in relation to the myocardial changes from anoxia. After the total repair, the ejection fraction parallels the clinical picture. When the patient is in cardiac failure (incomplete relief of the obstruc-

tion, residual large intraventricular shunt, large ventriculotomy),
the ejection fraction is greatly reduced, even below 20% while the
end diastolic volume of the right ventricle is not excessively increased.

i)   In restrictive cardiomyopathies (excluding infiltrative
disease of the heart) the reduction of the RV function is usually
associated to similar changes for the left ventricle. The pressures
are augmented in the right side of the heart, the right ventricular
volumes slightly augmented or normal, the compliance greatly reduced;
end systolic right ventricular volumes are normal, suggesting the
reduction of the distensibility of the right ventricle as the main
mechanism of the symptoms. Tricuspid insufficiency, when present, is
functional in its nature.

As general conclusions, the degree of accuracy of the functional
condition of the right ventricle can be evaluated, and seems satisfactory over a wide range of ages and different hemodynamic abnormalities. The comparison of data before and after major surgical
interventions in the same patient shows good correlation between the
functional conditions of the right ventricle and clinical status of
the patient. Tetralogy of Fallot, complete transposition of the
great vessels, pulmonic atresia with intact ventricular septum have
proved that the differences in the results of the surgery can be
explained much more clearly when a complete analysis of the anatomical details is associated to the study of the right ventricular
volumes. The natural history of some disease, as tricuspid insufficiency of the newborn or Ebstein's disease have a much closer
interplay with the different level of the right ventricular volumes
than with the anatomy of the lesion. The choice of the best surgical
treatment which can be offered to the individual patient seems to
benefit largely from the determination of the right ventricular
volumes, as the cases of tricuspid atresia and pulmonic atresia show.

The different age of the patients studied with this technique
does not seem to represent a major obstacle in the analysis of the
right ventricular volumes, as already stressed by Lange et al., one
year ago.

Taking into account the more compete data that can be obtained
by the study of the left ventricular function, this study emphasizes
the need of a complete study of the heart in all the congenital heart
disease. The study of the right ventricular function and the quan-
titation of the right ventricular volumes may be extremely helpful in
the management of these patients.

REFERENCES

1. R. Razzolini, R. Scognamiglio, L. Daliento, G. M. Boffa, F. Corbara, A. Ramondo, R. Chioin, and P. Stritoni, Valutazione in vivo dei metodi angiografici per la misurazione del volume ventricolare destro, Boll.Soc.It.Cardiol., 22:2271 (1977).

2. T. P. Graham, J. M. Jamarkani, G. F. Atwood, and R. V. Canent, Right ventricular volume determination in children, Circulation, 47:144 (1973).

3. I. Ferlinz, Angiographic measurement of right ventricular volume and ejection fraction, Cathet.Cardiovasc.Diagn., 2(1):5 (1976).

4. T. P. Graham, Right ventricular volume characteristics before and after palliative and reparative operation in tetralogy of Fallot, Circulation, 54:417 (1980).

5. P. E. Lange, G. W. Dietrich, G. W. Ormasch, A. Bernhard, and P. H. Heintzen, Left and right ventricular overload before and after surgical repair of tetralogy of Fallot, Am.J.Cardiol., 50:786 (1982).

6. G. M. Boffa, P. Stritoni, R. Chioin, R. Razzolini, L. Daliento, G. Fasoli, and S. Dalla-Volta, Transient tricuspid insufficiency of the newborn. Report of a case and further considerations, Giorn.It.Cardiol., 10:907 (1980).

7. R. Razzolini, R. Scognamiglio, L. Daliento, G. M. Boffa, F. Corbara, P. Stritoni, and R. Chioin, Emodinamica e angiografia quantitativa della stenosi polmonare valvolare isolata, Giorn.Emodinam., 1:59 (1981).

8. S. Dalla-Volta, L. Daliento, R. Scognamiglio, G. M. Boffa, P. Stritoni, A. Ramondo, M. Barbiero, and R. Chioin, Ventricular septal defect: medical aspects and surgical considerations, in: "Congenital Heart Disease in the First Three Months of Life," Bologna, p.403 (1981).

INVASIVE ASSESSMENT OF LEFT VENTRICULAR

FUNCTION IN ISCHEMIC HEART DISEASE

A. Chiddo, A. Gaglione, D. Quagliara and P. Rizzon

Division of Cardiology
University of Bari
Bari, Italy

Left ventricular function considerably affects prognosis and choice of treatment in patients with coronary artery disease (CAD). Nevertheless, a correct assessment of left ventricular (LV) performance, irrespective of the influence of pre- and after-load, is a hard and still unresolved problem, although solving it has been the aim of a number of clinical and experimental studies.

The indexes, mostly systolic, used in the past show well-known limitations such as 1) a considerable dependence on pre- and after-load variations; 2) a wide range of normal values which, in many instances, prevents the identification of contractile impairment in individuals. Yet, the reliability of the indexes obtained from the end-systolic force-length relationship, like Suga and Sagawa's Emax, proposed as "pure" inotropic indexes[1,2,3], has not yet been sufficiently demonstrated in humans; moreover, complexity of calculation hinders their routine use in cardiac carheterization laboratories. Recently, in studies performed on a wide series of patients with CAD, two indexes which are relatively simple to determine, have proved to be sensitive to latent impairments of LV function: the ejection fraction of the first third of systole (1/3 EF) [4,5,6,7,8,9], and left ventricular T (T) − a time constant expressing the rate of fall of left ventricular pressure during isometric diastole, active energy-dependent process[10,11]. In fact, it has been shown that left ventricular T index is the first to be impaired during transient ischemia induced by percutaneous transluminal angioplasty[12].

It has been the aim of our study to compare the reliability of the more commonly used systo-diastolic LV performance indexes with that of some recently proposed indexes like 1/3 EF and T, in a group of 48 patients, all with normal blood pressure and in sinus rhythm who were submitted to cardiac catheterization and angiography for diagnostic purposes. The patients were selected according to the presence of standard LV performance indexes within normal range (see Table 1).

According to clinical, ECG and angiographic findings, the patients were separated into three groups:

Group A: 30 patients with CAD (critical narrowing, $\geq$ 75%, of luminal diameter of at least one main coronary vessel), without previous myocardial infarction;

Group B: 11 patients with CAD and previous myocardial infarction, but without dyskinetic areas on left ventricularography;

Group C: 7 patients without cardiac disease.

Besides standard performance indexes, the following were evaluated: 1/3 EF, T, wall stress in the different phases of systole and diastole according to Mirsky's formula[13], volume stiffnes module Kp and wall stiffnes module Ks, aortic compliance according to Windkessel's model. All pressure recordings were obtained by Millar micromanometer-tipped catheters with two sensors (left ventricle and aorta), and were syncronized with cine-ventriculography frames.

Table 1.

	GROUP C	GROUP A	GROUP B
HR (beats/m)	80 ± 6	72 ± 13	76 ± 13
CI (ml/m^2)	3.2 ± 0.39	3.2 ± 0.65	3.1 ± 0.59
LVSP (mmHg)	125 ± 9	130 ± 10	127 ± 14
AoP (mmHg)	80 ± 9	78 ± 9	75 ± 7
Ao COMPLIANCE	1.3 ± 0.2	1.2 ± 0.3	1.3 ± 0.5
LVEDP (mmHg)	9 ± 2	11 ± 3	12.5 ± 1.9
EF (%)	63 ± 2.4	62 ± 7	60 ± 7
ESV (ml/m^2)	17.6 ± 4.8	29 ± 9	29.5 ± 10
EDV (ml/m^2)	65 ± 16	88 ± 16	90 ± 16
VCF (circ/sec)	1.8 ± 0.3	1.2 ± 0.2	1.2 ± 0.1

RESULTS AND DISCUSSION

All systolic and diastolic indexes (except global EF) showed
significant differences when comparing groups A and B with normal
subjects (see Table 2). In agreement with the data reported in
literature, 1/3 EF appeared to be more sensitive than global EF in
disclosing latent impairments of contractility. One possible hypothesis to explain this greater sensitivity takes in account wall
stress, which reaches its peak value early in systole. Still, while
this is true in normal subjects, it is also accepted that peak stress
is displaced towards mid-systole in CAD as a result of "tardokinesis"[14]. Indeed, no significant correlation was found in our CAD

Table 2.

	GROUP C	GROUP A	GROUP B
EF (%)	63 + 2.4	62 + 7 n.s.	60 + 7 n.s.
ESV (ml/m^2)	17.6 + 4.8	29 + 9 ($p < 0.01$)	29.5 + 10 ($p < 0.01$)
VCF (circ/sec)	1.8 + 0.3	1.2 + 0.2 ($p < 0.001$)	1.2 + 0.1 ($p < 0.001$)
EF 1/3 (%)	32 + 9	18 + 9 ($p < 0.001$)	15 + 4 ($p < 0.001$)
ES-STRESS (gr/cm^2)	207 + 37	328 + 135 ($p < 0.001$)	330 + 107 ($p < 0.001$)
LVEDP (mmHg)	9 + 2	11 + 3 ($p < 0.005$)	12.5 + 1.9 ($p < 0.001$)
EDV (ml/m^2)	65 + 16	88 + 16 ($p < 0.001$)	90 + 16 ($p < 0.01$)
T (sec)	35 + 1	43 + 11 ($p < 0.005$)	42 + 7 ($p < 0.001$)
ED-STRESS (gr/cm^2)	85 + 2	88 + 35 — * — ($p < 0.005$)	98 + 18 ($p < 0.05$)
Kp (10^{-3})	20 + 3	22 + 9 n.s.	24 + 7 ($p < 0.01$)
Ks	0.14 + 0.01	0.15 + 0.05 — ** — n.s.	0.19 + 0.09 ($p < 0.02$)

** $p < 0.02$

* $p < 0.05$

groups between 1/3 EF and wall stress of the first third of systole. Furthermore, wall stress in early systole was found to be comparable in the groups, whereas a significant difference between CAD patients and control group was found only in mid- and late-systole (see Table 3).

Left ventricular T was significantly prolonged in Groups A and B; no correlations appeared to exist between T and end-systole stress, and between T and end-systolic volume (ESV). Conversely, T shows significant correlation with heart rate (HR). These data are in agreement with Brutseart's hypothesis of a dual control of iso-metric diastole, the two processes involved being the load-effect, that is the effect of geometric deformation of the left ventricle at the end of ejection, and the inactivation of active processes con-nected with the action of the calcium-sequestering pump[15,16,17, 18,19]. When oxygen supply is inadequate, the latter mechanism becomes prevalent and masks load-dependence effects, the overall result being a prolonged isometric relaxation[20,21].

Among the considered systolic and diastolic indexes, end-diastolic stress and Ks alone were found to be able to distinguish CAD patients with previous myocardial infarction from those without myocardial infarction and thus proved to be the most sensitive to structural changes in a myocardium that is still efficient in terms of global performance[13,22,23].

SUMMARY

The main systolic and diastolic indexes of left ventricular function were evaluated in 41 subjects with coronary artery disease (CAD) selected according to absence of dyskinetic areas and of evident left ventricular dysfunction (EF - EDP - ESV - VCF within normal limits) and in a group of 7 normal subjects.

Table 3.

		SYSTOLIC TREND OF WALL STRESS	
	FIRST THIRD	SECOND THIRD	LAST THIRD
GROUP C	629 ± 66	503 ± 73	207 ± 37
GROUP A	665 ± 180	652 ± 207	328 ± 135
	n.s.	(P<0.005)	(P<0.001)
GROUP B	649 ± 112	629 ± 139	330 ± 107
	n.s.	(P< 0.005)	(P<0.001)

Amongst these, 1/3 EF and left ventricular T appeared to be the most sensitive indexes and those that are apt to detect latent left ventricular impairment.

Within the CAD groups only ED-stress and Ks were capable to separate patients with and without myocardial infarction, thus proving their sensitivity toward structural myocardial damage.

REFERENCES

1. K. Sagawa, H. Suga, A. Shoukas, and K. Bakalar, End-systolic Pressure-volume ratio: A new index of ventricular contractility, Am.J.Cardiol., 40:748 (1977).
2. K. Sagawa, The end-systolic pressure-volume relation of the ventricle: Definition, modifications and clinical use, Circulation, 63, No. 6:1223 (1981).
3. H. Suga and K. Sagawa, Instantaneous pressure-volume relationships and their ratio in the excised, supported canine ventricle, Circulation Res., 35:117 (1974).
4. A. Battler, R. Slutsky, J. Karliner, V. Froelicher, W. Ashburn, and J. Ross, Left ventricular ejection fraction and first third ejection fraction early after acute myocardial infarction: Value for predicting mortality and morbidity, Am.J.Cardiol., 45:197 (1980).
5. A. Cribier, J. Berland, F. Prigent, and B. Letac, Etude angiographique de la fraction d'éjection par tiers de systole chez le coronarien, Arch.Mal.Coeur., 6:641-652 (1982).
6. J. G. Dumesnil and R. M. Shoucri, Effect of the geometry of the left ventricle on the calculation of ejection fraction, Circulation, 65, No. 1, 91 (1982).
7. L. Johnson, K. Ellis, D. Schmidt, and P. J. Cannon, Volume ejected in early systole: A sensitive index of left ventricular performance in coronary artery disease, Circulation, 52:378 (1975).
8. R. Slutsky, D. Gordon, J. Karliner, A. Battler, S. Walaski, J. Vaba, M. Pfisterer, K. Peterson, and W. Ashburn, Assessment of early ventricular systole by first pass radionuclide angiography: Useful method for detection of left ventricular dysfunction at rest in patients with coronary artery disease, Am.J.Cardiol., 44:459 (1979).
9. R. Slutsky, J. S. Karliner, A. Battler, K. Peterson, and J. Ross, Comparison of early systolic and holosystolic ejection phase indexes by contract ventriculography in patients with coronary artery disease, Circulation, 61, No. 6:1083 (1980).
10. J. L. Weiss, J. W. Frederiksen, and M. L. Weisfeldt, Hemodynamic determinants of the time-course of fall in canine left ventricular pressure, J.Clin.Invest., 58:751-760 (1976).
11. W. Maughan, K. Sunagawa, and M. L. Weisfeldt, Time constant of relaxation: Influence of peripheral resistance, capacitance

and characteristic impedance, Circulation Abstr., 66,Suppl.II (1982).

12. U. Sigwart, M. Grbic, A. Essinger, A. Fischer, D. Morin, and H. Sadeghi, Myocardial function in man during acute coronary ballon occlusion, Circulation Abstr., 66, Suppl.II (1982).

13. I. Mirsky, P. F. Cohn, J. A. Levine, R. Gorlin, M. V. Herman, T. H. Kreulen, and E. H. Sonnenblick, Assessment of left ventricular stiffness in primary myocardial disease and coronary artery disease, Circulation, 50:128 (1974).

14. F. Cucchini, G. Baldi, A. L. Barilli, M. DiDonato, and O. Visioli, Tardokinesis in coronary artery disease: Evidence with instantaneous analysis of left ventricular ejection, Eur.J.Cardiol., 12:153-166 (1980).

15. D. L. Brutseart, P. R. Housmans, and M. A. Goethals, Dual control of relaxation: Its role in ventricular function in the mammalian heart, Circulation Res., 47,5:637 (1980).

16. J. D. Carrol, O. M. Hess, R. Widmer, H. O. Hirzel, and H. P. Krayenbuehl, The nature of inotropic stimulation and diastolic mechanics in man, Circulation, Abstr. 66, Suppl.II (1982).

17. M. M. Conhenye, N. M. DeClerck, M. A. Goethals, and D. L. Brutseart, Relaxation properties of Mammalian atrial muscle, Circulation Res., 48,3:352 (1980).

18. M. Hori, M. Inone, M. Fukunami, Y. Ishida, S. Nakajima, M. Kitakaze, A. Kitabatake, and H. Abe, Influence of ejection timing on left ventricular relaxation in isolated canine heart, Circulation Abstr., 66:Suppl.II (1982).

19. B. S. Lewis and M. S. Gotsman, Current concepts of left ventricular relaxation and compliance, Am.Heart J., 99,1:101 (1980).

20. J. D. Carrol, O. M. Hess, H. O. Hirzel, and H. P. Krayenbuehl, Exercise-induced ischemia: The influence of altered relaxation on early diastolic pressure, Circulation, 67,3,521 (1983).

21. H. Pouleur, M. Rousseau, W. Wijns, P. Mengeot, J. M. Detry, and L. Brasseur, Changes in left ventricular relaxation pattern during atrial pacing in ischemic heart disease, Circulation Abstr., 66:Suppl.II (1982).

22. J. Tyberg, J. S. Forrester, and W. W. Parmley, Altered segmental function and compliance in acute myocardial ischemia, Eur.J. Cardiol., 1/3:307-317 (1974).

23. S. A. Glantz and W. W. Parmley, Factors which affect the diastolic pressure-volume curve, Circulation Res., 42,2:171 (1978).

24. A. Blaustein, J. D. Carrol, O. H. L. Bing, and W. H. Gaasch, Preload does not alter myocardial relaxation rate: Studies in the intact heart and isolated muscle, Circulation, Abstr. 66, Suppl.II (1982).

25. A. P. Flessas and T. J. Ryan, Left ventricular diastolic capacity in Man, Circulation, 65,6:1197 (1982).

26. H. N. Sabbah and P. D. Stein, Incompatibility with the concept of passive left ventricular filling, Circulation Res., 48,3:357 (1981).

27. C. F. Sanford, M. L. Smucker, and K. Lipscomb, Dissociation of impaired left ventricular relaxation and diastolic filling during angina pectoris, Circulation Abstr., 66, Suppl.II (1982).

28. D. S. Thompson, C. B. Waldron, S. M. Juul, N. Nagui, R. H. Swanton, D. J. Coltart, B. S. Jenkins, and M. M. Webb-Peploe, Analysis of left ventricular pressure during isovolumic relaxation in coronary artery disease, Circulation, 65,4:690 (1982).

NON-INVASIVE ASSESSMENT OF LV SYSTOLIC FUNCTION AND DIASTOLIC

FILLING AT REST AND DURING EXERCISE IN CORONARY HEART DISEASE

P. Assennato, B. Candela, E. Hoffmann, G. Indovina,
L. Messina and A. Raineri

Cattedra di Fisiopatologia Cardiovasolare
Università di Palermo, Policlinico, Palermo, Italy

INTRODUCTION

In the assessment of the main determinants of cardiac function the study of the systolic phase and diastolic filling appear to be of paramount importance[1-3].

Nowadays such routine assessments can be carried out non-invasively by radiounclide angiography. This is an extremely important acquisition as abnormalities in left ventricular systolic and diastolic performance are common in coronary patients. Our work has the following objectives:

1) the assessment, by radionuclide angiography of the indices of left ventricular systolic function and diastolic filling, and any possible relation between the two;

2) the assessment of the effects of exercise on these indices both in normal and in coronary patients;

3) the relation between systolic and diastolic function and left ventricular functional reserve.

MATERIAL AND METHODS

The study was carried out on fifteen healthy male subjects with a mean age of 45, and 42 patients with ECG evidence of previous transmural myocardial infarction (Q wave>0.004 sec. of duration). 9 patients had had anterior infarction, 13 anter-lateral infarction, 11 inferior infarction, and 9 infero-lateral infarction.

245

At the time of the study all the patients were asymptomatic with normal artery pressure values and no clinical signs of heart failure or any valvular abnormalities.

The radiouclide angiography was carried out with a nonimaging nuclear probe, connected to a microprocessing computer which calculates cardiac volume (Nuclear Stethoscope Bios Inc.)[4].

The red blood-cells were labelled with a 99m technethium tracer dose by means of stannose pyrophosphate (Tc99m RBC). The dose was predetermined for each patient by an isotope calibrator (RAD/CAL II Victoreen) and ranged on average from 20 to 25 mCi[5]. After the injection of the radio-isotope, the probe was placed over the area of the left ventricle at an angle of 20° to 40° in an oblique left anterior position and a caudal angle of 5° to 10°. Since it is impossible to make use of anatomical pictures, the correct position of the probe is determined by the best combination between diastolic count and maximum stroke volume. This is obtained by locating the point where the greatest systo-diastolic flow can be detected, to which must correspond the highest end-diastolic count. This is signalled by the computer by means of a luminous sign on the monitor (bar); to be optimal the bar must reach the maximum length on the scale of those observed during the evaluation[6].

The next step is to locate the background area (BKG), by moving the detector into an immediately infero-lateral position, so that the curve is not periodic and the end diastolic count begins to drop.

Both the ventricular and BKG positions are marked on the patient's skin in order to facilitate the repositioning of the probe. After counting BKG activity the nuclear detector automatically acquires count rates by synchronizing them with the ECG signal and arranges the values in a left ventricular time-activity curve for many consecutive beats (Figure 1)[7].

Blood radioactivity is proportional to blood volume, so after the BKG correction the time-activity curve represents a measure of the change of relative left-ventricular volume with time.

The parameters used to describe the systolic function are the Ejection Fraction (EF) and the Peak Ejection Rate (PER). The parameters of diastolic function are: Peak Filling rate (PFR) and Time to Peak Filling Rate (TPFR). The EF is calculated by the formula CTD-CTS/CTD-CBKG. The CTD is given by the end-diastolic count, the CTS by the end-systolic count, and the CBKG by the background activity count[8]. The PER and PFR were found in the left ventricular time-activity curve (Figure 1). This graph turns out to be described by a third-degree polynomial function; the separate evaluation of the individual stages of the curve, the systolic and the diastolic, is automatically performed by the computer algorithm through the

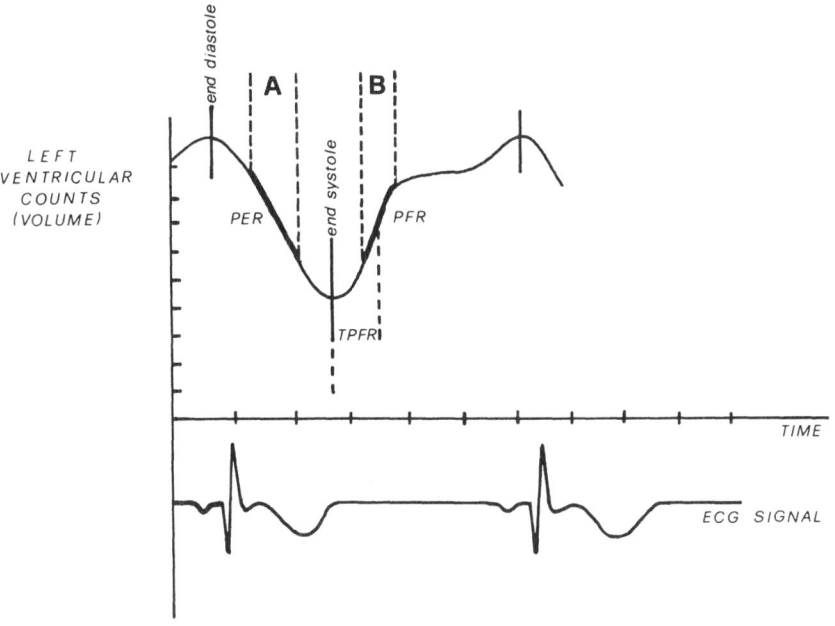

LEFT VENTRICULAR VOLUME CURVE

A: *SYSTOLIC EJECTION PHASE*
B: *DIASTOLIC FILLING PHASE*

Fig. 1. Left ventricular volume vs. time (time-activity) curve.

consecutive variations of the first derivative dv/dt, in a precise
area of the curve (Figure 1) which is composed of at least 8-10
consecutive points or 80-100 msec[9].

As regards th PER, if the proto-systole and the end-systole are
excluded the rest of the systolic phase has a linear pattern, so that
this value can be found along the descending phase of the curve as
shown in Figure 1 [9].

The diastolic phase of the volume curve and the relative dv/dt
cause greater problems of analysis than the pattern of the systolic
phase.

This is connected with the fact that diastole consists of three
clearly defined stages: rapid filling, diastasis, and atrial contraction. From the literature it emerges that the dv/dt for the diastolic phase or the PFR should be calculated in the rapid filling
stage, corresponding to the ascending phase of the curve (Figure 1B),
because it is then that approximately 80% of ventricular filling
occurs and there are the most important modifications affecting
muscular cell relaxation[10].

Both the PER and the PFR were calculated in left ventricular counts per second, and then expressed an end-diastolic counts per second (EDV/sec)[9].

The TPFR was calculated as the time intervening between the start of diastole and the highest point of PFR (Figure 1)[9]. These parameters were evaluated at rest, and during the fifth minute of a maximal or symptom-limited exercise, performed in the supine position, on a bicycle ergometer; all therapeutic treatment was suspended 48 hours before assessment.

Multiple variable analysis was applied in order to verify the reciprocal correlation on the data obtained. The analysis was carried out by means of multiple regression and Student "t", to compare the various samples. Calculations were worked out on an HP9826 computer.

Results

The systolic and diastolic function values at rest and during exercise in 15 normal subjects are given in Figure 2. The mean EF value at rest is 65±5%; during exercise this becomes 81±6%, with a +20 percent change.

The PER exhibits the same pattern: it moves from a mean value at rest of 3.17±0.13 EDV/sec to a mean value of 4.45±0.7 EDV sec during exercise, with a +28 percent change.

As regards diastolic filling the mean PFR value in normal subjects at rest is 3.08±0.5 EDV/sec; during exercise this becomes 5.48±1 EDV/sec. with a +43 percent change.

The TPFR shows a decrease during exercise. From a mean rest value of 144±20 msec it passes to a mean value of 103±10 msec during exercise, with a -27 percent change. The linear regression shows a multiple correlation of 0.26 for the rest values and of 0.50 for values during exercise and, as can be seen from the correlation matrix both at rest and during exercise, there is no reciprocal correlation between EF, PER. PFR and TPFR.

The values of systolic and diastolic function in 42 patients with a previous infarction and their pattern during exercise are shown in Figure 3.

In patients with previous infarction the mean EF at rest is 53±10%; this becomes 51±14% during exercise with a -4 percent change

The mean value of PER at rest is 2.52±0.4 EDV/sec; during exercise it is 2.64±0.8 EDV/sec, with +0.5 percent change.

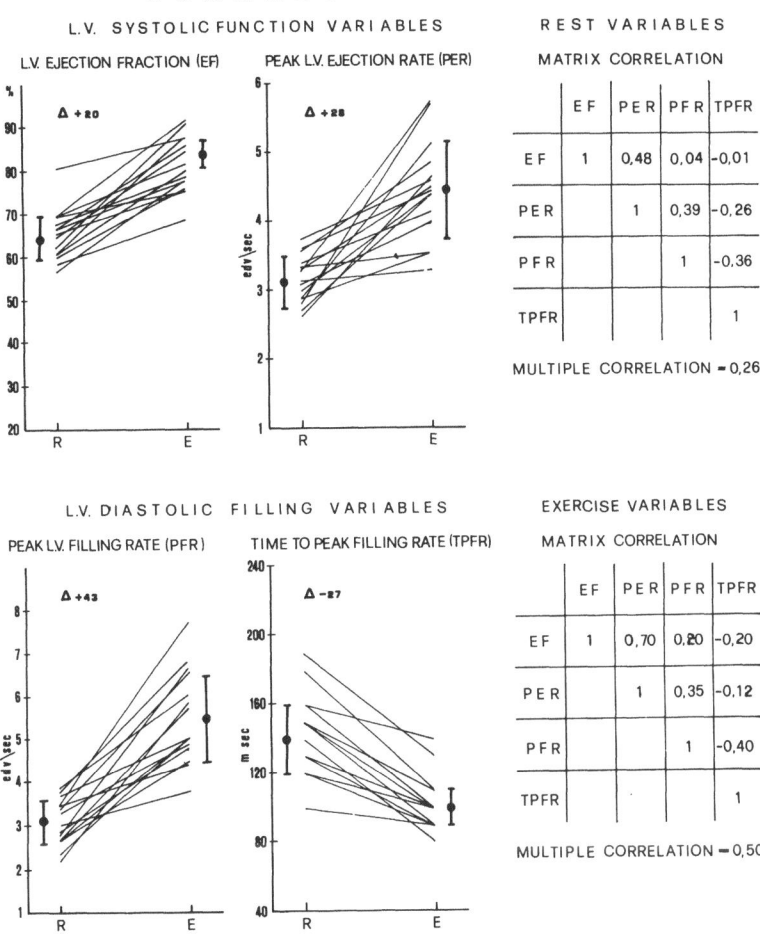

Fig. 2.   Systolic and diastolic function at rest (R) and during
          exercise (E) in 15 normal subjects.

The diastolic function in our patients shows the following
pattern: the mean PFR at rest is 2.42±0.6 EDV/sec, while during
exercise it becomes 3.12±0.9 with a +22 percent change. The mean
TPFR value at rest is 143±38msec, which becomes 128±26 msec during
exercise, with a -10 percent change.

The correlation matrix both at rest and during effort shows
coefficents between parameters of systolic and diastolic function
which are not indicative of any correlation. The coefficent of
multiple correlation is 0.70 at rest and 0.80 during exercise.

In relation to EF at rest and its variation during exercise, the
42 patients with previous infarction may be divided into three groups

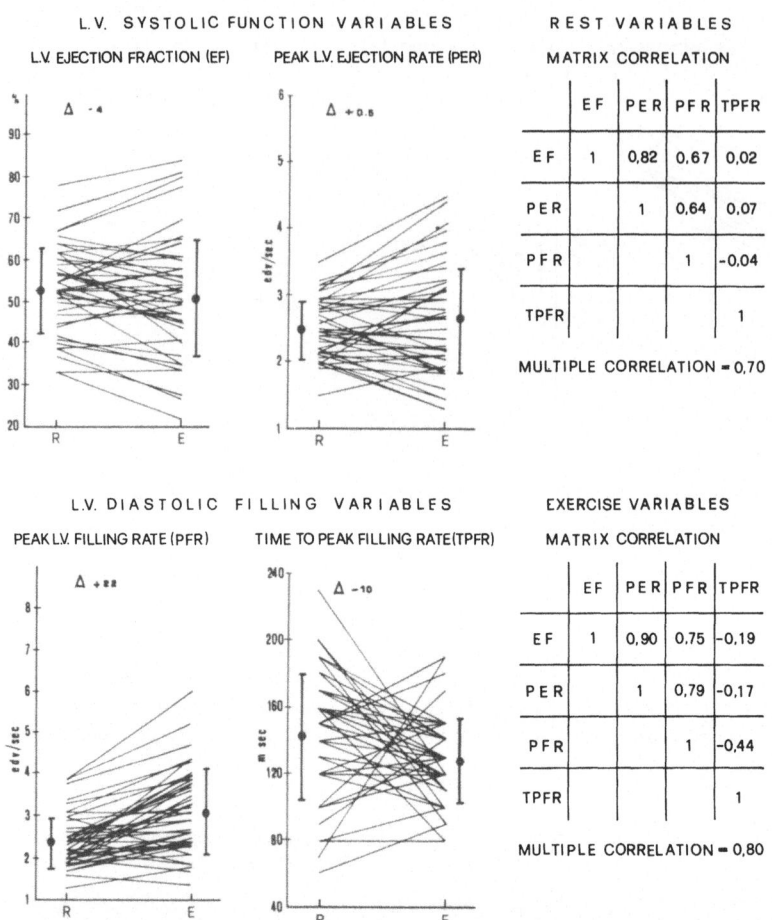

Fig. 3.   Systolic and diastolic values at rest (R) and during
          exercise (E) in 42 patients with previous myocardial
          infarction.

(Figure 4).   One group with a mean EF value at rest of 62±8% and
during exercise of 71±9% shows good left ventricular performance and
a pattern comparable to that of normal subjects.   Another group,
although having a EF rest value not unlike that of normal subjects
(59±4%), shows a decrease during exercise (52±6%), thus showing
reduced left ventricular functional reserve.   The last group consists
of patients who both at rest and during exercise show EF values
statistically lower than those of normal subjects (44±6% and 41±10%
respectively), thus showing poor left ventricular functional reserve.

Figure 5 shows the PER pattern in three groups of patients
differentiated with respect to their EF values.   The groups with good

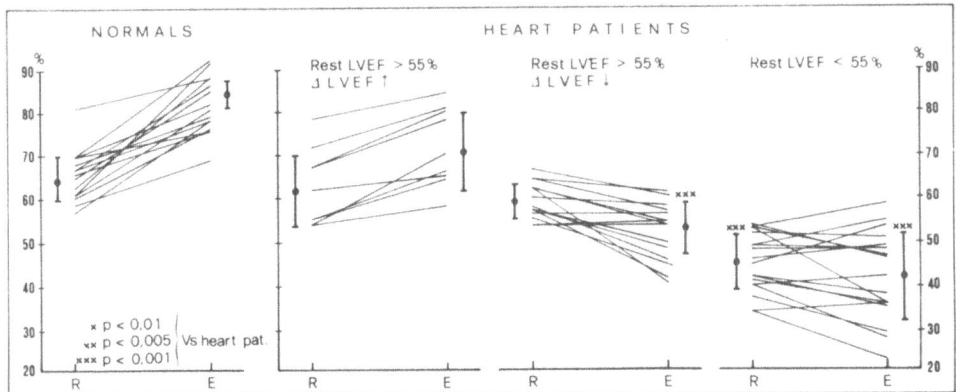

Fig. 4. Left ventricular ejection fraction (LVEF) variations during
exercise (E) in heart patients with normal rest (R) values
(LVEF>55%), and abnormal rest values (LVEF<55%). Among
patients with normal LVEF rest values, we can distinguish a
group with good (ΔLVEF), and a group with reduced (ΔLVEF↓)
left ventricular functional reserve.

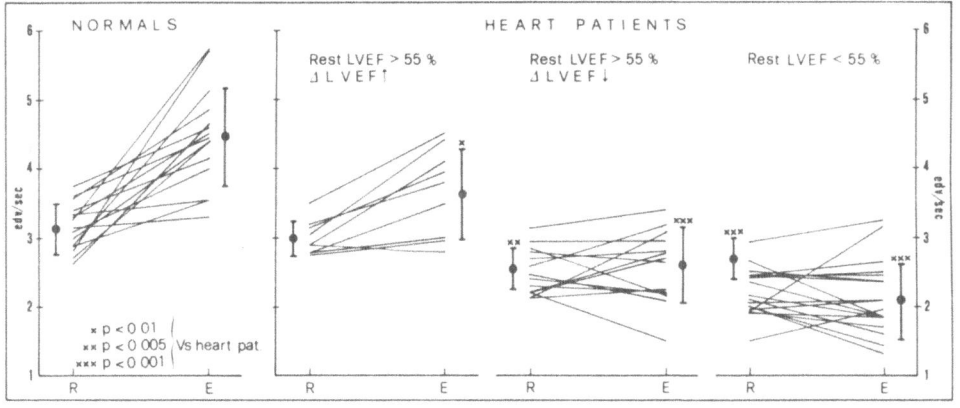

Fig. 5. Peak left ventricular ejection rate (PER) pattern at rest
(R) and during exercise (E), in patients with good (ΔLVEF↑),
reduced (ΔLVEF↓) and poor (LVEF<55%) left ventricular
functional reserve.

left ventricular performance shows a mean PER rest value comparable
to that of normal subjects (3.01±0.2 EDV/sec). During exercise all
the patients except one show an increase which reaches a mean value
of 3.66±0.6 EDV/sec., which is significantly lower that that of
normal subjects. The group with normal EF at rest and reduced EF
during exercise shows a mean value at rest of 2.57±0.3 EDV/sec.

which is significantly lower than that of normal subjects. During
exercise the PER shows an uneven pattern: in 6 patients there is an
increase, while in the rest there is a decrease; on average its value
remains approximately the same as the pattern at rest (2.60±0.5 EDV/
sec).

In patients with low left ventricular performance the PER is
significantly low at rest, in comparison with normal subjects, and
shows a further drop during exercise (2.20±0.3 EDV/sec and 2.12±0.5
EDV sec respectively). Only in three cases does the PER show any
increase during exercise.

Figure 6 shows the filling rate pattern in the three groups of
patients considered. The group with good left ventricular performance shows normal PER rest values (3.04±0.7 EDV/sec) with an increase
during exercise, which is nevertheless on average lower than that of
normal subjects (4.13±0.9 EDV/sec). Patients with normal LVEF at
rest and reduced LVEF during exercise show a mean PFR value that is
already significantly low at rest (2.29±0.3 EDV/sec). The PFR increases during exercise, reaching a mean value of 2.83±0.6 EDV/sec,
which is nevertheless significantly lower than that in normal subjects. Three patients even show a drop in PFR during exercise.

The group of patients with low left ventricular performance
display a similar pattern (PFR at rest 2.16±0.5 EDV/sec; during
exercise 2.77±0.8 EDV/sec). Two patients in this group show a decrease in PFR during exercise.

The mean TPFR values at rest of patients with good, reduced and
poor left ventricular performance are comparable to those of normal

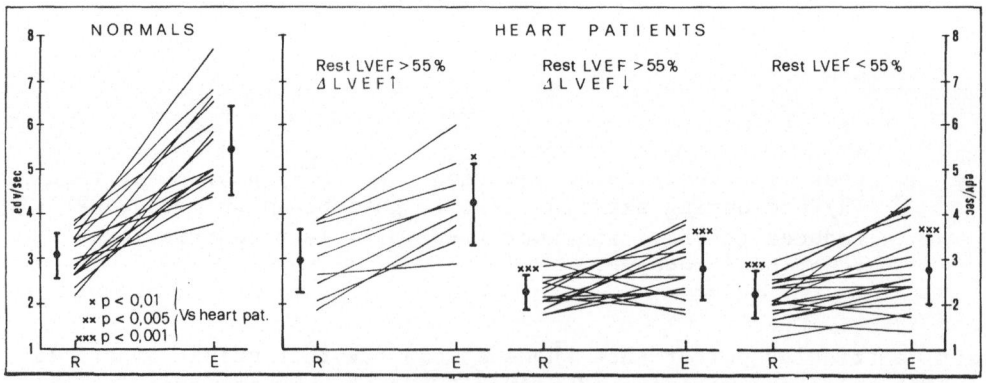

Fig. 6. Peak left ventricular filling rate (PFR) pattern, at rest
         (R) and during exercise (E) in patients with good (ΔLVEF↑),
         reduced (ΔLVEF↓) and poor (LVEF<55%) left ventricular
         functional reserve.

subjects (Figure 7). There is a decrease in TPFR in all three groups during exercise: the group with good left ventricular performance shows a mean TPFR value of 110±10 msec not unlike that of normal subjects, while two patients do not show any variation, and one shows an increase in TPFR during exercise.

The group with reduced left ventricular performance reaches a mean TPFR value of 142±29 msec; which is significantly higher than that of normals and four patients show an increase in TPFR under effort.

The low performance group has a TPFR pattern during exercise similar to that of the previous group (TPFR during exercise: 121±23 msec), and eight patients in this group show an increase under effort.

DISCUSSION

In normal subjects, multiple linear regression shows that both at rest and during exercise there is no reciprocal correlation between the parameters of systolic function and the parameters of diastolic function.

The patients with previous myocardial infarction show abnormalities in comparison with normal subjects both in systolic and diastolic filling, and show higher correlation coefficients especially during exercise, though not such as to bring out any clear relation between the parameters considered. Exercise proved to be a useful means of identifying abnormalities that cannot be detected at rest. In this connection the lack of any increase lower than 5% in its resting value is considered as an expression of left ventricular dysfunction[6-10].

This criterion enabled us to identify among our patients one group with good left ventricular performance, another group with diminished performance and a third with poor performance.

As a systolic index PER proved to be more sensitive than EF in detecting ejection abnormalities. In fact, its value in coronary patients is significantly lower even in those subjects whose EF both at rest and during exercise is normal (Figure 5).

The parameters of diastolic function are represented in our study by PFR and by TPFR. Alterations in diastolic filling were manifested in patients with normal left ventricular performance (Figure 6). In fact, in the group with normal left ventricular functional reserve only three patients out nine reach normal PFR values during exercise while in the group with normal EF at rest and reduced EF during exercise, the filling rate is already significantly reduced at rest.

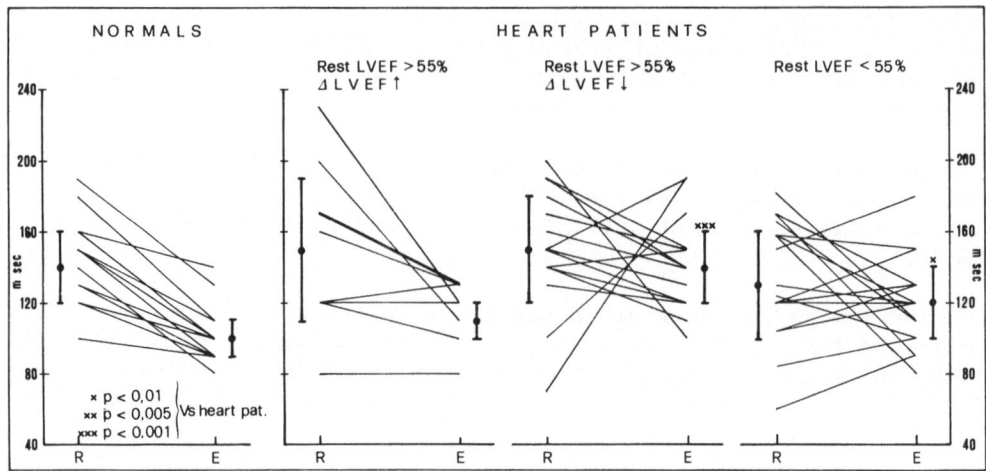

Fig. 7. Time to peak filling rate (TPFR) pattern at rest (R) and
         during exercise (E) in patients with good (ΔLVEF↑) reduced
         (ΔLVEF↓) and poor (LVEF<55%) left ventricular functional
         reserve.

These dates indicate that alterations in diastolic filling may be
present even in patients with normal left ventricular performance
both at rest and during exercise.

        Moreover, the drop in PFR during exercise in this group shows
that in several patients there may be a prevalence of abnormalities
in diastolic function in comparison with systolic function. In the
group with low left ventricular performance this is accompanied by
severe alterations both in systolic and diastolic function in those
cases in which the PFR drops during exercise.

        TPFR abnormalities could only be detected during exercise. At
rest three groups showed TPFR values comparable to those of normal
subjects (Figure 7).

        Hammermeister and Warbasse[10,11] were the first to demonstrate
the validity of the study of the systolic and diastolic phase by
means of the variations in left ventricular volume (dv/dt). They
used cineangiography to quantify variations in ejection volume durin
systole and variations in filling during diastole, by correlating
them respectively with systolic flow through the aortic valve and
with diastolic flow through the mitral valve, and by finding a
significant statistical coincidence. These authors compared the
diastolic filling rate of a group of heart patients with that of a
group of normal patients and discovered that in the first group ther
was a reduced diastolic filling.

Abnormalities in muscular relaxation[12], diminished compliance[13,15], and variations in the pressure-volume curve[16,17] at end-diastole have already been described in patients with previous myocardial infarction.

How and to what extent these different determinants cause alterations in diastolic filling has not yet been clearly explained. Myocardial relaxation is a complex interaction of a process that is both active and passive; the active O_2 dependent process affects the early phase of isovolumic diastole[18], while the passive process depends on the rigidity of the walls, and expresses compliance.

Gewirtz[19] has reported that after occlusion of the descending anterior artery in guinea-pig's hearts, there was a reduction in diastolic filling in comparison with controls. Such a reductions is concomitant with alterations in the active phase of muscle relaxation in the absence of compliance abnormalities. In fact ten minutes after the occlusion, he observed an extension of the duration of the T constant, which is the constant expressing the extent to which ventricular pressure drops in end-diastole. He observed no variation in the K constant which expresses the passive variations in wall rigidity, together with a small but significant increase in end-diastolic pressure. In a recent paper Carrol[20] reported that during exercise the T constant drops inadequately in coronary patients in comparison with control. Blaustein and Gaasch[12] have described this phenomenon as taking place even during temporal disruption of ventricular contraction experimentally induced in dog's hearts by ventricular pacing.

The mechanism by which a reduction in relaxation alters diastolic filling is not yet clear. The reduction in pressure gradient between the atrium and left ventricle induced by abnormal muscle relaxation might increase the resistance to filling and this might explain the reduction is diastolic filling rate[19].

On the other hand, experimental studies on animals suggest that active left ventricular muscle relaxation creates a suction effect which brains blood from the atrium to the ventricle.

Insufficent ischemia-induced diastolic relaxation reduces diastolic suction and hence there is a reduction in diastolic filling[22,23].

Our date agree with those obtained by other authors[24,25,26], and show that diastolic filling abnormalities are present even at rest in most of the patients under examination, even when there are no systolic function abnormalities.

In the assessment of these parameters radio-nuclide angiography presents limitations due to the lack of quantitive measurements of volume variations, and to the absence of a simultaneous analysis both of the pressure-volume ratio and the dynamics of mitral valve movement.

Nevertheless, all the indices obtained by this method are fairly reliable in the assessment of the parameters of diastolic and systolic function, and the possibility of making routine non-invasive assessments of these functions, provide us with a great deal of precious information on the overall performance of the left ventricle and is of great advantage in assessing the effects of therapeutic treatment whether medical or surgical.

REFERENCES

1. J. S. Borer, S. C. Bacharach, M. V. Green, K. M. Kent, S. E. Epstein, and G. S. Johnston, Real-time radionuclide cine-angiography in the non-invasive evaluation of global and regional left ventricular function at rest and during exercise in patients with coronary artery disease, N.Engl.J.Med., 296:839 (1977).

2. H. J. Berger, C. A. Reduto, D. E. Johnstone, H. Borkowski, J. M. Sands, P. F. Cohen, R. A. Langow, A. Gottschalk, and B. L. Zaret, Global and regional left ventricular response to exercise in coronary artery disease: assessment by quantitative radionuclide angiocardiography, Am.J.Med., 66:13 (1979).

3. J. W. Covell, J. Ross, Jr., Nature and significance of alterations in myocardiol compliance, Am.J.Cardiol., 32:455 (1979).

4. H. N. Wagner, The use of the nuclear stethoscope for temporal imaging of left ventricular function, in: "Nuclear Cardiology," selected computer aspects, p.45, Bachards, New York (1978).

5. D. G. Pavel, A. M. Zimmer, and V. N. Peterson, In vivo labeling of red blood cells with Tc 99m. a new approach to blood pool visualization, J.Nucl.Med., 18:305 (1977).

6. N. H. Wagner, P. Rigo, R. H. Baxter, P. O. Alderson, K. H. Douglas, and D. F. Housholder, Monitoring ventricular function at rest and during exercise with a non-imaging nuclear detector, Am.J.Cardiol., 43:975 (1979).

7. S. L. Bacharach, M. V. Green, and J. S. Borer, A realtime system for multi-image gated cardiac studies, J.Nucl.Med., 18:79 (1977).

8. M. V. Green, W. Brody, M. A. Douglas, J. S. Borer, H. G. Ostrow, B. R. Line, S. L. Bacharach, and G. S. Johnson, Ejection fraction by count rate from gated images, J.Nucl.Med., 19:880 (1978).

9. S. L. Bacharach, M. V. Green, J. S. Borer, J. E. Hyde, S. P. Farkas, and G. S. Johnston, Left-ventricular peak ejection

rate, filling rate and ejection fraction-frame rate requirements at rest and exercise: concise communication,
J.Nucl.Med., 20:189 (1979).

10. K. E. Hammermeister and J. R. Warbasse, The rate of change of
left ventricular volume in man: diastolic events in health
and disease, Circulation, 49:739 (1974).

11. K. E. Hammermeister, R. C. Brooks, and J. R. Warbasse, The rate
of change of left ventricular volume in man: validation and
peak systolic ejection rate in health and disease,
Circulation, 49:729 (1977).

12. D. G. Gibson, T. A. Prewitt, and D. J. Brown, Analysis of left
ventricular wall movement during isovolumic relaxation and
its relation to coronary artery disease, Br.Heart J., 38:1010
(1976).

13. W. Grossman and L. P. McLaurin, Diastolic properties of the left
ventricle, Ann.Intern.Med., 84:316 (1976).

14. G. Diamond and J. S. Forrester, Effect of coronary artery
disease and acute myocardial infarction on left ventricular
compliance in man, Circulation, 45:11 (1972).

15. J. Mirsky, P. F. Cohen, J. A. Levine, R. Gorlin, A. V. Herman,
T. H. Krenlen, and E. H. Sonnenblick, Assessment of left
ventricular stiffness in primary myocardial disease and
coronary artery disease, Circulation, 50:158 (1976).

16. J. D. Bristow, B. E. Van Zee, and M. P. Judkins, Systolic and
diastolic abnormalities of the left ventricle in coronary
artery disease: studies in patients with little or no
enlargement of ventricular volume, Circulation, 42:219
(1970).

17. W. H. Gaasch, H. J. Levine, M. A. Quinones, and J. K. Alexander,
Left ventricular compliance: mechanisms and clinical implications, Am.J.Cardiol., 38:645 (1976).

18. P. E. Pool and E. H. Sonnenblick, The mechanochemistry of
cardiac muscles. I. The isometric contraction,
J.Gen.Physiol., 50:951 (1967).

19. H. Gewirtz, W. Ohley, J. Walsh, D. Shearer, M. Sullivan, and A.
S. Most, Ischemia-induced impairment of left ventricular
relaxation. Relation to reduced diastolic filling rates of
the left ventricle, Am.Heart J., 105:72 (1983).

20. J. D. Carrol, O. M. Hess, H. O. Hirzel, and H. P. Krayenbuchil,
Exercise-induced ischemia. The influence of altered relaxation on early diastolic pressures, Circulation, 67:521
(1983).

21. A. S. Blaustein and W. H. Gaasch, Myocardial relaxation. VI. The
effects of left ventricular synchrony and beta adrenergic
tone on the load dependency of isovolumic relaxation rate,
Am.J.Cardiol., 47:515 (1981).

22. H. N. Sabbah and P. D. Stein, Pressure diameter relations during
early systole in dogs. Incompatibility with the concept of
passive left ventricular filling, Circ.Res., 45:357 (1981).

23. J. K. Tyberg, W. J. Keon, E. H. Sonnenblick, and C. W. Urschel,

Mechanics of ventricular diastole, Cardiovasc.Res., 4:423 (1970).

24. R. O. Bonow, S. L. Bacharach, M. V. Green, K. M. Kent, D. R. Rosing, C. C. Lipson, M. B. Leon, and S. E. Epstein, Impaired left ventricular diastolic fillings in patients with coronary artery disease. Assessment with radionuclide angiography, Circulation, 64:315 (1981).

25. L. A. Reduto, W. J. Wichermeyer, J. B. Young, L. A. Del Ventura, J. W. Reid, D. H. Glaeser, M. A. Quinones, and R. R. Miller, Left ventricular diastolic performance at rest and during exercise in patients with coronary artery disease, Circulation, 63:1228 (1981).

26. G. B. Mancini, R. A. Slutsky, S. L. Norris, V. Bhargava, W. L. Ashburn, and C. B. Higgins, Radionuclide analysis of peak filling rate, filling fraction, and time to peak filling rate, Am.J.Cardiol., 51:43 (1983).

GLOBAL AND REGIONAL FUNCTION AFTER BY-PASS SURGERY

S. Caponneto,* F. Bruzzone,* S. Borziani,* C. Pastorini*
P. Fiorio,** and D. Ravera**

*Cattedra di Malattie Apparato Cardiovascolare
 Università di Genova
**Servizio di Medicina Nucleare
 Ospedale di Sampierdarena, Genova, Italy

INTRODUCTION

Coronary artery by-pass graft (CABG) surgery relieves chest pain
due to coronary artery disease (CAD) and improves quality of life in
60-90% of patients, while the effect of this surgical procedure on
ventricular performance requires further elucidation[1,2,3]. Because
of the regional nature of CAD, assessment of global and regional
ventricular function is important in evaluation and prognosis of
patients with CAD[4]. Thus, the present investigation is designed to
assess left ventricular ejection fraction (LVEF) and regional wall
motion (RWM) by non invasive radionuclide angiography (RNA) in
patients with CAD before and after CABG.

METHODS

Twenty-two patients with CAD, with a mean age of 58, 8±6 (range
45-65), have been studied before and 2 months after CABG. Twenty one
patients were male, one was female. Eight patients had previous
myocardial infarction (PMI). Eighteen patients have been studied six
months after CABG (17 male, 1 female) with a mean age of 59±3; eight
of them had PMI.

Consent was obtained from each patient. The diagnosis of myocardial infarction was based on the following criteria: typical
prolonged chest pain, development of pathologic q waves 0.04" in
duration in at least two electrocardiographic leads and characteristic serial elevations of serum-enzymes.

Complete revasculatization was performed with 52 (mean 2.4) saphenous vein aortocornary by-passes (range 1-5) during moderate hypothermia. However, graft patency was not assessed in these patients. In our study, patients with perioperative myocardial infarction, documented by development of q waves and a significant increase in CK-MB, were not included. Each patient was considered to be an operative candidate because of persistence of refractory angina pectoris. RNA in equilibrium was performed before and later (mean two months, mean 6 months) after CABG at rest and during isometric exercise (HG). All patients had no heart failure, important arrhythmias and therapy were withheld for at least 24 hours before testing. Premedication with 4 mg of stannons pyrophosphate intravenously was given to permit "in vivo" erytrocytes labeling RNA in equilibrium was performed after i.v. bolus injection of 25 mCi of 99 mTc as pertechnetate. RNA equilibrium imaging was performed in the best septal LAO projection (35°-45°) with 10° of caudal tilt using the incore gate synchronized mode of computer acquisition.

We used a single crystal gamma camera (ELSCINT U.F) with computer ELSCINT DIMAS 80. The cardiac cycle was divided into 12 frames. Images at rest were acquired for a period of time resulting more than 100,000 counts per frame.

To perform HG the patients were asked to squeeze a hand dynamometer (Vigometer Manometer) to the maximum extent possible and then maintain contraction at one third of the predetermined maximum for 4'-4' 30", RNA data were recorded in the last 2' 30" of HG exercise.

As a global index of left ventricular function we considered LVEF calculated as endiastolic counts - endsystolic counts divided by background corrected end-diastolic counts. To assess left ventricular regional function we performed a semiquantitative analysis, using Fourier analysis, by dividing the left ventricle (LV) in 45° LAO view into five segments: proximal and distal postero-lateral segments, infero-apical segment was assigned a score with 3 representing normal wall motion, 2 hypokinesis, 1 akinesis and 0 dyskinesis. The regional wall motion index (RWMi) was obtained by totalling the segmental scores. Therefore 15 (5 x 3) was the highest possible normal RWMi score.

The data were reviewed and interpreted independently by at least 2 experienced observers, and the results were averaged. These methods of global and regional LV function assessment were validated by comparison against contrast angiography in our laboratory[5,6,7].

Statistical Analysis

We compared data by Student's t-test for significance of paired data. Data are expressed as mean ± standard deviation.

RESULTS

Six months after operation 13 of 18 patients (72%) become asymptomatic, and an additional 3 patients reported that they could perform a greater level of activity with fewer symptoms than before CABG.

Left Ventricular Ejection Fraction

At rest, mean LVEF was preoperatively 51.4%±9.3, two months post CABG was 52.1±10.8 (increase of 1.3%, NS), six months post CABG was 50.6±10.1 (decrease of 1.5%, NS). (Figure 1). Comparing individual LVEF values pre-op versus six months post-op we observed that: of the twelve patients who had a normal LVEF remained unchanged and three got worse; for the six patients who had a low LVEF three improved to a normal value and three didn't change.

If we consider separately patients with PMI we obtain the following data: at rest mean LVEF was pre-op 43.7±9.9, two months post-op was 47.5±13.4 (increase of 8.6%, NS), 6 months post-op was

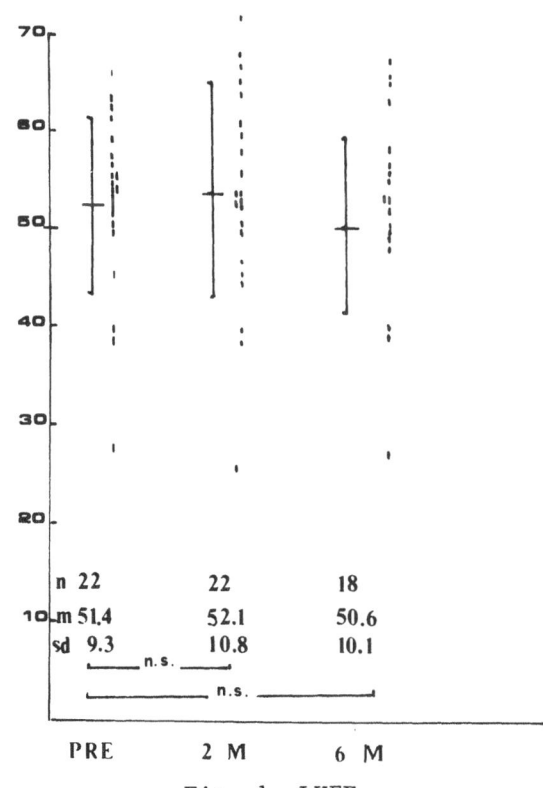

Fig. 1. LVEF

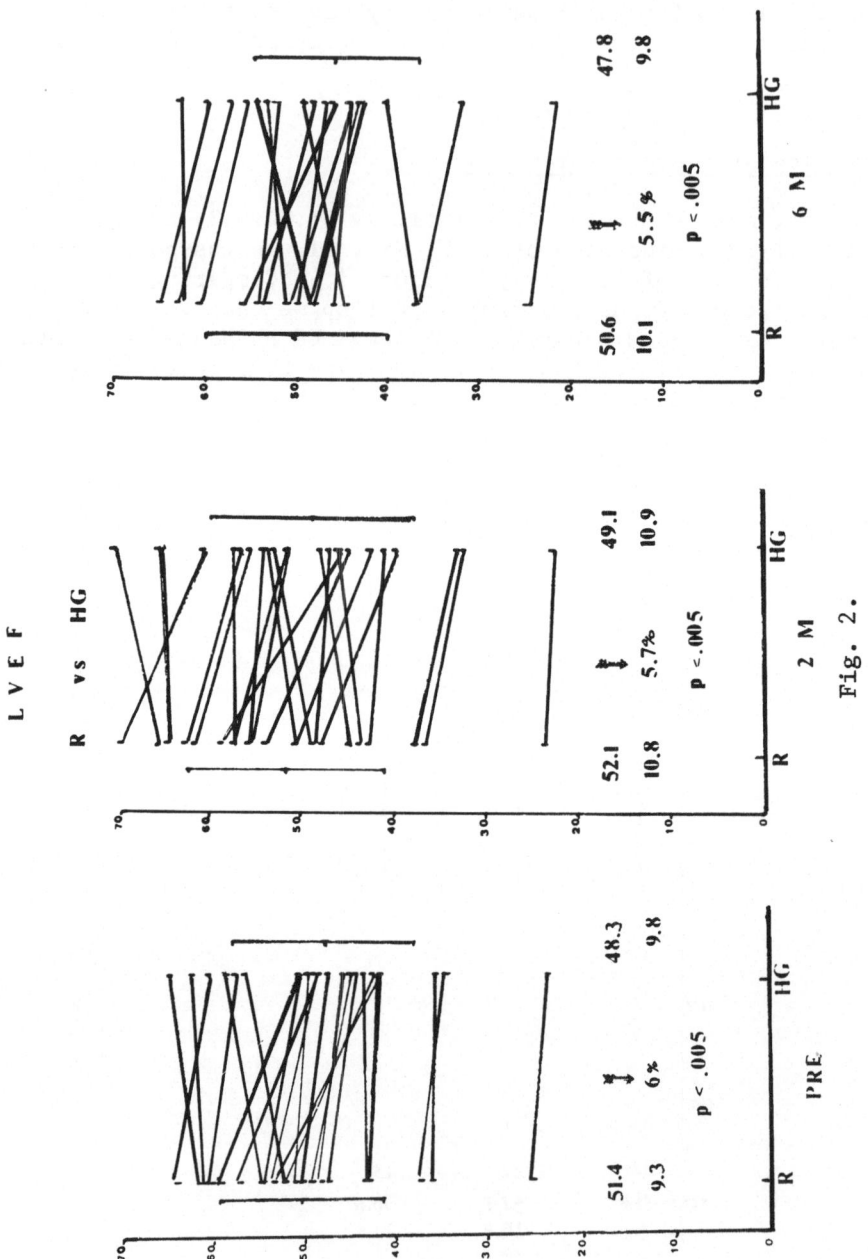

Fig. 2.

48.1±11.8 (increase of 10%, NS). During preoperatively HG, in all
patients, we had a reduction of mean LVEF from 51.4±9.3 to 48.3±9.8
(decrease of 6%, p=0.005); two month post-op the reduction was from
52.1±10.8 to 49.1±10.9 (decrease of 5.7%, p=0.005); six months
post-op from 50.6±10.1 to 47.8±9.8 (decrease of 5.5%, p=0.005).
(Figure 2). All patients performing HG had a decrease of LVEF except
four cases preoperatively (not one had PMI), six cases two months
(one had PMI) and four cases six months after surgery (not one had
PMI) who improved their LVEF. During pre-op HG, patients with PMI
had a reduction of LVEF from 43.7±9.9 at rest to 40.5±8.2 (-7.3%,
NS); two months post-op from 47.5±13.4 to 43.7±12.1 (-8%, NS); six
months post-op from 48.1±12.9 to 42.8±10.5 (-11%, NS).

Regional Wall Motion

At rest the RWMi decreased from 12.7±1.3 pre-op to 11.8±1.5 two
months post-op (decrease of 7%, p=0.05) and to 12.05±1.6 six months
post-op (decrease of 5.5%, p 0.05) (Figure 3).

In patients with PMI at rest RWMi was pre-op 12±0.9; two months
post-op was 11.6±1.5 (decrease of 7%, p=0.03); 6 months post-op was
11.2±1.5 (decrease of 5.5%, NS). During pre-op HG, in all patients,
the results show a reduction of RWMi from 12.7±1.3 to 11.1±1.8
(-12.5%, p=0.001); two months post-op from 11.8±1.5 to 11.1±1.8
(-5.9%, p=0.025); 6 months post-op from 12±1.6 to 11.7±1.8 (-2.5%,
NS) (Figure 4). During pre-op HG in patients with PMI we had a
reduction of RWMi from 12±0.9 to 9.7±1.4 (-19.1%, p=0.02); two months
post-op from 11.6±1.5 to 10.7±2.2. (-7.7%, NS); six months post-op
from 11.2±1.5 to 10.7±1.3 (-4.4%, NS).

The post-op response at rest of individual segments was also
evaluated (Figure 5). Before surgery 63 segments (57%) showed a

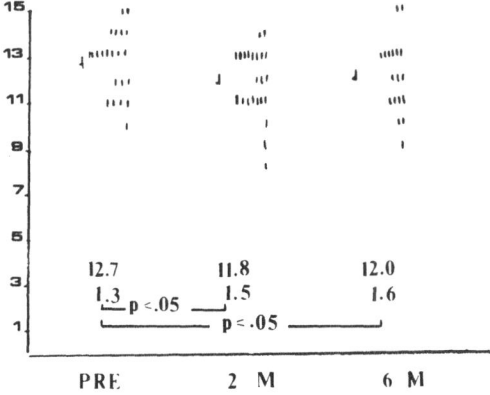

Fig. 3. Regional wall motion index.

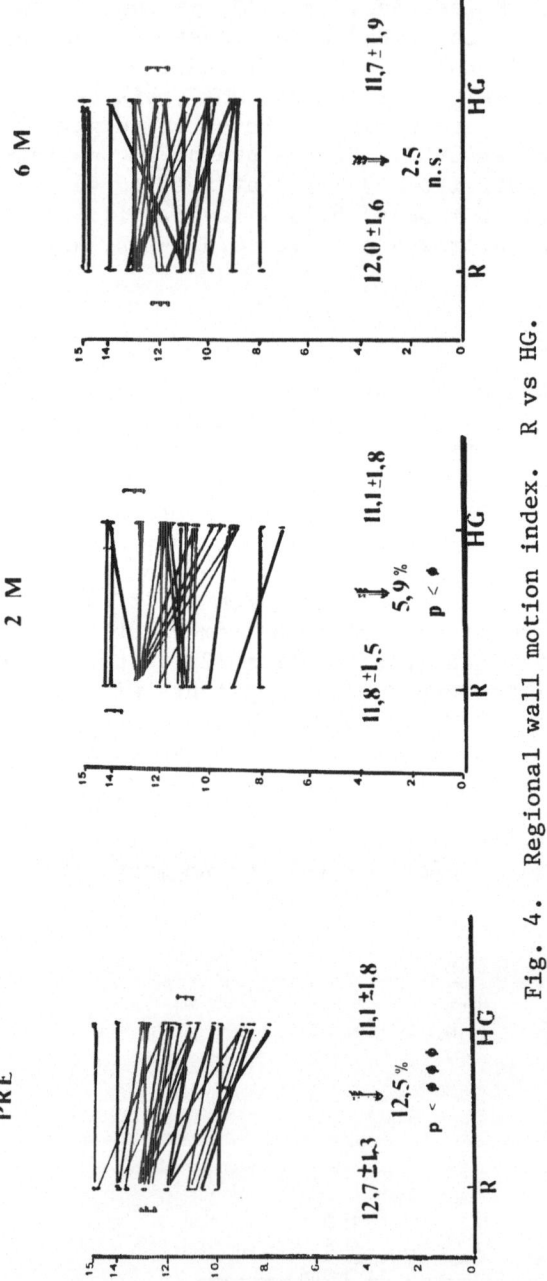

Fig. 4.   Regional wall motion index. R vs HG.

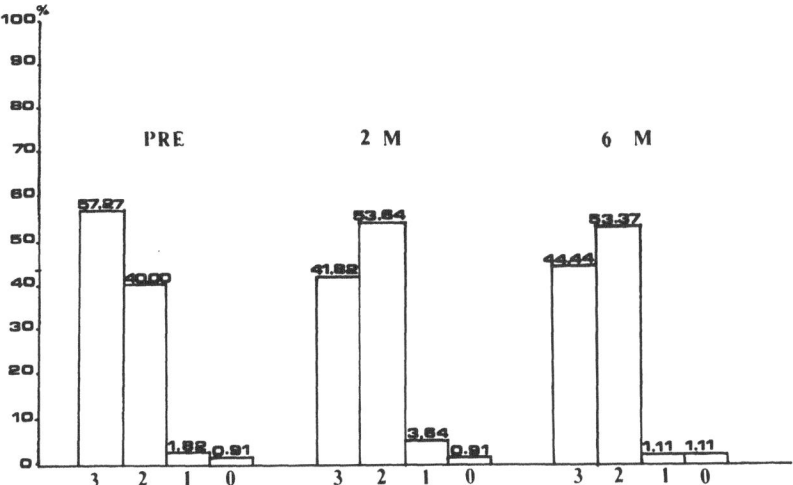

Fig. 5.   Distribution of RWMI before and late CABG

normokinesis, 44 segments a hypokinesis (40%), 2 segments an akinesis
(1.8%) and one segment a dyskinesis (0.9%). Two months post-op there
were 46 segments normokinetics (41%), 59 segment hypokinetics (53%),
1 segment akinetic (3.6%) and 1 segment dyskinetic (0.9%). Six
months post-op (considering 18 patients) there were 40 segment normokinetic (44%), 48 segments hypokinetics (53%), 1 segment akinetic
(1.1%) and 1 segment dyskinetic (1.1%). Two months post-op, of 110
segments (5 segments x 22 patients) 66 did not change (34 normokinetics, 30 hypokinetics, 1 akinetic and 1 dyskinetic), 31 worsened
(28 from normokinetics to hypokinetics to akinetic). Thirteen segments improved (12 from hypokinetics to normokinetics and 1 from
akinetic to hypokinetic). Of the 31 segments which got worse 17
(55%) were septals. Six months post-op of 90 segments (5 x 8
patients) 53 did not change (26 normokinetic, 26 hypokinetic and 1
dyskinetic); 22 segments got worse (21 from normokinetics to hypokinetics and 1 from normokinetic to akinetic; 15 segments improved
(14 from hypokinetics to normokinetics and 1 from akinetic to hypokinetic. Of 22 segments which got worse 13 (59%) were septals.

DISCUSSION

     It's well known that CABG produces a marked symptomatic improve-
ment and ameliorate quality of life in patients with CAD [1,2,3].
This effect could be determined by increased perfusion of ischemic
areas, perioperative myocardial infarction of previous ischemic
segments or to surgical neural differentiation of the ischemic
segment[8]. In agreement with other studies we found a regression of
angina in the 72% of patients[1,2].

Global LVEF is an index of ventricular function, however this global assessment might include both regions with normal function and regions with depressed function due to ischemia and scar formation, Hence, LVEF can mask regional disorders detectable with regional wall motion analysis[9,10,11]. At rest we didn't find significant variation after surgery LVEF in early (2 months) and late (6 months) follow-up, in agreement with most previous studies[12,13,14]. Several studies showed that HG is a valuable method to detect haemodynamics abnormalities in patients with CAD[14,15,16,17]. In normal patients HG doesn't change significatively LVEF, on the contrary, in patients with CAD, LVEF generally decreases and areas of dissynergy can be unmasked or worsened[4,18,19]. In our study considering all patients, exercise LVEF had a significant reduction before and after surgery. Considering patients with PMI the reduction exercise LVEF was significant only preoperatively but after CABG it didn't show a significant worsening. This HG response would demonstrate the ability of revascularization to diminished exercise induced ischemia in the presence of PMI.

Regional wall motion: our data show that the post surgery RWMi decreases at rest in all patients; this could be ascribed, mainly, to the deterioration of interventricular septal segments. In fact, out of 31 segments which got worse 2 months post CABG and 22 six months post CABG, 17 and 13 were respectively interventricular septal segments. This result agrees with previous studies[20-21]. Besides, other reasons that can explain post-op deterioration of same segments might be: by-pass grafts closure and progression of CAD[14].

The response of segmental wall motion to HG exercise differs before and after surgery. Before surgery in all patients we had a significant reduction of exercise RWMi (p=0.001); two months after CABG we found a poorly significant decrease (p=0.025); 6 months after CABG we hadn't any significant change of exercise RWMi. This suggests that reperfusion of ischemic muscle leads to an improvement of regional wall motion response to isometric exercise. In fact by-pass improves the capacity of coronary arteries to augment blood glow during period of increased metabolic demand like during HG exercise[13-22].

In conclusion this study shows that myocardial revascularization leads to a regression of angina in the largest number of patients. After CABG global left ventricular function doesn't change; regional wall motion gets worse at rest but RWM response to HG exercise improves.

REFERENCES

1. W. C. Sheldon, G. Rincon, D. B. Effer, V. L. Proudfit, and F. M. Sones, Jr., Vein graft surgery for coronary artery disease:

Survival and angiographic results in 1000 patients, Circulation, 48:(Suppl.III) 111-184 (1973).

2. E. L. Alderman, H. J. Mattlof, L. Wexler, N. E. Shumway, and D. C. Harrison, Results of direct coronary artery surgery for the treatment of angina pectoris, N.Engl.J.Med., 288:535 (1973).

3. H. D. McIntosh and J. A. Garcia, The first decade of aorto-coronary by-pass grafting 1967-1977, Circulation, 57:405-415 (1978).

4. M. M. Bodenheimer, S. V. Banka, C. M. Fooshee, M. S. Gillespie, and R. H. Helfant, Detection of coronary heart disease using radionuclide determined regional ejection fraction at rest and during handgrip exercise: Correlation with coronary arteriography, Circulation, 58(4):640-648 (1978).

5. C. Pastorini, P. Muttini, S. Mazzantini, P. Fiorio, M. Vischi, D. Ravera, and F. Bruzzone, Affidabilità e riproducibilità della frazione di eiezione ventricolare sinistra studiata con la 99mTc cineangiografia radioisotopica, Giornale di Emodinamica 2(4):165-172 (1982).

6. S. Borziani, F. Bruzzone, D. Ravera, P. Fiorio, S. Mazzantini, and C. Pastorini, Correlazioni nella valutazione della motilità regionale di parete del ventricolo sinistro fra cineangiografia contrastografica e cineangiografia radio-isotopica in equilibrio, Atti del XLIV Congresso Nazionale della Società Italiana di Cardiologia, Torino (1983).

7. F. Bruzzone, P. Muttini, P. Fiorio, D. Ravera, S. Caponnetto, and C. Pastorini, Correlazioni inter ed intra osservatore fra due programmi computerizzati nel calcolo delle frazioni di eiezione del ventricolo sinistro mediante cineangiografia radioisotopica in equilibrio, Atti del XLIII Congresso Nazionale della Società Italiana di Cardiologia, Roma (1982).

8. R. S. Ross, Ischemic heart disease: An overview, Am.J.Cardiol., 36:496 (1975).

9. N. Schad, Non traumatic assessment of left ventricular wall motion and regional stroke volume after myocardial infarction, J.Nucl.Med., 18:333 (1977).

10. N. Schad and O. Nickel, Radionuclide angiography in coronary artery disease: Where do we stand? Cardiovasc.Radiol., I:27 (1978).

11. M. M. Bodenheimer, V. S. Banka, and R. H. Helfant, Nuclear cardiology: Radionuclide angiographic assessment of left ventricular contraction: Uses, limitations and future directions, Am.J.Cardiol., 45:661 (1980).

12. L. J. Mintz, N. B. Ingels, Jr., G. T. Daughters II, B. E. Stinson, and E. L. Alderman, Sequential studies of left ventricular function and wall motion after coronary artery by-pass surgery, Am.J.Cardiol., 45:210 (1980).

13. K. Kent, J. Borer, V. Green, S. Bacharach, C. McIntosh, M. Conkle, and S. Epstein, Effect of coronary artery by-pass on global and regional left ventricular function during exer-

cise, N.Engl.J.Med., 298:1434-1439 (1978).

14. M. Freeman, R. Gray, D. Bernan, J. Maddahi, M. Raymond, G. Forrester, and G. Matloff, Improvement in global and segmental left ventricular function after coronary artery by-pass surgery, Circulation, 64(Suppl.II) (1981).

15. P. Rosselli, V. Martini, M. Iannetti, C. Pastorini, M. A. Masperone, and S. Caponnetto, Comportamento dei tempi della sistole ventricolare sinistra in pazienti con cardiopatia ischemica sottoposti ad esercizio isometrico, Giorn.Ital. Cardiol., V(3):364 (1975).

16. S. Caponnetto, M. Iannetti, C. Pastorini, M. A. Masperone, P. Perugini, and G. Oriani, Effects of sustained isometric handgrip on ventricular systolic time intervals in patients with ischemic heart disease, Jap.Heart J., 18(4):434 (1977).

17. M. M. Bodenheimer, V. S. Banka, C. M. Fooshee, G. A. Hermann, and R. H. Helfant, Comparison of wall motion and regional ejection fraction at rest and during isometric exercise: Concise communication, J.Nucl.Med., 20:724-732 (1979).

18. R. Slutsky, Response of the left ventricular to stress: Effect of exercise, atrial pacing, afterload stress and drugs, Am.J.Cardiol., 47:357 (1981).

19. W. Savin, E. Alderman, and W. Haskell et al., Left ventricule response to isometric exercise in patients with denervated and innervated hearts, Circulation., 61:897-904 (1980).

20. J. Lindsay, N. Nolan, E. Kotlyarov, S. Goldstein, and J. Bacos, Radionuclide evaluation of the interventricular septum following coronary artery by-pass surgery, Radiology, 142: 489-493 (1982).

21. P. A. Vignola, C. A. Boucher, and G. D. Curfman, Abnormal interventricular septal motion following cardiac surgery: Clinical, surgical, echocardiographic correlates, Am.Heart J., 97:27-34 (1979).

22. P. Chesebro, Left ventricular performance before and after aortocoronary artery by-pass surgery, Circulation., 65(Suppl.II) (1982).

THE VALUE OF SYSTOLIC TIME INTERVALS AS

PROGNOSTIC INDICATORS IN CORONARY ARTERY DISEASE

A. M. Weissler and H. Boudoulas*

Prof. of Medicine, University of Colorado
Chairman of Medicine, Rose Medical Center
Denver, Colorado
*Prof. of Medicine, The Ohio State University
College of Medicine, Columbus, Ohio

"I hold that it is an excellent thing for a physician
to practice forecasting. For if he discover and declare
unaided by the side of his patients the present, the past
and the future, and fill in the gaps in the account given
by the sick, he will be the more believed to understand the
cases, so that men will confidently entrust themselves to
him for treatment."   (Hippocrates 460-375BC) [1]

As evidenced in this quotation from Hippocrates' treatise on
Prognostics, interest in predicting the course of disease dates
from the very beginnings of scientific medicine. With respect to
coronary artery disease, the dire prognosis of patients with angina
pectoris was uncovered by the very physician who described the
symptom, William Heberden. In 1772 he wrote: "The termination of
the angina pectoris is remarkable. For if no accident intervene, but
the disease go on to its height, the patients all suddenly fall down,
and perish almost immediately."[2]. It was not until the 19th century, however, when quantitative methods for expressing mortality in
large populations of patients emerged, that the earliest scientific
studies on the prognosis of patients with varying signs and symptoms
of coronary artery disease were published. It is notable, in this
regard, that the first actuarial study of angina pectoris was reported by Herrick and Nazum[3] only a few years after Herrick had
described the clinical picture of sudden obstruction of the coronary
arteries. These studies, followed by those of Richards, Bland,
and White[4] and Block and co-workers[5], established that among
patients with angina pectoris a poor prognosis was associated with

269

the presence of cardiac enlargement, hypertension, an abnormal
electrocardiogram, a past history of myocardial infarction and the
presence of clinically evident congestive heart failure. The more
recent Framingham study confirmed the poor prognosis of patients with
manifest clinical heart failure after myocardial infarction[6].

The advent of coronary arteriography and contrast ventriculography uncovered additional subsets of patients with a high mortality
risk. Early attention was focused on the extent of coronary arterial
occlusion as the primary prognostic variable[7-11]. More recent
investigations[12-15] have emphasized the critical prognostic influence of left ventricular performance and have supported the view
that the extent of left ventricular dysfunction is a more dominant
risk determinant than the extent of coronary artery disease.

Previous studies from this laboratory, outlined earlier in this
symposium, have documented a high of abnormal global left ventricular
performance among patients who have recovered from a previous episode
of acute myocardial infarction[16]. On the basis of these studies,
and knowledge that it is the presence of left ventricular dysfunction
per se which presents a major mortality risk factor in coronary
artery disease, we hypothesized that noninvasive measurements of left
ventricular performance in patients with established coronary artery
disease might provide useful prognostic indicators which could be
applied on repeated testing. It was for this reason that we launched
the studies that I will elaborate herein on the potential value of
the systolic time intervals as risk determinants in patients with
coronary artery disease. The methodologic approach to the measurement of the systolic time intervals (STI) has been outlined in a
previous paper in this symposium[17]. The PEP/LVET measurement has
been established as the most accurate discriptor of global left
ventricular performance by the STI method. Among normal individuals
the PEP/LVET averages .34 with an SD of 0.04. Left ventricular
dysfunction is designated by an increase in PEP/LVET to >0.42.

The studies were performed on 136 patients observed consecutively, all of whom had recovered by an average of 14 months from a
previous myocardial infarction[18]. The patients were studied pros-
pectively with a minimum follow-up of two years and an average
follow-up of 44 months. During the period of study, 32 of the
patients succumbed, 21 died suddenly, and the remaining patients died
of recurrent myocardial infarction or congestive heart failure. In
order to test the prognostic potency of the noninvasive estimate of
left ventricular performance relative to that for the extent of
coronary arterial occlusive disease, only patients who had coronary
arteriography immediately prior to their entry into the study were
included. Patients with left bundle branch block, which obscures the
accuracy of the PEP/LVET measurement, were excluded. The cumulative
five-year survival curves relative to the extent of coronary arterial
disease among the patients with one-, two-, and three-vessel disease

are shown in Figure 1. It is clear that with increased extent of
coronary arterial disease there was increasing mortality. The
patients in the study hence followed a pattern much like that
reported in previous studies. When the cumulative five-year survival
for the same patients were grouped according to the presence of a
normal PEP/LVET (n=65) or an abnormal PEP/LVET (n=71), a clear
separation into low and high mortality risk was observed (Figure 2).
On further segregation of the patients into three subgroups, those
with normal PEP/LVET (< 0.42) and those with moderate (0.42<0.50) and
severe (>0.50) abnormality of PEP/LVET, there was clear stratification of the cumulative survival curves with decreasing survival as
left ventricular function diminished (Figure 3). An insight into the
relationship of the extent of coronary arterial disease and the
abnormally in ventricular performance was provided when the level of
left ventricular performance among the patients with one-, two-, and
three-vessel disease was compared (Table 1). Thus, with one-, two-,
and three-vessel disease respectively, the mean PEP/LVET ratio increased significantly. Similarly, the percentage of patients with
abnormal PEP/LVET increased with increasing severity of coronary
arterial occlusive disease. Clearly, there is a close relationship
between the extent of coronary arterial disease and left ventricular
impairment in patients with coronary artery disease.

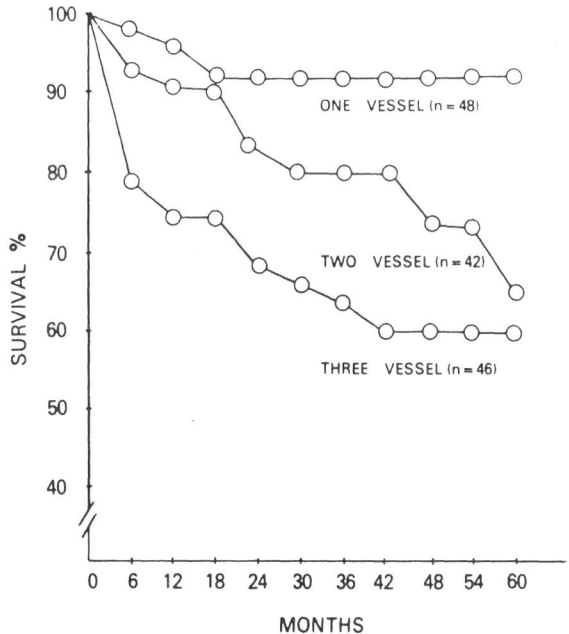

Fig. 1. Cumulative 5-year survival for the 136-patient cohort
         classified according to the presence of one-, two- or three-
         vessel disease. The trend for decreasing survival with
         increasing extent of coronary artery disease is significant
         (p<0.01). (Reprinted with permission).

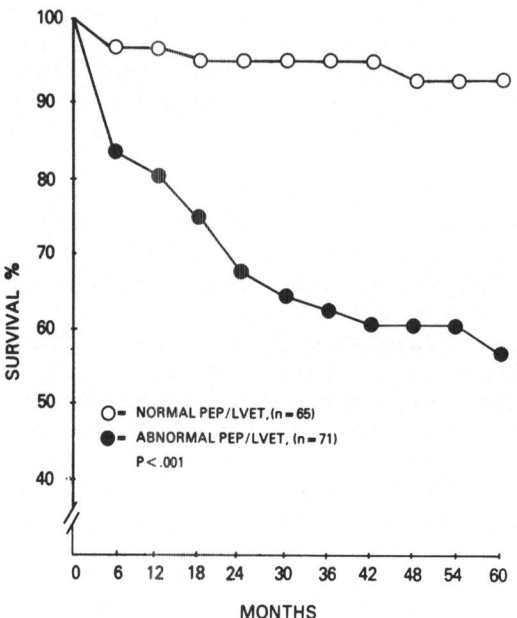

Fig. 2.   Cumulative 5-year survival for patients with a normal ratio
          of preejection period to left ventricular ejection time
          (PEP/LVET) (0.42 or less) and an abnormal PEP/LVET ratio
          (greater than 0.42).   The difference in survival is
          significant (p<0.001).   (Reprinted with permission)

        If, as we hypothesized, the STI could provide a practical non-
invasive indicator prognosis among patients with coronary artery
disease, it would have to be demonstrated that the data could be
interpreted without the need to know the extent of coronary arterial
occlusive involvement.   To test this thesis, we compared the cumulative survival curves among patients with normal and abnormal PEP/
LVET for each level of coronary arterial disease.   The cumulative
survival curves for patients with normal PEP/LVET and one-, two-, and
three-vessel disease are illustrated in Figure 4.   Despite the presence of advanced coronary arterial vessel disease, the patients with
normal PEP/LVET after recovery from acute myocardial infarction
retained a relatively low risk of mortality.   In contrast, as is
illustrated in Figure 5, the presence of abnormal PEP/LVET was associated with a high mortality risk even among the patients with
one-vessel disease.   Thus, among patients with a normal PEP/LVET the
cumulative five-year survival was 97%, 92%, and 87% (mortality rate
3%, 8% and 13% respectively) for patients with one-, two-, and three-
vessel disease.   Among patients with abnormal PEP/LVET, the cumulative five-year survival was 82%, 51%, and 41% (mortality rate 18%,
49%, and 59% respectively) for patients with one-, two-, and three
vessel disease.   Thus only among patients with abnormal left ven-

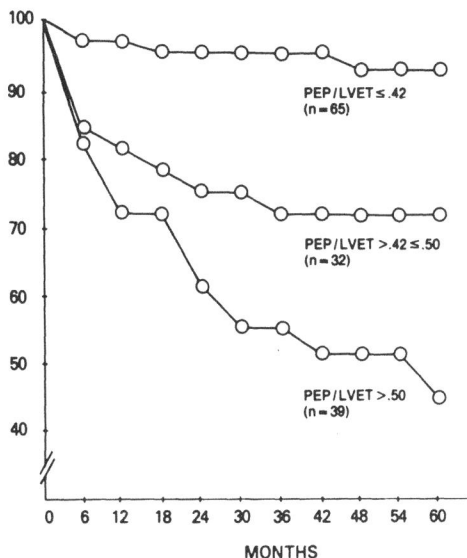

Fig. 3. Cumulative 5-year survival for patients subgrouped according
to the presence of a normal ratio of preejection period to
left ventricular ejection time (PEP/LVET) (0.42 or less), a
moderately abnormal PEP/LVET ratio (0.43 to 0.50) or a
severely abnormal PEP/LVET ratio (greater than 0.50). The
trend for diminishing survival with increasing left
ventricular dysfunction is significant (p<0.05). (Reprinted
with permission)

Table 1. Ventricular Performance in Coronary Artery Disease*
(mean ± standard deviation)

Vessels With Coronary Disease	Patients (n)	Mean PEP/LVET Ratio	Patients With Abnormal PEP/LVET Ratio (%)
One	48	0.40 ± 0.06	35
Two	42	0.46 ± 0.10	55
Three	46	0.49 ± 0.11	67

* p <0.001 for differences in ratio of preejection period to left
ventricular ejection time (PEP/LVET) and percent with an abnormal
PEP/LVET ratio for patients with one, two and three vessel disease
(analysis of variance).

tricular function did the extent of coronary artery disease add to
the prognostic strength of the PEP/LVET. On multivariate proportional hazards analysis (a computerized discriminant function analysis) for risk descriptors that have been previously reported
(i.e. presence of hypertension, diabetes. hypercholesteremia, family

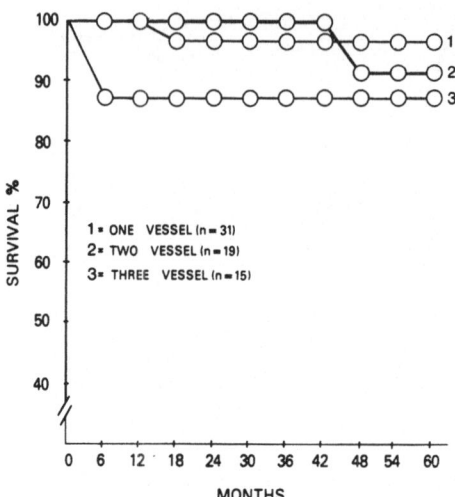

Fig. 4.   Cumulative 5-year survival for patients with a normal ratio
          of preejection period to left ventricular ejection time
          (PEP/LVET) (0.42 or less) with one-, two- or three-vessel
          coronary artery disease. The differences in survival with
          increasing extent of coronary artery disease are not
          significant. (Reprinted with permission)

history of coronary artery disease, history of smoking, history and
site of previous myocardial infarction, presence of angina pectoris
or dyspnea, and presence of cardiomegaly as reflected in a cardiothoracic ratio of < 0.5), the PEP/LVET determination proved to be a
far more significant determinant of mortality risk. Among the risk
discriptors only a cardiothoracic ratio (CTR) of >0.5 added significantly to the prognostic information provided by PEP/LVET. Clearly,
the noninvasive measurement of STI offers a prognostic indicator
which stratifies patients with coronary artery disease into high and
low risk groups at a level of prognostic potency greater than previously reported clinical risk indicators. In addition, by similar
comparative analysis for risk potency, the noninvasive assessment of
risk by the PEP/LVET determination proved to be superior to coronary
arteriography. The previously reported stratification of mortality
risk by extent of coronary arterial disease was shown to pertain only
among patients with established left ventricular dysfunction.

     In considering the overall problem of mortality artery disease,
it is clear that the most common cause of death among patients with
previous myocardial infarction is sudden death. Approximately 60% of
the more than 700,000 deaths from coronary artery disease per annum
in the United States occurs suddenly outside of a hospital. The
primary disorder leading to sudden death for the most part is ven-

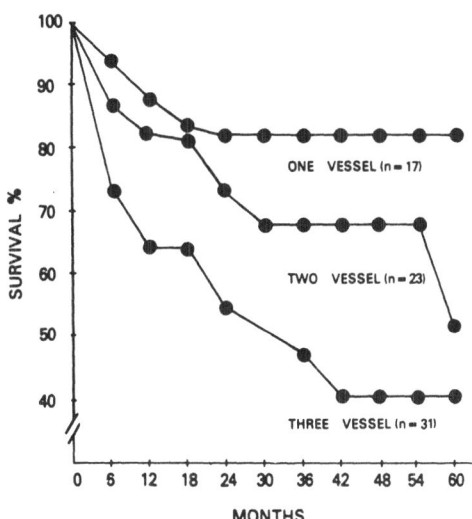

Fig. 5.   Cumulative 5-year survival for patients with an abnormal
ratio of preejection period to left ventricular ejection
time (PEP/LVET) (0.42 or greater) with one-, two- or three-
vessel coronary artery disease.   The trend for decreasing
survival with increasing extent of coronary artery disease
is significant (p<0.01).   (Reprinted with permission)

tricular fibrillation, an electrophysiologic calamity which accounts
for some 65% to 85% of all such episodes.   Profound bradycardia or
asystole and ventricular tachycardia are electrophysiologic events
that account for the remaining episodes of sudden death out of the
hospital.   Although once thought to be a common contributing event,
the occurrence of acute myocardial infarction immediately prior to
sudden death is now known to be relatively infrequent as a precipitating factor.   It is clear from the observations reported above that
not all patients at risk following recovery from a previous myocardial infarction are detected by the noninvasive assessment of left
ventricular performance.   Our attention in recent years has, hence,
been focused on additional means of detecting the patient at risk of
cardiac death.   It was our thesis, in pursuing these studies, that
the use of an additional noninvasive measure for the detection of the
patient vulnerable to lethal cardiac arrhythmia might sensitize our
capacity to uncover the patient at high risk.

The additional method, which we have chosen to pursue, represents an extension of the systolic time intervals concept.   Based
on knowledge that a high incidence of sudden death occurs among
patients with congenital prolongation of electrical repolarization[19,20], and other studies suggesting a higher sudden death risk

among patients with previous myocardial infarction when electrical
repolarization is prolonged relative to heart rate[21-23], our group
has focused on the QT interval as a predictor of high risk of ventricular arrhythmia in patients who have recovered from a previous
myocardial infarction. It is of interest in this regard that, in
previous studies on patients who had recovered from acute myocardial
infarction, the QT interval, measured electrocardiographically and
corrected for heart rate, was prolonged in a significantly higher
proportion of survivors of ventricular fibrillation than among those
without a history of ventricular fibrillation[22].

   Our studies have applied knowledge of the relationship between
the QT interval and the duration of electromechanical systole, the
QS_2 interval. Among a normal population, the QT interval terminates
at an average of 26 msec prior to the end of mechanical systole (as
determined by the timing of the second heart sound) with a standard
deviation of 13 msec[24]. Thus, prolongation of the QT relative to
the QS_2 is considered to be abnormal. The cumulative five-year
survival rate was studied among 100 patients who had recovered from a
previous myocardial infarction an average of 14 months previously and
who were followed for an average of 43 months. A comparison of the
cumulative survival among patients with a normal QT interval relative
to QS_2 (n=80) and those with prolongation of QT relative to QS_2
(n=20) is illustrated in Figure 6. The difference in five-year
cumulative survival rate of 58% is highly significant (p<.001).
Among the same cohort group the QT interval, corrected by the conventional Bazzett formula for heart rate (QTc), did not significantly
stratify the survival curves.

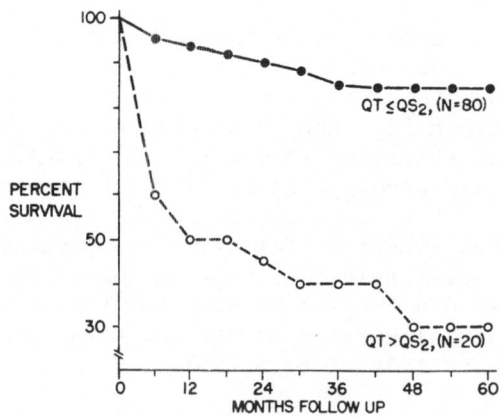

Fig. 6. Survival following QT, QS_2 determination. Cumulative 5-year
         survival rate for patients with and without QT > QS_2. The
         difference in cumulative 5-year survival is significant
         (p<0.001). (Reprinted with permission)

We next tested the hypothesis that the use of noninvasive measures designed to evaluate both left ventricular performance and vulnerability to ventricular arrhythmia by noninvasive measure may prove to be of greater sensitivity in detecting the patient at high risk than single measure alone[25].   This was accomplished by comparing the cumulative survival curves of groups of patients with no abnormality, a single abnormality, and combined abnormalities. Cumulative five-year survival among 100 patients with stable angina post myocardial infarction-grouped according to the presence of no, one, or two abnormalitites in PEP/LVET and QT relative to QS_2-was 98%, 73%, and 25% respectively.   Among 20 of the patients who died, 16 died suddenly.   Of these patients with sudden death, 11 had both abnormal PEP/LVET and $QT>QS_2$, while the remaining 5 had one or the other of the abnormalities.   Of 43 patients with no abnormality in PEP/LVET or QT relative to QS_2 only one succumbed during the follow-up period.   The four distribution of the patients relative to the absence or presence of an abnormality in PEP/LVET and QT relative to QS_2 is shown in Figure 7.   The clustering of patients with sudden death in the quadrant containing patients with both abnormalities is apparent.   The low incidence of sudden death among patients with neither abnormal PEP/LVET nor $QT>QS_2$ is also evident.   Thus, the application of combined noninvasive measures in the detection of left ventricular dysfunction and vulnerability to arrhythmia among patients with previous myocardial infarction adds considerably to the capacity to stratify patients into high and low risk groups.

It is appropriate to ask why such seemingly disparate phenomena as the occurrence of left ventricular dysfunction and the propensity to ventricular arrhythmia bear such a close relationship in the patient prone to death from coronary artery disease.   The studies of Lie and Titus are pertinent in this regard[26].   In a histopathologic

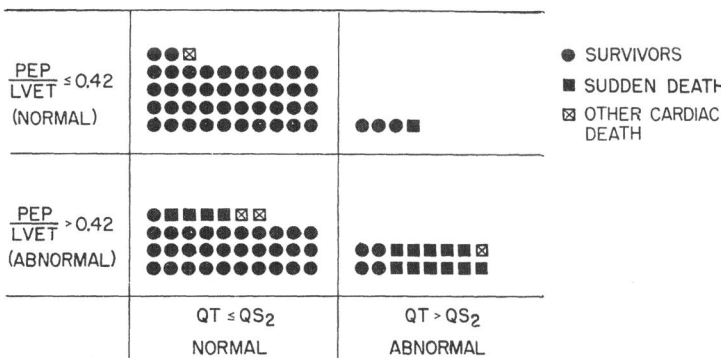

Fig. 7.   Distribution of survivors and non-survivors.   Distribution of 100 patients with previous myocardial infarction for presence of normal and abnormal PEP/LVET and for $QT \leq QS_2$ and $QT \geq QS_2$.   Survivors ●, sudden death ■ and other cardiac death ⊠ are shown.

analysis of the conduction system among patients dying suddenly,
either with or without an acute myocardial infarction, a scarcity of
acute lesions in the conduction system per se was noted. In contrast
52% to 81% of victims of sudden death showed at least one of three
histologic markers of ischemic myocardial injury or acute or chronic
type. This study corroborates several others demonstrating the high
incidence of old myocardial infarction, healing myocardial infarction, and/or myocardial fibrosis in patients with coronary artery
disease and sudden death. It would appear as if electrophysiologic
instability has its origin, not in the central components of the
ventricular conduction system, but rather in the peripheral myocardium, the same site as the disordered contractile mechanism.

It is our impression that the identification of the patient at
high risk by noninvasive methods has two important practical advantages. First, risk identification allows for better separation of
patients into management subgroups. Thus, the patient at low risk
may be advised to reestablish normal activity without the need for
intensive pharmacologic management, On the other hand, the patient
at high risk should have close clinical monitoring and probably
should be assigned a higher priority when considering antiarrhythmic
therapy and coronary bypass surgery. A second practical advantage of
risk identification relates to the planning and design of clinical
investigative studies on the treatment of patients with coronary
artery disease. In the design of studies on the influence of pharmacologic agents and/or surgical intervention in patients with coronary
artery disease relatively little attention has been given to the
influence of risk. Thus, the inclusion of large numbers of patients
at low risk is likely to yield little evidence of improvement from
any maneuver when a clinical trial is conducted over periods of five
or even ten years. The reason for this relates to the fact that
among low mortality risk groups it is difficult statistically to
identify a variable which can improve life expectancy, even when the
study is carried through several decades. On the other hand, among
patients at high risk, such interventions may demonstrate effectiveness in reducing mortality over relatively brief study periods.
Thus, the degree to which low risk patients are interspersed among a
study population diminishes the capacity to identify the salutory
influence of therapeutic maneuvers. In future studies on the potential beneficial effects of such maneuvers in coronary artery disease,
the randomization of patients at high risk by noninvasive measures of
ventricular function is likely to yield more definitive results in
shorter study periods than has heretofore been accomplished.

As our understanding of the fundamental mechanisms underlying
mortality risk in coronary artery disease evolves, it is anticipated
that the application of noninvasive methods will allow for a high
level of accuracy in predicting not only the subgroups vulnerable to
lethal cardiac arrhythmia but those individuals subject to left
ventricular failure and recurrent acute myocardial infarction as

well. As these methods are refined, it is anticipated that the improved precision in defining low- and high-risk subgroups will permit better identification of the high risk patient for special therapy and the implementation of more critical clinical trials on the effectiveness of medical and surgical procedures.

REFERENCES

1. R. H. Major, A history of medicine, Charles C. Thomas, Springfield, IL, p.132 (1954).
2. W. Heberrden, Some account of a disorder of the breast, Med.Tr. Royal Coll.Phys., London, 2:59-67 (1772).
3. J. B. Herrick and F. R. Nazum, Angina pectoris; clinical experience with 200 cases, JAMA, 70:67-70 (1918).
4. D. W. Richards, E. F. Bland, and P. O. White, A completed 25 year follow-up of 456 patients with angina pectoris, J.Chron. Dis., 4:423-33 (1956).
5. W. J. Block, E. L. Crumpacker, T. J. Dry, and R. P. Gage, Prognosis of angina pectoris. Observations in 6,882 cases, JAMA, 155:259-64 (1952).
6. W. B. Kannel, P. Sorlie, and P. M. McNamara, Prognosis after initial myocardial infarction: The Framingham Study, Am.J.Cardiol., 44:53-9 (1979).
7. A. V. G. Bruschke, W. L. Proudfit, and F. M. Sones Jr., Progress study of 590 consecutive nonsurgical cases of coronary disease followed 5-9 years, Circulation, 47:1147-53 (1973).
8. G. W. Burggraf and J. O. Parker, Prognosis in coronary artery disease. Angiographic, hemodynamic and clinical factors, Circulation, 51:146-56 (1975).
9. T. J. Reeves, A. Oberman, W. B. Jones, and L. T. Sheffield, Natural history of angina pectoris, Am.J.Cardiol., 33:423-30 (1974).
10. A. Oberman, W. B. Jones, C. P. Riley, T. J. Reeves, and L. T. Sheffield, Natural history of coronary artery disease, Bull.NY Acad.Med., 48:1109-25 (1972).
11. J. O. Humphries, L. Kuller, R. S. Ross, G. C. Friesinger, and E. E. Page, Natural history of ischemic heart disease in relation to arteriographic findings, Circulation, 49:489-97 (1974).
12. G. R. Nelson, P. F. Cohn, and R. Gorlin, Prognosis in medically-treated coronary artery disease, Circulation, 52:408-12 (1975).
13. R. E. Vliestra, J. L. Assad-Morell, and R. L. Frye, Survival predictors in coronary artery disease. Medical and surgical comparisons, Mayo Clin.Proc., 52:85-90 (1977).
14. H. Gross, A. K. Vaid, and M. V. Cohen, Prognosis in patients rejected for coronary revascularization surgery, Am.J.Med., 64:9-20 (1978).
15. A. M. Weissler, R. S. Stack, and Y. H. Sohn, Systolic time

intervals, in: "Heart Failure," A. P. Fishman, ed.,
Hemisphere Publishing, Washington, DC, 203-14 (1978).

16. A. M. Weissler, R. S. Stack, and Y. H. Sohn, Global left ven-
    tricular performance in chronic stable angina pectoris: the
    critical role of myocardial infarction, Coeur Med.Interne.,
    19:143-52 (1980).

17. A. M. Weissler and H. Boudoulas, Systolic time intervals in
    coronary artery disease, in: "Assessment of Ventricular
    Function," to be published by Progress in Scientific Culture
    (1983).

18. A. M. Weissler, W. W. O'Neill, Y. H. Sohn, R. S. Stack, P. C.
    Chew, and A. H. Reed, Prognostic significance of systolic
    time intervals after recovery from myocardial infarction,
    Amer.J.Cardiol., 48:995-1002 (1981).

19. A. Jervell and F. Lange-Nielsen, Congenital deaf mutism,
    functional heart disease with prolongation of QT interval an
    sudden death, Am.Heart J., 54:59 (1957).

20. C. Romano, G. Gemme, and R. Pongiglione, Aritmie cardiache rare
    dell'eta pediatrics, La Clinic Paeditr., 45:656 (1963).

21. P. J. Schwartz and S. Wolf, QT interval prolongation as
    predictor of sudden death in patients with myocardial infarc
    tion, Circulation, 57:1074 (1978).

22. R. E. Haynes, A. P. Hallstrom, and L. A. Cobb, Repolarization
    abnormalities in survivors of out-of-hospital ventricular
    fibrillation, Circulation, 57:654 (1978).

23. S. Ahme, C. Halmers, T. Lundman, N. Rehnquist, and A. Sjogren,
    QTc intervals in acute myocardial infarction: first year
    prognostic implications, Clin.Cardiol., 3:303 (1980).

24. H. Boudoulas and A. M. Weissler, The QT > QS$_2$ syndrome: a new
    mortality risk indicator in coronary artery disease, Amer.
    J.Cardiol., 50:1229 (1982).

25. H. Boudoulas, A. M. Weissler, Y. Sohn, and W. W. O'Neill, The
    detection of the high mortality risk by combined noninvasive
    measures of mechanical and electrical dysfunction in coronar
    artery disease, Clin.Res., 30:456A (1982).

26. J. T. Lie and J. L. Titus, Pathology of the myocardium and the
    conduction system in sudden coronary death, Circulation,
    52(Supp.III):41 (1975).

COMMENTS ON CHAPTER III:

A. Reale

Cattedre di Cardiologia
Università di Roma
Italy

After having listened to the lecturers and the discussion of the preceding sessions, so I shall be brief to leave as much time as possible to the discussion.

But first allow me to make a few general considerations. It appears to me that we are talking about ventricular performance, ventricular function, contractility and so on. We are not always talking the same language. Our own personal understanding of the concepts is often quite different from that of others and this is probably the reason why we really have no clear cut definitions of the terms we are using. Defining complex entities such as performance, behavior or contractile state, is indeed a very difficult task and if you think that we do not even have a generally accepted and comprehensive definition of the end results of abnormalities of the above mentioned variables, that is of heart failure, this gives you an idea of the difficulties.

There is an editorial in the July issue - I think July - of the European Heart Journal by Professor Denolin et al., attempting to define heart failure. We thought that it was rather a good one, but it immediately received, in the same issue of the Journal more disagreeing than favorable comments.

So it is understable that we are and shall be discussing how to evaluate the ventricle by different methods which represent the ultimate goal. As was repeatedly said in this meeting our aim is to relate these indeces to the pratical clinical problem of the individual patient in terms of a quantative diagnosis, prognosis and therapeutic indications. In the past few days we have heard some excellent presentations on the clinical applications of more or less

sophisticated ventricular function studies in different forms of
heart disease.

Since the beginning of aorto-coronary by-pass surgery the
question was raised whether myocardial revascularization besides
reducing angina would improve overall cardiac performance, thus
contributing to the prolongation of survival, the issue is really not
yet settled.

My questions are, for instance: if such an improvement does not
occur - as we have seen in Doctor Caponnetto's paper - does it mean
that revascularization was ineffective or unnecessary?

What happens to medically treated impaired ventricular
functions? Does it deteriorate progressively over the same observation period whereas ventricular behavior in surgically treated
patients remains unchanged? This is another unanswered or only
partly answered question.

Speaking of surgical treatment of heart conditions in general
and of ischemic heart disease in particular, the problem is of course
what happens after surgery, but mostly what has happened before the
operation. All our efforts should be directed at sorting out the
best time for surgery, so as to avoid going beyond the point of no
return in an individual patient. Demonstration that this is possible
in some patients has been given by Dr Mattioli in aortic valve
disease. Doctor Peterson and Rothlin in marked incompetence, Mariani
in mitral stenosis and Sergio Dalla Volta in congenital heart
disease. In ischemic heart disease the problem is still more complex, since ventricular function is one of the major prognostic
determinant of survival, but it is certainly not the only one. And
this leads us to Dr Weissler's presentation and there is very little
I can say about it, except that his results are impressive and that
it is somewhat ironic that prognosis in coronary artery disease can
be evaluated with such a high degree of accuracy by the simplest and
expensive of non-invasive techniques. It could well be that in the
near future most of our present problems could be solved by 2 sets of
investigations: systolic time intervals and computer enchanced
digital angiographic imaging of the coronary arteries. Those will
probably be sad times for the man in the invasive lab but easier
times for the patient with ischemic heart disease!

CHAPTER III:

DISCUSSION

DALLA VOLTA

In clinical practice tricuspid insufficiency associated with
initial stenosis is frequently observed. We found that right
ventricular function is important in determing the post-surgical
outcome in mitral stenosis. On the other hand I wonder if the
right ventricular ejection fraction, in the presence of tricuspid
insufficiency, is the ideal parameter for the evaluation of right
ventricular function. Perhaps it would be necessary to consider
volumetric parameters performing at the same time as a quantative
analysis of tricuspid insufficiency.

MARIANI

I think that the evaluation of the right ventricular function
nowadays gives even less reliable results in comparison with the
evaluation of left ventricular function. Nevertheless, the majority of the patients with a bad post-surgical outcome are those
with a low ejection fraction. The patients with a good outcome
are those with a higher ejection fraction. I'd like to add something about the treatment of tricuspid insuffiency. From the hemodynamic point of view when there is a relevant regurgitation due to
valvular damage, there is an indication for surgery. When there
is a high degree of pulmonary hypertension it is possible that the
regression of pulmonary pressure due to mitral surgery, results
in a reduction of tricuspid insufficiency. In any case I think
that cardiologists and cardiosurgeons definitely disagree about
this problem.

RAINERI

With the development of chronic mitral regurgitation therapy,
firstly fiber slippage, and then as a chronic response, hypertrophy
develops. After surgery you discussed regression of hypertrophy.
What happened to the fiber slippage, do you know?

PETERSON

I don't know. I think this is an extremely interesting question I tried to approach that question when I was on a Sabbatical Year in 1979 and I had a surgeon in Geneva who was taking biopsies for me, during surgery, hopefully to correlate with some later myocardial biopsies that we could take serially over a period of time. The problem is that you can never assess sarcomere links or sarcomere slippage unless you catch the ventricle at end diastole.

The study by Ross and Sonnenblick was a successful study because they quickly potassium arrested the heart at end diastole and unfortunately when you take a normal endomyocardial biopsy or a normal operative biospy, you see a mish-mash, you see sarcomeres in all ranges of shortening.

WEISSLER

In any evaluation employing end-diastolic volume as a measurement of ventricular function, be it the volume-pressure loop or the volume-stress loop, or the Starling approach, unless we know what the length of the sarcomeres are and the lining up of the sarcomeres, whether there is slippage, we'll never be able to evaluate ventricular function from any hemodynamic measure.

GAASCH

Ten years ago, Doctor Israel Mirsky calculated an index of preload based on the idea that the force acting to stretch the fiber was end-diastolic stress, that preload was analogous to the force stretching an isolated muscle, and that the force resisting stretch at end-diastole was end-diastolic muscle stiffness. Theoretically, it should be possible to get a handle on end-diastolic fiber stretch, and in fact with his calculations he did feel that he could actually correlate calculated indeces of fiber stretch with sarcomere links in end-diastole.

Unfortunately those measures were based on some very crude values for stress and muscle stiffness at end-diastole. I think we are in a position now to better estimate preload, end-diastolic stress and muscle stiffness at end-diastole; and we should be able,

at least, to deal with acute changes in sarcomere link - in terms of
end-diastolic muscle stretch.

PETERSON

The problem is once a ventricle dilates in response to a volume
overload and you are getting this fiber slippage, and the sarcomeres
are all staying 2.2 microns, then you are beginning to use a volume
for calculation of end-diastolic stress that does not necessarily
relate to what the sarcome link is, and in my opinion, preload is
sarcomere link, not end-diastolic stress.

RAZZOLINI

I agree that preload should reflect sarcomere length not volume,
so I believe that strain measurements should be an adequate measure
for preload. We have tried to measure these parameters in some
patients. Of course we have been forced to make very crude approxi-
mations and assumptions because of the difficulty to make such
measurements in vivo, in humans, We think we have encouraging results. For instance in aortic stenosis, diastolic strain always
seems to be lower than normal and in mitral regurgitation always
much more elevated than normal.

In view of the fact that mitral valve replacement in miral
regurgitation certainly induces an increase in total outflow
impedance to the left ventricle and probably an increase in afterload, too, do you think that a conservative treatment which usually
allows some small leakage through the mitral valve is less risky
even though it induces a less abrupt increase in impedance?

PETERSON

Mitral valve reconstructive procedures today are providing a
much better actuarial curve than is mitral valve replacement. One
of the reasons could be, just as you said, that you are partially
correcting the volume overload but still leaving some low impedance
pathways into the left atrium. So there is a physiological unloading
of the ventricle. Now it is hard for me to believe that hemodynamically that patient is still better off than he would be if the
leakage was completely closed off. It is possible that his ejection
fraction might be better, but that does not mean that his symptoms
are better.

RAINERI

I would like to ask Dr. Leachman his opinion about medical and surgical therapy of IHSS.

LEACHMAN

The treatment for IHSS at the present time is a very important question. I believe that probably if we look at the vast majority of patients with obstructive myopathy they need little if any treatment. I am not sure that medical treatment itself improves these patients. We have some patients who seem to feel better when they take propanolol than when they don't take it, perhaps the appreciation of his own symptoms is the best guide to each individual ventricular function.

So we offered these patients medicine, if they improved with the medicine then we just observed them, nothing more. There are about 15%, of those patients who seemingly continued with their symptoms or had progression of symptoms in spite of no treatment or as a result of medical treatment, and those are people that come to surgery and, as I alluded to, I believe many of those become worse as they develop progressive mitral valve regurgitation.

Now, if we look at the type of surgical procedure the tendency in the younger patients is to do a myomectomy. If there is major mitral regurgitation, then in addition to the myomectomy the current tendency for our surgeons is to do mitral valve replacement as well as myomectomy. I think clinical results are similar with better hemodynamic results following mitral valve replacement or mitral valve replacement in combination with myomectomy, than with myomectomy by itself.

DALLA VOLTA

I would like to refer to an important outcome which is sudden death. You have a rather and unusually high number of cases for Europe of hypertrophic cardiomyopathy with coronary vessel stenosis. Do you think that this condition might be important for sudden death?

LEACHMAN

There are many publications in the literature taking about deaths from ventricular arrhythmias and yet all of the information with regard to the diagnosis is echocardiography. We don't have systematic data in this group of patients with coronary disease to know whether they had bigger degree of arrhythmia time than those

without coronary disease. But I would suspect that they did, as we
know, that patients with coronary disease with normal hearts have a
higher incidence of arrhythmias.

The long term rate for mortality of medically treated patients
with IHSS with unknown coronary disease, with or without, the rate is
about 3% to 5% a year, the surgical survivors are about 1.6 per year.
We have now numbers which would correspond to 1.6 per year. We had 2
deaths among a group of 30 patients that we had long term data on,
the ones that I showed, some myomectomy, some valve and myomectomy
and some valve alone. And it turns out that there were 2 deaths in
that group, both women, both with extremely large hearts, both with
chronic atrial fibrillation and both with pulmonary hypertension, and
in fact we have since come to recognize that there is a hemodynamic
combination associated with a very high surgical mortality. But
there is no question that there has to be some increase in danger of
embolus, of malfunction of the valve.

ROTHLIN

I believe that for the survival with or without an operation it
is not possible yet to give an answer because there is not any study
where comparable patients were studied with or without surgical
treatment. We have long term experience, up to 15 years of total
series of 70. The group of patients with no limiting symptoms for
surgery all survived up to 15 years, while in the group of patients
with limiting symptoms there was a mortality which gave us after 10
years a 85% survival. Now I want to ask Dr. Leachman another
question. You gave evidence of decrease of left ventricular hypertrophy in a echocardiographic measurement of the ventricular wall,
and I have seen on your slides that you measured the left ventricular
septum. Because the left ventricular septum in these conditions is
not probably contributing very much to the left ventricular contractions, I would be interested to know if you also had measurements
of the left ventricular free wall, and whether you could show a
decrease of the left ventricular free wall dimension, because in our
experience this has not yet decreased, although in the myomectomy
operations performed in Zurich, we have the same hemodynamic results
as you presented for the mitral valve replacement, practically all
gradients were abolished and there was a significant drop in left
ventricular end-diastolic pressure.

LEACHMAN

I think, it is sometimes very difficult by a M-mode echo to get
what you really feel as a reliable dimension of the left posterior
wall. We tried to look at those in the the echos, but we had trouble
what the exact thickness of the posterior wall was. We thought that

these numbers are particularly significant because the septum was
not touched in those cases, the surgeon did not do anything to the
septum, but there is another potential explanation, and that is if
the ventricle is increased in its diastolic volume, the thinning of
the septum may be partly related just to the diastolic dimension
increase, so that even if you looked at the systolic dimension as
well, you would hope that it would be thinner than preoperatively.

I appreciate the fact that myomectomy does not improve the
gradient as well as changing of the valve alone, and we have had a
similar experience, as I indicated. The only operation that I know
of that has been 100% uniform in eliminating both the gradient and
the gradient induced by the infusion of catecholamines in our
experience, has been by replacing the valve. It seems that if you
leave the valve you end up with the possibility of inducing some
obstruction.

ROTHLIN

I would like to ask Doctor Dalla Volta, he mentioned three main
factors which are important for ventricular function in congenital
heart disease. By reviewing our transposition patients who had been
operated on and corrected in the early '60's, we had found a number
of patients who survived well but some of them are already with
atrial fibrillation; and some of them have just started with inter-
mittent dual arrhythmias, and from echocardiographic studies it
appears that these patients have right ventricular dilatation and
decreased shortening of the right ventricle, and one of the ideas
which was put forward to explain this was that these children survived for 4 to 7 years with extreme arteriolar desaturation and that
this is also a factor which could be responsible for right ventricular dysfunction in their later life.

The long survival with severe desaturation could be a reason to
operate on these patients much earlier.

DALLA VOLTA

I am not surprised, the atrial arrhythmias and the nodal
arrhythmia are one of the main causes as well as the closure of the
pulmonary vein or inferior vena cava, of the bad outcome of patients
with transposition of the great arteries. We must not forget one
very important thing: that in spite of the fact that we are changing
the mode of circulation and enabling the unoxigenated blood to come
from the two vena cavas and enter the pulmonary circulation, the
tricuspid valve must sustain the burden of very high pressure of the
right ventricle and this is totally unphysiological.

I think it is one of the major causes of late failure of
patients who underwent successful operation of either Mustard or
Senning.

LEACHMAN

I was just curious to ask Doctor Dalla Volta whether they had
any information to show a change in the actual ventricular performance in patients with extreme cyanotic tetralogy who underwent
surgical correction with normal oxygenation of the ventricle after-
wards.

DALLA VOLTA

In our experience the pre-operative level of th left ventricle
is not a major predictor of the outcome of the patient after operation.

KELLERMANN

I have one general question for Dr Chiddo and Dr Assennato.
How long was your assessment done after the acute onset of the
infarction?

ASSENNATO

This evaluation was done in patients that were discharged from
our unit care therapy 3 weeks after the acute onset of MI and the
evaluation decided about the rehabilitation program and other treat-
ments.

CHIDDO

Six months after MI.

DENOLIN

Mr Chairman, I would like to ask a question to both speakers.
I would like to know how they use these very interesting pathophysiological data to make their judgement on therapeutic indications and
prognosis. I would like to see exactly how they use all this information in a specific case, how they decided on the basis of this
sophisticated approach, that surgery is possible, considering at the
same time the clinical situation and the coronary situation.

ASSENNATO

It is very useful to have data which give us information on the diastolic and systolic function, especially if we consider that nowadays there are drugs that improve or decrease both the functions.

CHIDDO

Certainly the functional state of the left ventricle influences the prognostic assessment of patients with ischaemic cardiopathy. Obviously two-vessel disease patients, with an ejection fraction lower than 50% and a high degree of stress do not have the same prognosis as a two-vessel disease patient who is normal with respect to this index.

As regards indication and therapeutic method there is - unfortunately, in my opinion - an aspiration on the part of the cardiologist that the surgeon should take into account the evaluation of the functional state of the left ventricle. But I don't believe this happens in practice, that is, we have some patients where, even in the presence of serious impairment to the left ventricle, from our point of view, and therefore with an ejection fraction of 40%, the surgeon operates just the same, sometimes with good results and above all at low risk.

RIZZON

In this special selected group of patients these data are not important for surgical decision but we selected these patients to study latent impairment of left ventricular function, and to see which of these indexes is the best. And from our results we saw that Ks and left ventricular T are far worse in infarction patients when compared with patients without myocardial infarction.

DALLA VOLTA

I have a short question for Dr Chiddo. Do you think that the change in geometry of the ventricle in patients who had myocardial infarction and in some who had a change in the stiffness of the left ventricle can explain your data which at the end are based on two main data?

CHIDDO

Certainly modifications in the geometry of the left ventricle are very important. In effect the stress levels are decidedly high,

and besides, Ks, that is the elastic rigidity module - as you have
seen - distinguished within a single group subjects with infarction
from subjects without previous infarction, and this is a sure sign,
I would say, that geometry is very important. And it is of such
importance that I have been to stress that in our patients T is
lengthened, despite the fact that there are important modifications
in geometry that should have produced, in a non-ischaemic condition,
a different T result. That is, if the geometry is greatly altered, T
shortens, and I say this because I think this type of alteration is
very important.

PETERSON

     This is an interesting paper - I think - and the first thing
that occurs to me is how complex these analyses are and how important
it is to look at the methodology. The margin of identification on the
ventricle is very important. It happens that the beginning of the
second third of systole, is just when the trabeculae are beginning to
infold during systole, and you can get quite significant differences
in your first third ejection fraction depending upon where you draw
the outline of the ventricle. Secondly, with respect to the calculation of the time constant T, you can get quite different answers
depending what model and what mathematical approach you use. If you
differentiate the equation and then plot dP/dT versus P, and do a
linear regression, you then get a time constant which is independent
of what your pressure calibration is. And that is another extremely
important point. Doctor Weissler in his initial publications made a
mistake of not taking into account what the pressure calibration
constant might do to the actual calculation of T. So this is another
very important thing to examine.

     Thirdly, when you go to analyse diastole and calculate the
elastic constant, either on the pressure volume curve or on the
strain curve, you can get quite different answers whether you use
the three constant model or the mono-exponential fit; and again
whether you use a two constant model I think we would have to
examine features to know how valuable these data are to us.

     So I am not sure if we have time to go into all those points
but I merely wanted to highlight that these are very complex analyses
and if I were you and I had to write a paper with this kind of data
in, I would want very clear definition of what the methods were.

CHIDDO

     Well in connection with the questions, Dr Peterson asked
about methodology: first of all, the ejection fraction of the
first third was calculated by dividing the systole exactly into

three parts, from the end-diastolic volume, taken as the largest
volume corresponding to the R wave, to the first of the end-systolic
volumes, that is to the first of the smallest volumes, and by dividing it exactly into three parts. We did not eliminate the pre-
ejection period, as has been done by other authors.

Regarding the second questions about ventricular T: we calculated T by the semilogarithmic method.

Actually, a great step forward has been taken in the calculation
of T, by using a new and much more complex method which also takes
into account the possibility of extrapolating the pressure pattern
from the active to the passive phase of filling.

BUZZONE

I would like to ask Dr Assennato a question.

As you know by using the nuclear stethoscope we cannot assess
regional performance because we have not the functional imaging of
the left ventricle. Regarding the global function, it's very difficult to have a correct background subtraction without the functional imaging of the left ventricle, because of the overlapping of
the left atrium and the left ventricle in LAO projection. About
background subtraction, did you find the same problems in the
positioning of the nuclear stethoscope especially during exercise
because it is very important to have a right acquisition of the raw
material.

ASSENNATO

When carrying out an evaluation with a nuclear stethoscope it
is extremely important that it should be used properly. The nuclear
stethoscope doesn't yeald regional evaluations in since it doesn't
produce any images, but the curves of systolic and diastolic function
are valuable in assessing left ventricle performance. Background
scanning is not critical, as the instrument is capable of identifying
the exact values, which can be obtained without any difficulty even
during exercise.

DALLA VOLTA

Dr Chiddo, there was a point you emphasized correctly in my
opinion, in your paper about the particular nature of monovascular
pathology. I should like to know if in this connection you
think collateral flow is important. I believe that in Padua

Dr Scognamiglio has shown that in cases with the same degree of
coronary stenosis when there is a collateral flow it is easier to
find a hypokinetic area than an akinetic one, and this might be
important for those indexes which you examined.

CHIDDO

Just now I can't give you any data concerning the presence or
not of the collateral vessels in these patients, but I remember that
the collateral vessels were present in all of them and - as I said
before - our infarctions were mostly related to occlusion of the
right coronary artery in relatively young subjects. I think that if
studied six months after infarction, the collateral vessels may be
formed.

WEISSLER

I was wondering whether you have data on the frequency of
regional contraction abnormality in your group A, to explain whether
or not there was a higher incidence of ventricular fibrosis in that
group as one would expect in the usual group of patients with
coronary disease without infarction.

Furthermore I do understand that your patients with previous
infarction were selected out of a group of 11 patients with low end
diastolic pressure. Hence in discriminating any measure of ventricular function one reduces his capacity to differentiate which
measure is better than another because all of them have such mild
degrees of abnormal function.

CHIDDO

The patients were not only selected on the basis of end-
diastolic pressure but also, of course, on the basis of volumes.
That is, they had analogous and, above all, normal-range values for
end-systolic volume, end-diastolic volume, VCF and overall ejection
fraction. Therefore not only end-diastolic pressure.

As regards the distribution of the areas, in group A, that
is in the group of patients without infarction, there were 12
hypokinetic areas, six of the diaphragmatic wall, 4 of the medial
anterior segment and 2 of anterior apical segment. On the other
hand, in the infarction group we had 9 hypokinetic areas, the same
number (6) of the diaphramatic wall, 3 of the medial anterior segment and 4 akinetic areas of the diaphragmatic segment of the
posterior wall.

RUBINO

I have a question for Dr Caponnetto. Would you agree that for
the analysis of the regional dynamics of the left ventricle, to
detect any improvement due to by-pass, a projection should be made in
all the regions and, therefore in the right oblique, as well?

CAPONNETTO

As you know, by using equilibrium radionuclear angiography, one
can only evaluate the function of the left ventricle in the left
anterior oblique or, at most, in antero-posterior. We have carried
out an anlogous study with the first pass, which enables us to study
the left ventricle in right anterior oblique to obviate this problem.
The advantage of equilibrium lies in the evolution of the function of
the septal region, something which is impossible with the first pass.
Therefore, when it was possible, that is in the first pass, we
analyzed that the right anterior oblique, which naturally has the
advantage of having the same projection as the angiography.

LEACHMAN

Dr Weissler, I was wondering, did you look at the QT/QS_2 in
coronary patients who had not had an infarction but presumably had
still a normal ventricle?

Was there any profitable data that could come out of that
particular group?

WEISSLER

Remember that all the data I'm talking about are patients who
had previous infarction: I selected those because this is the only
way in 5 years one can demonstrate the difference in mortality.

So, in fact once PEP/LVET is normal, one already has a low
mortality rate in the low risk group, if at the same time we have a
normal QT/QS_2 that makes them even less vulnerable to death.

The group of patients - I think - are the most important ones.
In fact, the greatest challenge comes from the group I did not study.
The patients with angina pectoris who did not have an infarct and
died suddenly. In order to do that study I am going to have to come
back here when I am 60, 70 years old or so, and have all the data on
a large group of patients, clearly about 500 patients, because the
mortality rate in that group is going to be very low. The indicators
in patients - I believe - who have angina pectoris are going to be

ischemic indicators, not functional indicators, that is, individuals
who develop severe ischemia with exercise and the individuals who
develop vulnerability to arrhythmia.

RAINERI

Doctor Weissler, can you give us some data on incidence of
serious ventricular arrhythmia in the group with prolonged QT/QS_2
index, in the group with sudden death and patients who did not have
sudden death?

WEISSLER

Unfortunately the only written figures we have are of the
systolic time interval measurements, and we have not published those
data because I don't think it is a reliable indicator of the presence
of arrhythmia. I can tell you that among the patients with prolonged
QT there was a slightly higher incidence of PVCs during their
electrocardiogram and during the recording of systolic intervals.

But most of the patients, 80% of the patients, at least, who had
QT lenthening had no arrhythmia during the recording of the systolic
intervals which are all arbitrary, you could have about 1 or 2
minutes of full recording, but that's about it.

CHAPTER IV

PHARMACOLOGICAL MANIPULATION OF VENTRICULAR FUNCTION

CHAPTER IV:

INTRODUCTORY REMARKS

A. Raineri

Cattedre di Fisiopetologie, Cardiovascoliere
Universitè di Palermo, Italy

Over the past two decades, there have been great advances in our understanding of the cardiac physiology, and in particular, of the factors governing myocardial performance and energetics.

In the intact circulation, cardiac function is regulated by four major parameters: 1) contractile state; 2) preload; 3) afterload; 4) heart rate.

The test stand of these parameters is heart failure.

The primary derangement in myocardial failure is impairment of contractile performance. In this setting a number of compensatory mechanisms become operative.

The heart will dilate, thereby increasing its preload, and according to the Frank-Starling principle, its stroke volume. The circulatory adjustments which characterize heart failure appear to have evolved in order to protect two critical physiological parameters, initially stroke volume and cardiac output and, ultimately, arterial pressure.

The previously presented background indicates the rationale employment of different drugs to treat failure, and among these vasodilators.

This can be appreciated by recognizing that, while in normal subjects cardiac performance represents an interplay between the loading conditions of the left ventricle and its contractile state, in patients with myocardial failure contractility is essentially fixed and cardiac performance is more directly determined by the peripheral circulation.

299

The parameters which determine the symptomatic status of the patients, the ventricular filling pressure and cardiac output, can therefore be affected by drugs which act directly on the peripheral blood vessels.

The therapeutic implications of these consideration are clear. At present is is impossible to reverse the basic molecular fault responsible for the loss of myocardial contractile activity.

However, it is logical to attempt to correct this defect by drugs with positive inotropic activity.

It is equally logical to attempt to offset the biophysical disadvantage of ventricular dilatation by reduction of the blood volume with diuretics and counter the deleterious effects of increased impedance with vasodilator drugs.

On the basis of this view I have the pleasure of introducing the speakers who will continue this last session of our Course.

LEFT VENTRICULAR UNLOADING WITH CALCIUM ANTAGONISTS

M. D. Guazzi

Cattedra di Cardiologia, Instituto Ricerche
Cardiovascolari "G. Sisini", Centro Ricerche
Cardiovascolari del Consiglio Nazionale delle Ricerche
University of Milan, Milan, Italy

SUMMARY

Calcium channel blockers reduce the arterial smooth muscle tone and lower blood pressure. They may be regarded as left ventricular (LV) unloading agents.

LV unloading efficacy of nifedipine (15 cases) and verapamil (14 cases) was tested in hypertensive decompensated patients, through a one-month treatment period.

Nifedipine persistently reduced systemic vascular resistance (SVR), mean arterial pressure (MAP), mean pulmonary wedge pressure (PWP) and LV diastolic diameter (DD) and improved cardiac index and Vcf. All of the patients had relief from dyspnea and reduction in heart size. The only side effect was ankle edema in 6 cases.

Verapamil reduced SVR and MAP was not effective on PWP, LV DD and Vcf. The drug was discontinued in 2 patients who developed severe dyspnea at rest after a 3-4 day continued oral treatment. Clinical symptoms and signs did not improve in the remaining subjects despite persistent pressure reduction.

A less potent vasodilating action of verapamil and a prominent depression in cardiac contractility may account for the different results with the two compounds, in spite of a shared vasodilating antihypertensive effect. These findings prove that functional changes in the failing hypertensive heart may differ from one calcium blocker to another as a result of interaction and relative preponderance of influences on afterload and contractility.

Circulatory load impinging on LV during systole depends upon
duration of contraction, ventricular pressure and dimensions, and
wall thickness. Resistance to peripheral runoff is a major determinant of the impedance that blood faces as it is ejected from the
left ventricle and importantly influences the velocity and the extent
of fiber shortening, ejection fraction and, probably the diastolic
and the end-systolic ventricular volume. Therefore, afterload may
vary in parallel with changes in vascular resistance and an over-
loaded left ventricle benefits from systemic vasodilatation. The
functionally impaired hypertensive heart is such an example. In it,
cardiac enlargement[22,23] adds to the elevated impedance to left
ventricular ejection (vascular resistance) and intraventricular
pressure, resulting in an extremely high load impinging on the left
ventricle. Vasodilating antihypertensive compounds are drugs of
choice in these cases. Since calcium channel blockers possess these
properties, they may be regarded as potentially effective ventricular
unloading agents.

This possibility was tested[1] in a population of 29 hospital-
ized untreated primary hypertensive patients whose supine diastolic
pressure was >120 mm Hg, who were free of valvular, ischemic or
idiopathic myocardial disease, arrhythmias or conduction disorders,
segmental left ventricular contraction abnormalities. The majority
of these patients presented ECG signs of left ventricular strain,
variable degree of lung congestion on chest roentgenogram, mild
dyspnea at rest. These patients were assigned randomly to two groups
and treated with, respectively, nifedipine (Group I, 15 cases) and
verapamil (Group 2, 14 cases). Circulatory measurements were performed in the control state and repeated 90 min after the first oral
dose of the drug, and at the end of one month continued oral treatment. Nifedipine was given at the dosage of 20 mg q.i.d. and
verapamil at the dosage of 160 mg t.i.d.

Average of the circulatory values in the control condition
(period A) after acute (period B) and prolonged (period C) calcium
channel blockade in the two groups are reported in Figure 1. Groups
were homogeneous with regard to age; baseline heart rate, arterial
pressure, diastolic left ventricular minor dimension (echocardiography), systemic vascular resistance, cardiac index, pre-ejection
and end-systolic wall stress and Vcf were also similar in the groups;
wedge pulmonary pressure was slightly higher in Group 1.

Acute administration of nifedipine (20 mg) in Group 2 influenced
all of the examined hemodynamic functions. At 90 min after the drug
the average fall in mean arterial pressure from control was 41 mm Hg,
cardiac output was invariably augmented and peripheral vascular
resistance reduced. Additional changes were a decrease of the
wedge pulmonary pressure, of both pre-ejection and end-systolic wall
stress, and an increase of the left ventricular mean circumferential
fiber shortening rate. At the end of one month continued oral treat-

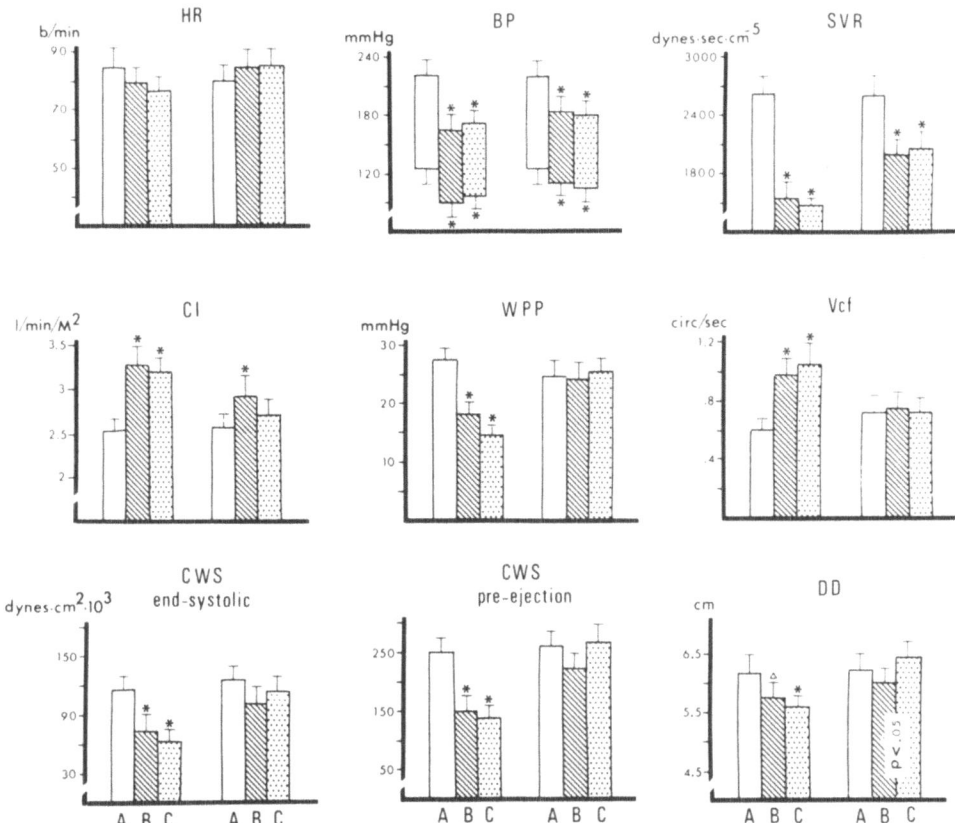

Fig. 1. Hemodynamic measurements in Group 1 and Group 2 patients in
the control state (A), after acute calcium channel blockade
(B) and after one month continued treatment (C). Bars
represent the means and SD. HR, heart rate; BP, aortic
pressure (systolic and diastolic); SVR, systemic vascular
resistance; CI, cardiac index; WPP, mean wedge pulmonary
pressure, Vcf, LV mean velocity of circumferential fiber
shortening; CWS, end-systolic and pre-ejection LV circumferential wall stress; DD, end-diastolic LV minor diameter.
Δ and * indicate p values of <0.05 and <0.01, respectively,
for differences between control and drug periods; p values
for differences between drug periods are also shown in the
figure.

ment (period C) the averages of the circulatory variables were
similar in quantity and quality to those at period B.

The circulatory response to a 160 mg oral dose of verapamil in
Group 2 (period B) was characterized by a decrease in systolic and
diastolic arterial pressure (the average fall below control mean

arterial pressure was 29 mm Hg), associated with reduction in peripheral vascular resistance and increase in cardiac index. Wedge
pulmonary pressure and left ventricular mean velocity of circumferential fiber shortening were unchanged, ventricular diastolic
minor axis and wall stress were slightly diminished. At period C,
although blood pressure and vascular resistance remained reduced to
the same extent as at period B, there was an increase in the end-
diastolic left ventricular minor axis and reversion of pre-ejection
wall stress towards the baseline levels. Averages at period C in
Group 2 were derived from 12 cases, two patients having been with-
drawn from the trial during the first 3-4 days of verapamil treatment. Percentage variations from control during period B and C in
arterial pressure, cardiac output and peripheral vascular resistance
were greater after nifedipine than after verapamil.

All patients in Group 1 experienced progressive disappearance of
dyspnea after nifedipine was started in parallel with the circulatory
improvement. Reduction of heart size was documented by chest roent-
genograms; the electrocardiogram reverted to normal in five cases;
signs of left ventricular strain were inequivocally reduced in seven
cases.

In the verapamil group (Group 2), two patients developed severe
nocturnal dyspnea that required interruption of the drug and prompt
use of digoxin and furosemide. In the remaining cases of this group
cardiac symptoms and signs persisted substantially unchanged.

These findings support two points: that both nifedipine and
verapamil have a vasodilating antipertensive efficacy in man; that as
far as these effects are concerned, prolonged administration is not
associated with drug resistance. Referring to the clinical applic-
ability of the two calcium channel blockers as ventricular unloading
agents, patients were purposely selected in whom each of the variables which are positively related to the stress (afterload) of the
ventricular wall augmented: high blood pressure, elevated impedance
to ejection (peripheral vascular resistance), enlarged ventricular
cavity. It is obvious that ventricular wall stress is not static
during contraction and that pre-ejection and end-systolic stresses
are single frames of a continuously varying phenomenon; however,
pre-ejection and end-systolic stresses may have a special conceptual
value as afterload markers, as they reflect the maximal levels of the
instantaneous force (pressure, wall tension or stress) for the existing instantaneous myocardial length or volume during the isovolumic
and the ejection phases[2].

Symptoms, signs and hemodynamic data (high wedge pulmonary
pressure, low cardiac output and Vcf) document that in these patients
the excessive afterload was associated with advanced impairment in
the performance of the heart. The fact that performance varied with
treatment and proportionally to changes in afterload supports the

interpretation that a cause-effect relation existed between excessive afterload and ventricular dysfunction. Three facts are proved for nifedipine: 1) left ventricular preload was reduced (lower filling pressure and end-diastolic volume); 2) pre-ejection and end-systolic afterload were also reduced; 3) the performance of the left ventricle was improved (diminished pulmonary artery wedge pressure associated with increased cardiac output). It is presumable that in consequence of the vasodilatation, impedance to left ventricular ejection was reduced. This effect was such as to probably improve the myocardial oxygen balance, to facilitate emptying of the left ventricle, to augment ejection velocity and to facilitate venous return from the lungs. Information about systemic venous circulation and the possible contribution of venodilatation to reduction of preload are not available; however, venous fooling need not to be postulated for this effect. Changes in Vcf would suggest[3] that the inotropic state of the heart was improved by the nifedipine administration; it cannot be excluded, however, that they were simply the consequence of the consistent variations in both preload and afterload[4] with the drug.

Verapamil shared with nifedipine a vascular smooth muscle relaxant action which, although less pronounced, resulted in a definite reduction in peripheral vascular resistance and, conceivably, in impedance to LV ejection. In spite of this, the performance of the left ventricle was almost unaffected in the acute study and tended to become depressed further with continued treatment. Two patients developed paroxysmal nocturnal dyspnea that required withdrawal from the trial. Persistance of enhancement of an elevated wedge pulmonary pressure after a drug, verapamil, whose dilating action involves the arterial and possibly the venous side of systemic circulation[5], suggests the intervention of a negative inotropic effect. Regarding the relation between left ventricular afterload and function, both pre-ejection and end-systolic afterload were poorly reduced at period B and even less at period C compared to control state; the performance of the left ventricle (as indicated by the Vcf and the wedge pulmonary pressure-cardiac index relation) varied in the same direction. The beneficial effects on afterload of the arterial pressure reduction were almost entirely offset by the concomitant enlargement of the ventricular cavity both in diastole (Figure 1) and at end-systole (Figure 2). This pattern reinforces the possibility of a predominant negative inotropic action of verapamil.

Differences between the two calcium channel blockers in their influences on myocardial contractility are impressively shown in Figure 2 by the end-systolic stress-left ventricular end-systolic diameter relation, which was taken as an index[6] of the end-systolic force-length relationship[2,7,8,]. Leftward movement of the stress-length values following nifedipine might reflect either a shift to a different point in the same stress-length line (what would indicate no change in contractility) or a shift to the left of the stress-length line itself (what would indicate improved contractility); on

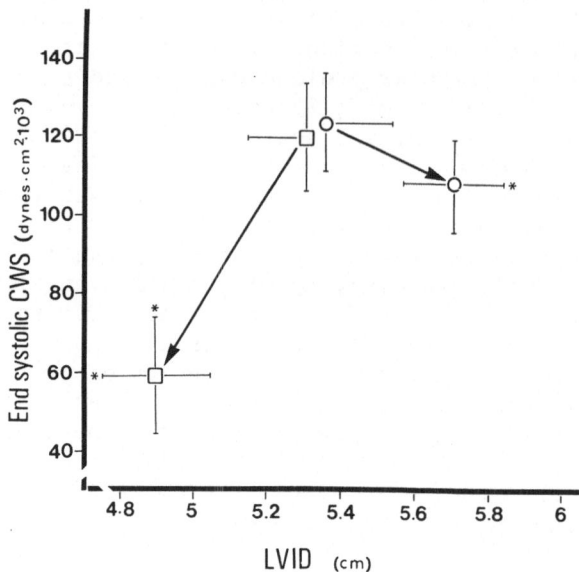

Fig. 2. Relation between LV end-systole wall stress and LV end-
         systolic internal diameter (LVID) in Group 1 (open squares)
         and in Group 2 (open circles) in the baseline and after
         one-month continued treatment with nifedipine and, respec-
         tively, verapamil. Symbols represent means and SD.
         * indicates p values of <0.01 for differences in wall stress
         and diameter, respectively, between control and treatment.

the contrary, rightward shift of the same values with verapamil must
reflect depressed contractility.

     Findings discussed here indicate that drug-induced functional
changes in the dilated hypertensive heart may vary from one calcium
channel blocker to another as a result of the interplay and the
relative preponderance of different influences in afterload and
contractility. They suggest a clinical role for nifedipine (and its
derivatives) in unloading the enlarged hypertensive heart, and appear
to negate the value of verapamil and verapamil-like compounds in this
clinical setting.

REFERENCES

1. M. Guazzi, F. Fabbiocchi, C. Cipolla, P. Montorsi, and P.
    Sganzerla, Disparate unloading effects of verapamil and
    nifedipine on the decompensated hypertensive left ventricle,
    Am.Heart J., in press.
2. K. T. Weber, J. Janicki, and L. L. Hefner, Left ventricular
    force-length relations of isovolumic and ejecting contrac-
    tions, Am.J.Physiol., 231:337 (1976).

3. N. J. Fortuin, W. P. Hood, Jr., and E. Craine, Evaluation of
    left ventricular function by echocardiography, Circulation,
    46:26 (1972).
4. E. Braunwald, Determinants and assessment of cardiac function,
    N.Engl.J.Med., 296:86 (1977).
5. A. Zanchetti, Perspectives in antihypertensive treatment, In:
    "Calcium Antagonism in Cardiovascular Therapy: Experience
    with Verapamil," A. Zanchetti and D.M. Krikler, eds.,
    Excerpta Medica, Amsterdam, p.292 (1981).
6. N. Reichek, J. Wilson, M. S. J. Sutton, T. A. Plappert, S.
    Goldberg, and J. W. Hirshelfd, Noninvasive determination of
    left ventricular end-systolic wall stress: Validation of the
    method and initial application, Circulation, 65:99 (1982).
7. K. T. Weber and J. J. Janicki, Instantaneous force-length re-
    lations: Experimental findings and clinical correlates,
    Am.J.Cardiol., 40:740 (1977).
8. W. Grossman, E. Braunwald, T. Mann, L. P. McLaurin, and L. H.
    Green, Contractile state of the left ventricle in man evalu-
    ated from the end-systolic pressure volume relations,
    Circulation, 56:845 (1977).

NEW AND OLD INOTROPIC DRUGS:

CONTROVERSIES AND CHALLENGES

G. Cocco and P. Jerie

Klinik Barmelweid, CH-5017 Barmelweid, Switzerland
Cardiological Ambulance, Solbadklinik
CH-4310 Rheinfelden, Switzerland

SUMMARY

Many studies, using different parameters for contractility, have
established that some inotropic reserve exists even in severly failing myocardium. Thus, various agents (such as digitalis, sympathomimetic drugs etc.) can augment the myocardial inotropic state in
heart failure to various degrees. The aim of this paper is to describe available positive inotropic drugs, classified according to
their mechanism of action. Drugs with established positive inotropic
effect are described first. In addition, new agents with sufficiently high oral bioavailability and long duration of action are
presented. Of increasing importance are drugs which combine a positive inotropic and a vasodilator effect (e.g., methylxanthines, some
β-agonists, amrinone etc.). Indeed, it appears that one of the most
favorable effects of positive inotropic drugs is the simultaneous
direct or reflex peripheral vasodilation. Lastly, it remains to be
determined whether or not pronounced, sustained inotropic stimulation
of the myocardium can produce a potentially deleterious imbalance in
the myocardial energy supply-demand relationship, or whether the
potential exists for arrhythmia induction by some of these agents.
Based on these considerations, it seems likely that inotropic drugs
alone will not represent the most favorable therapy for the severely
failing heart.

INTRODUCTION

In recent years the management of heart failure has improved.
In some cardiac diseases treatment is now possible, e.g., with reduction of high blood pressure in hypertensive cardiomyopathy; or with

309

surgery in patients with valvular malfunction. In the majority of
cases, however, etiology is uncertain and specific and radical treatment is not yet possible. Also, unloading therapy with diuretics
and/or vasodilators is an effective approach. By reducing either
preload or afterload, these agents place the myocardium in a better
hemodynamic and metabolic position in the Starling curve. Unloading
drugs, however, do not exert any relevant inotropic action on the
heart itself. On the other hand, many studies using different parameters for contractility have established that some inotropic reserve
exists even in severely failing myocardium[1]. Indeed, many drugs
can augment the inotropic state of the failing heart to various
degrees. Unloading agents are reviewed elsewhere in this book, and
we will not discuss these drugs. The aim of our paper is to review
the most important positive inotropic drugs and to estimate their
therapeutic value according to their mechanism of action (Table 1).

1. CARDIAC GYLOSIDES

Cardiac glycosides (generally called 'digitalis') are still the
most important agents, and the only inotropic drugs recognized as
effective in the long-term treatment of chronic congestive heart
failure[2-8]. Cardiac glycosides play a major role in the treatment
of the failing heart with atrial fibrillation. There are, however,
controversial opinions about the efficacy of digitalis in patients
with cardiac failure who are in sinus rhythm. Digitalis was introduced in an era when strict proof of efficacy had not been required.
In recent years some studies were not able to demonstrate a sustained
effect of digitalis in patients with chronic treatment[9-12]. Other
investigators, however, have confirmed the sustained effect of digitalis as maintenance therapy in patients with unquestionable diagnosis
of congestive heart failure and sinus rhythm[13-17]. Assessment of
the effect of digitalis on maintenance therapy in the clinical setting may be difficult. Prejudices about digitalis may also play a
role[2,8]. We believe, that whenever the myocardium is compromised,

Table 1. Positive Inotropic Drugs

---

1. Cardiac glycosides (digitalis)
2. Beta-adrenergic agonists
3. Alpha-adrenergic agonists
4. Methylxanthine (e.g., theophylline)
5. Glucagon
6. Varia: – histamine
         – amrinone
         – sulmazol
         – Ro 13-6438
         – inosine

---

it is logical to consider digitalis therapy. Most clinicians would employ digitalis in this condition unless contraindications exist. With other forms of heart failure, other therapeutic measures and unloading drugs (especially diuretics) may be preferable.

The exact mechanism of action of digitalis is still unknown. It is generally accepted that digitalis leads to an increase in the amount of intracellular Ca^{2+} which reacts with the contractile proteins, but it is unclear how this increase in Ca^{2+} is achieved[2, 3,18]. There is no doubt that digitalis reacts with the (Na^+-K^+) -ATPase[2,3]. Some authors[2,3,19] argue that inhibition of (Na^+-K^+)-ATPase results in an increase in intracellular Ca^{2+}, via an effect on transsarcolemmal Na^+-Ca^{2+} exchange. Other authors[2,3,19] assume that such an inhibition in (Na^+-K^+)-ATPase is only important during 'toxic' actions of cardiac glycosides, whereas 'therapeutic' concentrations would lead to an increased Ca^{2+} release from the sarcolemmal stores during excitation.

Whatever the mechanism of action of digitalis may be, the net hemodynamic effect is characterized by a marked increase in cardiac output, by a reduction in heart rate and, especially if elevated, by a decrease of left ventricular filling pressure, thus reducing oxygen consumption[2,4]. Table 2 summarizes the cardiocirculatory effect of digitalis in the failing heart. Unfortunately cardiac glycosides have low therapeutic/toxic index. Furthermore, many patients remain symptomatic despite therapeutic levels of digitalis and are thought to have 'refractory' heart failure[2,5] and sensitivity to digitalis is enhanced in the elderly[2]. Lastly, many drugs may interfere with

Table 2. Cardiocirculatory Effect of Digitalis in
          Congestive Heart Failure

_____

        DIRECT: positive inotropic effect
        INDIRECT: - arteriolar vasodilation
                    - systemic venous dilation
                    - hepatic venous dilation
                    - sympathetic diminution

        Consequences of these effects:
                    - cardiac output   ↑↑
                    - arterial pressure   ↑↓
                    - venous pressure   ↓
                    - portal venous pooling   ↓
                    - venous return   ↑↑
                    - blood volume   ↓
                    - heart rate   ↓
                    - myocardial O_2 consumption   ↓
_____

↑↑: marked increase; ↑↓: increase or decrease; ↓: decrease

the pharmacokinetics of digitalis and induce toxicity[20,21]. The
pharmacokinetic properties of cardiac glycosides in patients with
normal[20,21] or impaired[22,23] renal function are important and the
use of these drugs may be difficult. For these reasons, there is
considerable effort being put into the development of various
inotropic drugs, possibly with fewer side effects and a low toxicity.

2. β-AGONISTIC ADRENERGIC DRUGS

The β-adrenoreceptors are localized on the external surface of
the sarcolemma[24]. Following stimulation of the β-adrenoceptor, the
increase in contractile force (inotropic effect) develops rapidly and
is accompanied by an increase in the rate of force development
(klinotropic effect), and by a decrease in the duration of contraction. This latter action is often referred to as the 'relaxant
effect' of β-agonists, and it is shared only by agents which increase
the myocardial cAMP content, such as methylxanthines[3]. The relaxant effect is hemodynamically important in facilitating ventricular
filling in the face of an increased heart rate. It is accepted that
β-agonists increase the Ca^{2+} concentration in the vicinity of the
contractile proteins as a consequence of an increase in myocardial
cAMP level and that both Ca^{2+} and the cAMP system activate each
other[3]. The β-agonists increase the Ca^{2+} uptake into the sarcoplasma, thus increasing the amount of releasable Ca^{2+} within, they
also increase the slow inward Ca^{2+} current during the action
potential[3]. The increased Ca^{2+} uptake by the sarcoplasma not only
contributes to the positive inotropic effect, but also explains the
relaxant effect. The relaxant effect, however, might also be due to
a decreased Ca^{2+} sensitivity of the contractile proteins[3]. The
above effects are not direct, but are the result of stimulation of
adenylcyclase, with a subsequent increase in cAMP levels. It follow
an increased cAMP phosphorylation of the contractile proteins[3].
The β-agonist is the 'first messenger' and the cAMP and the Ca^{2+} the
'second' and 'third' messenger, respectively.

2a. β₁-Adrenergic Drugs

Stimulation of β₁-adrenoceptors results in an increase in
cardiac contractility and heart rate, and a decrease in atrioventricular condition time. The most important β₁-agonists are shown
in Table 3. Adrenaline (epinephrine), noradrenaline (norepinephrine
isoprenaline (isoproterenol), dopamine and dobutamine stimulate the
cardiac β-adrenoceptors and thus increase myocardial contractility,
usually improving hemodynamics in patients with heart failure[2,7,
28,29]. Despite relevant hemodynamic differences, these drugs have
the common disadvantage of requiring slow intravenous infusion and
careful titration, and are not suitable for long-term treatment of
chronic heart failure.

Table 3. Beta-adrenergic Agonists

A. NON SPECIFIC (beta$_1$ + beta$_2$) AGONISTS
- adrenaline (epinephrine)
- noradrenaline (norepinephrine)
- isuprenaline (isoproterenol)
- dopamine
- dobutamine

B. BETA$_1$-AGONISTS
- prenalterol
- xamoterol

C. BETA$_2$-AGONISTS
- salbutamol (albuterol)
- fenoterol
- terbutaline
- pirbuterol
- metaproterenol (orciprenaline)

The list does not pretend to be complete. It summarizes
the most important agents with β-agonistic activity. Some
drugs possess additional relevant effects, e.g., dopamine
stimulates dopaminergic receptors and, at higher concentrations, it may stimulate α-adrenoceptors.

Prenalterol is a new β$_1$ selective adrenoceptor agonist. The
l-isomer is biologically active. The drug is active in the paren-
teral and oral form[30]. Its pharmacologic profile is: a) inotropic
β-agonist, b) inotropic effect >> chronotropic effect, c) β$_1$ selective, d) no α-adrenergic effect, and e) partial β-agonist. Prenalterol does not exert any effect on blood glucose, free fatty acids,
triglycerides, lactate and pyruvate[30]. The bio-availability of a
20 mg sustained-release oral preparation is about 45%, and the elimination half-life around 2 hours[31]. Hjalmarson et al.[32] have
treated patients with moderate to severe congestive heart failure
with prenalterol, using up to 225 ug/kg body weight intravenously.
The cardiac index and the heart rate increased significantly, while
the pulmonary artery end-diastolic pressure and the calculated systemic vascular resistance decreased significantly; the effect on
heart rate was variable in individual patients. The positive hemodynamic effect was also evident during exercise[32]. The same investigators[32] and our group[33] have also treated patients with
moderate to severe cardiac failure with oral prenalterol who were
'resistant' to digitalis and diuretics given at therapeutic doses.
The basic therapy was left unchanged and, according to a double-blind
randomization schedule, either prenalterol or placebo was added. The
daily dosage of prenalterol was 20-200 mg p.o., the dose being care-
fully titrated so that the heart rate and blood pressure was not

allowed to increase more than 10% over the resting baseline, the
daily dosage being divided into 4 equal parts. Hjalmarson et al.[32]
treated 16 patients up to two weeks: the stroke volume and cardiac
output increased by 16 and 21% respectively, both changes being
significant; the ejection fraction (by echocardiography) and the
systolic blood pressure decreased by 10 and 4% respectively, and the
heart rate increased by 10%. The pre-ejection period and the PEP/
LVET (systolic time intervals) decreased significantly and no arrhythmias were detected by ECG dynamic monitoring. The hemodynamic
effect lasted for the 2-week observation period and no tachyphylaxis
was observed. Nine out of 17 patients experienced less dyspnea and
tiredness, and the effort tolerance increased; five out of 16
patients felt no difference and another patient worsened because of
mitral regurgitation. The main side effects were tachycardia and
palpitation. We have treated 14 patients during 2-4 weeks (average
2, 3 weeks)[33]; the therapy was stopped according to the instruc-
tions of the manufacturer, due to severe changes in the long-term
animal toxicology. In the prenalterol-group, stroke volume and
cardiac output (first-pass technetium-99m) increased by 13 and 19%
respectively, while in the placebo-group they were unchanged. The
end-diastolic and end-systolic left ventricular volumes were decreased by 10 and 8% respectively. The positive hemodynamic effect
was maintained during exercise and no signs of tachyphylaxis were
observed. In two out of 14 patients (both with ischemic heart
disease), cardiac output decreased and angina pectoris worsened, the
regional scintigraphy showed a left ventricular dyssynergy during
exercise. The dynamic ECG-monitoring did not detect any worsening of
the pretreatment arrhythmias, but the average heart rate increased by
10% and the episodes of spontaneous sinus tachycardia augmented by
20%, while in the placebo-group both heart rate and the episodes of
sinus tachycardia remained unchanged.

Xamoterol (ICI 118,587, Corwin®) is another cardioselective
β-adrenergic blocker with partial $β_1$-agonistic activity[34]. Its
partial agonistic activity is equal to 43% of the maximal response of
isoprenaline and this should prevent any further increase in potentially harmful sympathetic stimulation[35,36]. It is expected to be
devoid of any $β_2$-activity and metabolic effect[37]. Xamoterol has
been given i.v. to patients with cardiac failure and coronary artery
disease: global systolic and diastolic left ventricular function
improved and myocardial O_2 consumption did not increase[37]. It
seems to be a promising drug for the i.v. treatment of patients with
coronary artery disease and heart failure. It is unclear, however,
if it will be valuable for the maintenance oral treatment in patients
with chronic heart failure.

2b. $β_2$-agonists

These drugs are widely used for the treatment of asthma. $β_2$-
adrenoreceptors are present in smooth muscle, bronchi and uterus; the

myocardium should have mainly β_1-receptors. It appears, however, that β_2-agonists exert powerful effects on the heart. The reasons are unclear: either the available β_2-adrenergic agonists are not specific, and possess some β_1-agonistic activity as well, or the classification into β_1- and β_2-adrenoreceptors has clinical limitations and it might be impossible that the simple concept of only β_1-adrenoceptors in the heart is untrue. Alternatively, due to reduced systemic resistance (direct effect of β_2-stimulation), a reflex release of catecholamines follows the administration of β_2-agonists and thus stimulates the cardiac β_1-adrenoceptors.

Salbutamol albuterol[38,41], fenoterol[37,42], terbutaline[37, 43] and pirbuterol[37,41,44,49] are selective β_2-agonists. In addition to a vasodilating effect (stimulation of vascular β_2-adrenoceptos), these agents induce a positive inotropic effect[37, 41,43-49], whereas the positive chronotropic effect is relatively small[3,37,41]. Salbutamol and pirbuterol have been found to improve hemodynamics when given either i.v. or orally to patients with severe heart failure[41,50]. The β_2-angonists have also several metabolic effects: glycogenolysis is increased in the liver and skeletal muscle so the hyperglycemia and lactic acidosis may occur[50]. Serious cardiac arrhythmias are rare in patients treated with β_2-agonists, but they have been observed[41,50]. On the other hand, cardiac arrhythmias with β_2-agonists are neither more frequent nor worse than with other inotropic drugs e.g.m digitalis or amrinone[50]. β_2-agonists may also increase myocardial oxygen consumption. Preliminary studies with pirbuterol are encouraging[50], but conclusive data are unavailable and some caution is necessary.

3. α-ADRENERGIC AGONISTS

The positive inotropic response to catecholamines is mediated predominantly by β_1-adrenoreceptors. However, α-adrenoreceptors exist in the myocardium[3,51]: their stimulation increases the contractile force[3]. The inotropic effects of α- and β-adrenergic agents are qualitatively different: the α-adrenergic response develops relatively slowly and is accompanied by a prolongation of the contraction, while the effect of the β-agonists is fast and there is a relaxant effect (shortening of contraction)[3]. The α-adrenergic inotropic effect is strongly dependent on the frequency of stimulation, being greater at low frequencies[3]. Furthermore, the α-adrenergic positive inotropic effect is more pronounced in hypothyroidism and in hypothermia, and does not lead to marked positive chronotropic effects[3,51].

The mechanism of action of α-adrenergic agonists is unclear. There is no increase in cAMP level and no relaxant effect, cGMP level and (Na^+-K^+)-ATPase activity remain unchanged, and changes in slow Ca^{2+} current are of very little magnitude[3].

At present α-agonists are not routinely used in the therapy of
heart failure, neither α_1 nor α_2. On the other hand, since the
effect of α-adrenergic stimulation is different from that of β-
stimulation, α-agonists might be of some interest in the treatment of
cardiac failure with low heart rate or hypothyroidism.

4. METHYLXANTHINES

These agents, e.g., theophylline, have positive inotropic and
vasodilating effects[3,41]. The positive inotropic effect is not due
to either a release of catecholamines or stimulation of adrenergic
receptors, but is secondary to an increase in Ca^{2+} uptake from the
extracellular space. The positive inotropic effect can be altered by
slow-channel blocking agents. The effect of these agents develops
within seconds and is of the same amplitude as that of β-agonists[3].
The inotropic effect is due to increased cAMP levels, secondary to an
inhibition of the degradation of cAMP[3,52]. Aminophylline is the
best studied and largely used methylxanthine: 240 mg, given parenterally, orally or rectally, increases cardiac output and heart rate
and induced a powerful vasodilation. Problems could arise, however,
with the use of aminophylline as an inotropic agent, especially
because of its positive chronotropic effect. Furthermore, to avoid
toxicity, monitoring of serum concentration is necessary, particularly in the elderly, in patients with liver insufficiency and in
smokers[52]. Cardiac arrhythmias and central nervous side effects
(tremor, nervousness, insomnia, convulsions, coma etc.) may follow
toxicity. Theophylline and caffeine are two well known methylxanthines. They induce a positive inotropic effect, however, they
prolong the duration of contraction, an effect which is opposite to
the of β-agonists[3]. The prolongation of contraction is cAMP independent and does not belong to the essential action of all methylxanthines. For example, 3-isobutyl-1-methylxanthine is a potent inhibitor of phosphodiesterase and thus inhibits the degradation of cAMP,
nevertheless it decreases the duration of contraction[3]. Some
experimental phosphodiesterase inhibitors are at present under investigation, hoping that they will induce fewer side effects. Preliminary results are available for buquineran (Pfizer), BDPU (UK 14,275
Pfizer), carbazeran, D4975 (Degussa), HA542, HA543 and HA550 (Asahi
Chem), agents with phosphodiesterase inhibiting effect, but these
agents are not yet available for routine clinical use.

5. GLUCAGON

The i.v. administration of the pancreatic hormone glucagon
induces considerable inotropic effect[53,3,41] which is the result of
an increased cAMP level, secondary to stimulation of adenylcyclase.
The effect is not mediated, however, by β-adrenoreceptors and thus
cannot be antagonized by β-antagonists[3,41]. Glucagon must be given

parenterally, induces hyperglycemia and other relevant metabolic effects and thus cannot be used for long-term treatment.

6. OTHER DRUGS

6a. Histamine

The myocardium contains H-receptors, especially of the H_2-type[3]. H-agonists exert a direct positive inotropic effect, due to an increased cAMP level secondary to a stimulation of adenylcy-clase[3]. H-agonists are not available for clinical management of heart failure, but are theoretically interesting.

6b. Amrinone (Inocor)

Amrinone is a bipyridine derivative with positive inotropic and chronotropic effect, and with a vasodilating effect[3,4,41,54-56]. Amrinone does not effect adrenoreceptors or histamine-receptors and does not change phosphodiesterase and (Na^+-K^+)-ATPase activities or cAMP levels[3,4,55]. Amrinone potentiates the effect of β-agonists and histamine and increases the slow Ca^{2+} inward current[3]. At high concentrations it prolongs the duration of contraction, similarly to theophylline and is therefore reasonable to assume that amrinone might indeed be a phosphodiesterase inhibitor[3]. Following i.v. administration, the inotropic effect begins within a few minutes and lasts approximately 90 minutes. The positive inotropic effect of oral doses can be maintained for up to 24 hours and the drug could therefore be used for long-term treatment of heart failure. However, severe side effects such as thrombocytopenia, cardiac arrhythmias, hypotension, jaundice, abdominal discomfort, hypogeusia, dysosmis, fever etc. have been reported[57] and there is a subgroup of patients with congestive heart failure who do not benefit from this drug[54]. Although there is no doubt that amrinone is a positive inotropic agents acutely[3,4,41,54-57], the question arises to whether it is an improvement in the long-term therapy of heart failure. Despite the initial enthusiasm, amrinone has not proved to be suitable for routine clinical use.

6c. Sulmazol (Vardax)

Sulmazol (ARL 115 BS) is a phenyl-imidazo-pyridine derivative. Chemically it has no similarities with digitalis, catecholamines, phosphodiesterase inhibitors, glucagon or amrinone. Its mechanism of action is unclear, but it has positive and vasodilating effects[3,4, 41,58]. The inotropic effect is not mediated by stimulation of adrenoreceptors or release of endogenous catecholamines and the drug is not bound to the (Na^+-K^+)-ATPase[3]. The effect resembles that of

theophylline. The drug increases the duration of contraction, potentiates the inotropic effect of noradrenaline and inhibits phosphodiesterase activity. It also increases the Ca^{2+} sensitivity of
contractile proteins, the mechanism of action being probably complex[3]. Given i.v. sulmazol improves myocardial function by decreasing the pre- and afterload, together with a positive inotropic
effect[58]. The safety of this agent is, however, as yet unknown.

6d. Ro 13-6438

Ro 13-6438 (Roche) is a derivative of imidazoquinoline, whose
pharmacologic properties resemble that of amrinone[59]. Its mechanism of action is unclear, being probably complex. During early
clinical testing, relevant tachycardia and marked orthostatic hypotension were limiting side effects[60]. It has been given to
patients with moderate cardiac failure: cardiac contractility has
increased together with a positive chronotropic effect and vasodilation[61]. At this stage the drug is merely experimental, but it
preliminary results are not encouraging.

6e. Inosine

The endogenous nucleoside inosine has been found to exert impor
tant cardiovascular actions. It possesses marked inotropic properties, dilates coronary arteries and reduces free fatty acid concentrations[62]. Inosine stimulates myocardial glucose uptake. It may
also stimulate insulin release and may favorably influence the metab
olism of the myocardium, especially in the presence of ischemia[62].
Inosine is an interesting agent, in the treatment of heart failure i
the presence of myocardial ischemia, but clinical experience is not
yet available.

DISCUSSION

If heart failure is dominated by systolic dysfunction, agents
that enhance inotropism may be central in the therapeutic program.
On the other hand, with low cardiac output and impaired filling,
unloading agents are more important. In the majority of patients
with severe heart failure, however the combination of diuretics,
unloading measures and positive inotropic drugs will be required.

When noradrenaline was first introduced in the therapy of car-
diogenic shock after myocardial infarction, it was not clearly under
stood whether the blood pressure increase was either the result of
perypheral vasoconstriction or of positive inotropic effect or even
of both factors. Experimental studies and clinical results in the
last decade have clarified the mechanisms and definitely confirmed

the usefulness of β_1-adrenoceptor stimulating agents in the treatment of heart failure. The positive intropic effect of these agents results from an increase in slow Ca^{2+} inward current and in the uptake of Ca^{2+} into the sarcoplasmatic reticulum. This, in turn is probably due to an increased adenylcyclase activity and cellular cAMP level. β_1-adrenoreceptor stimulants are useful in patients with severe and refractory heart failure, especially in combination with vasodilators. Their use is, however, limited because they require i.v. administration and careful titration. Glucagon and histamine stimulate adenylcyclase by a different mechanism than adrenergic agonists. Glucagon has been repeatedly recommended in patients with heart failure: it must be given parenterally and induces marked metabolic effects and thus cannot be used for long-term treatment. Prenalterol can be given orally and its chronotropic effect is not too marked, however its usefulness in chronic heart failure is as yet unknown. Furthermore, severe changes in the long term animal toxicology are discouraging. β_2-agonists have vasodilating properties and possibly some direct inotropic effect. They improve hemodynamics in patients with heart failure, but conclusive data about long-term treatment of chronic heart failure is not yet available. Some caution is necessary, since acidosis and hyperglycemia may occur, precipitating arrhythmias in predisposed patients. Methylaxanthines are effective inotropic agents, but their positive chronotropic effect and the need for careful titration do not allow their routine clinical use in patients with chronic heart failure. Amrinone, a bipyridine derivative, probably a phosphodiesterase inhibitor, has a positive inotropic effect and vasodilating properties. It induces favorable hemodynamic effects in patients with heart failure and it can be given i.v. and orally. However, severe side effects make this drug not suitable for routine clinical use.

The critical approach to the treatment of heart failure has contributed to a better understanding of the role of digitalis. Its positive inotropic effect was confirmed by several authors and its indications were revised. But other conditions modified the therapeutic approach with cardiac glycosides. The clinical diagnosis of heart failure was better defined. Highly effective cardiac glycosides were introduced, the galenics were improved and the bioavalability was standardized. Very cautious prescription of digitalis, due to safety problems, and some scepticism concerning its efficacy, led to the 'digitalis controversy'. Nevertheless, digitalis still remains the drug of choice for life-long treatment of heart failure in patients with atrial fibrillation or tachycardia.

Since the use of alternative newer drugs has been disappointing, there is a revival in recommending salt restriction, limiting physical activity and mental stress and classic unloading agents in patients with severe heart failure. These measures should be considered in all patients with heart failure: they may restore the sensitivity to medicaments and are a cornerstone in the complex therapy of heart failure.

The 'drug of the future' for the treatment of chronic heart failure will combine the positive inotropic effect with vasodilating properties, it will be free of any secondary contraregulatory mechanisms and systemic adverse reactions, it will not have serious side effects. Of course, none of the available drugs has such a pharmacologic profile.

REFERENCES

1. J. Ross, Mechanisms of cardiac contraction. What roles for pre-load, afterload and inotropic state in heart failure? Eur.Heart J., 4(suppl.A):19-28 (1983).
2. L. Storstein, Clinical and circulatory aspects of digitalis in heart failure, Eur.Heart J., 3(suppl.D):59-64 (1982).
3. H. Scholz, Pharmacological actions of various inotropic agents, Eur.Heart J., 4(suppl.A):161-172 (1983).
4. K. Weber, V. Andrews, and J. S. Janicki, Cardiotonic agents in the management of chronic heart failure, Am.Heart J., 103:639-649 (1982).
5. P. Jerie, Behandlung der therapieresistenten Herzinsuffizienz, Schweiz.Rundschau.Med., 72:15-21 (1983).
6. M. Telerman, P. Decoodt, and B. Peperstraele, The rationale of the treatment of congestive heart failure, Acta Cardiol., (Suppl.28):77-83 (1982).
7. M. Artman, M. D. Parrish, and T. P. Graham Jr., Congestive heart failure in childhood and adholescence: recognition and management, Am.Heart J., 105:471-480 (1983).
8. H. Holzgreve, Glykosidtherapie: Suche nach alternativen Behandlungsmöglichkeiten, Münch.med.Wschr., 124:14-16 (1982).
9. A. M. Weissler, J. R. Syder, C. D. Schoenfeld, and S. Cohen, Assay of digitalis glycosides in man, Am.J.Cardiol., 17:768-780 (1966).
10. C. Davidson and D. Gibson, Clinical significance of positive inotropic action of digoxin in patients with left ventricular disease, Brit.Heart J., 85:970-976 (1973).
11. V. Cohn, A. Selzer, E. Kersh, L. S. Kerpmann, and N. Goldschleger, Variability of hemodynamic response of acute digitalisation in chronic cardiac failure due to cardiomyopathy and coronary artery disease, Am.J.Cardiol., 35:461-468 (1975).
12. G. D. Johnston and D. G. McDevitt, Is maintenance digoxin necessary in patients with sinus rhythm? Lancet, 1:567-570 (1979).
13. N. H. Carliner, C. A. Gilbert, A. W. Pruitt, and L. I. Goldberg, Effect of maintenance digoxin therapy on systolic time intervals and serum digoxin concentrations, Circulation, 50:94-98 (1974).
14. M. H. Crawford, J. S. Karliner, and R. A. O'Rourke, Favourable effect of oral maintenance digoxin therapy on left ventric-

ular performance in normal subjects. Echocardiographic study, Am.J.Cardiol., 38:843-848 (1976).

15. J. H. Kleinmann, N. B. Ingels, G. Daughters, E. B. Stinson, E. L. Aldermann, and R. H. Goldmann, Left ventricular dynamics during long-term digoxin treatment in patients with stable coronary artery disease, Am.J.Cardiol., 41:937-943 (1978).

16. S. B. Arnold, C. B. Randolph, W. W. Parmley, and K. Chatterjee, Long-term digitalis therapy improves left ventricular function in heart failure, New.Engl.J.Med., 303:1443-1448 (1980).

17. R. G. Murray, A. C. Tweddel, W. Martin, D. Pearson, J. Hulton, and T. D. Lawrie, Evaluation of digitalis in cardiac failure, Br.med.J., 284:1526-1528 (1982).

18. H. Lüllmann, T. Peters, and J. Preuner, Mechanism of action of digitalis glycosides in the light of new experimental observations, Eur.Heart J., 4(suppl.D):45-51 (1982).

19. T. Godfraind, The biphasic action of cardiac glycosides on the Na^+-K^+ pump and its relevance in the treatment of heart failure, Eur.Heart J., 3(suppl.D):53-57 (1982).

20. U. Peters, Pharmacokinetic review of digitalis glycosides, Eur.Heart J., 3(suppl.D):65-78 (1982).

21. K. Kochsiek, Pharmacokinetic review of digitalis review, Eur.Heart J., 3(suppl.D):79-86 (1982).

22. G. Cocco and A. Audibert, Probleme der digitalistherapie bei patienten mit niereninsuffizienz, Schweiz.Rundschau.Med., 39:1242-1245 (1975).

23. G. Cocco and A. Audibert, Digitalistherapie und niereninsuffizienz, Mediz.Monatschrift., 1:22-25 (1976).

24. E. G. Krause and A. Wollenberger, Cyclic nucleotides and heart, in: "Cyclis 3', 5'-nucleotides Mechanisms of Action," H. Cramer and J. Schultz, eds., London (1977).

25. G. I. Drummond and D. L. Severson, Cyclic nucleotides and cardiac function, Circ.Res., 44:145-153 (1979).

26. H. Scholz, Effects of beta- and alpha-adrenoceptor and activators and adrenergic transmitter releasing agents on the mechanical activity of the heart, in: "Handbook of Experimental Pharmacology," Part I, L. Szekers, ed., Springer Verlag, Berlin - Heidelberg - New York, 651-733 (1980).

27. H. Reuter, Localization of beta adrenergic receptors and effect of noradrenaline and cyclic nucleotides on action potentials, ionic currents and tension in mammalian cardiac muscle, J.Physiol. (London), 242:429-451 (1974).

28. S. I. Rajfer and L. I. Goldgerg, Dopamine in the treatment of heart failure, Eur.Heart J., 3(suppl.D):103-106 (1982).

29. K. Chatterjee, A. Bendersky, and W. W. Parmley, Dobutamine in heart failure, Eur.Heart J., 3(suppl.D):107-114 (1982).

30. H. Mattson, A. Hedberg, and E. Carlsson, Basic pharmacological properties of prenalterol, Acta Med.Scand., 659(suppl.):9-37 (1982).

31. O. Ronn, E. Fellenius, C. Graffner, G. Johnsson, P. Lundborg, and L. Svensson, Metabolic and hemodynamic effects and pharmacokinetics of a new selective β_1-adrenoceptor agonist, prenalterol, in man, Eur.J.Clin.Pharmacol., 17:81-86 (1980).

32. A. Hjalmarson, N. Abelardo, and E. Waagstein, Effects of prenalterol in congestive heart failure, Eur.Heart J., 3(suppl.D):115-121 (1982).

33. G. Cocco and C. Strozzi, Oral treatment with the beta-agonist prenalterol in patients with congestive heart failure. Unpublished data. Ciba-Geigy Basle files.

34. E. J. Ariens, Molecular pharmacology. The mode of action of biologically active compounds, Academic Press, New York, 148-170 (1964).

35. J. D. Harry, H. F. Marlow, A. G. Wrdleworth, and J. Young, The action of ICI 118,587 (a beta-adrenoceptor partial agonist) on the heart rate response to exercise in man, Br.J.Clin. Pharmacol., 12:266-267 (1981).

36. H. F. Marlow, A. G. Wardleworth, and L. M. Booth, The haemodynamic effects of oral doses of ICI 118,587, a beta-adrenoceptor partial agonist, in healthy volunteers, Br.J.Clin.Pharmacol., 13:269-270 (1982).

37. H. Pouleur, M. F. Rousseau, P. Mengeot, C. Veriter, M. F. Vincent, and L. A. Brasseur, Improvement of global and regional left ventricular function in patients with previous myocardial infarction by a new β_1-adrenoceptor partial agonist, ICI 118,587, Eur.Heart J., 3(suppl.D):123-127 (1982).

38. B. Sharma and J. F. Goodwin, Beneficial effect of salbutamol on cardiac function in severe congestive cardiomyopathy. Effect on systolic and diastolic function of the left ventricle, Circulation, 58:449-460 (1978).

39. C. H. Kerr, Hemodynamic effects of oral salbutamol in patients with severe left ventricular failure, Circulation, 60(suppl.II):11-41 (1979).

40. P. D. V. Bourdillon, J. R. Dawson, R. A. Foole, A. D. Timmis, P. A. Poole-Wilson, and G. C. Sutton, Salbutamol in treatment of heart failure, Brit.Heart J., 43:206-210 (1980).

41. C. S. Maskin, Th. Le Jemtel, J. Kugler, and E. H. Sonnenblick, Inotropic therapy in the management of congestive heart failure, Cardiovasc.Rev.and Reports, 3:837-846 (1982).

42. R. Slutsky, W. Hooper, K. Gerber, G. Cartis, J. Karliner, and W. Ashburn, The effect of terbutaline on left ventricular function and size, Am.J.Cardiol., 45:412-419 (1980).

43. M. Irmer, H. Wollschläger, and H. Just, Behandlung der schweren herzinsuffizienz mit dem beta-stimulator fenoterol, Klin.Wschr., 59:639-645 (1981).

44. W. S. Colucci, R. W. Alexander, and G. H. Mudge Jr., Acute and chronic effects of pirbuterol on left ventricular ejection fraction and clinical status in severe congestive heart failure, Am.Heart J., 102:564-568 (1981).

45. J. R. Dawson, R. Canepa-Anson, P. Kuan, N. H. G. Whitaker, J. Carnie, C. Warnes, S. R. Reueben, P. A. Poole-Wilson, and G. C. Sutton, Treatment of chronic heart failure with pirbuterol: acute haemodynamics responses.

46. R. E. Rude, Z. Turi, E. J. Brown, B. H. Lorell, W. S. Colucci, G. H. Mudge Jr., C. R. Taylor, and W. Grossman, Acute effects of oral pirbuterol on myocardial oxygen metabolism and systemic hemodynamics in chronic congestive heart failure, Circulation, 64:139-145 (1981).

47. B. Sharma, J. Hoback, G. S. Francis, M. Hodges, R. W. Asinger, J. N. Cohn, and C. R. Taylor, Pirbuterol: a new oral sympathomimetic amine for the treatment of congestive heart failure, Am.Heart J., 102:533-541 (1981).

48. C. R. Taylor, J. R. C. Baird, K. J. Blackburn, D. Cambridge, J. W. Constantine, M. S. Ghaly, M. L. Hayden, H. M. McIhenny, P. F. Moore, A. Y. Olukotun, L. G. Pullman, D. S. Salsbury, C. A. P. D. Saxton, and S. Shevde, Comparative pharmacology and clinical efficacy of newer agents in treatment of heart failure, Am.Heart J., 102:515-532 (1981).

49. J. R. Dawson, J. Bayliss, M. S. Norell, R. Canepa-Anson, P. Kuan, S. Reuben, P. A. Poole-Wilson, and G. C. Sutton, Clinical studies with β_2-adrenoceptor agonists in heart failure, Eur.Heart J., 3(suppl.D):135-141 (1982).

50. R. Canepa-Anson, J. R. Dawson, W. S. Frankl, P. Kuan, G. C. Sutton, S. Reuben, and P. A. Poole-Wilson, β_2-Adrenoceptor agonists; pharmacology, metabolic effects and arrhythmias, Eur.Heart J., 3(suppl.D):129-134 (1982).

51. G. Cocco, F. Burkart, D. Chu, and F. Follath, Intrinsic sympathomimetic activity of beta-adrenoceptor blocking agents, Europ.J.clin.Pharmacol., 13:1-4 (1978).

52. W. S. Hillis and M. Been, Phosphodiesterase inhibitors: haemodynamic effects related to the treatment of cardiac failure, Eur.Heart J., 3(suppl.D):97-101 (1982).

53. M. Robert and L. Humair, Cardiotone wirkung vom glucagon. Klinische prüfung und therapeutische indikationen, Schweiz.med Wschr., 100:1345-1352 (1970).

54. E. L. Kinney, B. Carlin, J. O. Ballard, J. M. Burks, W. F. Hallahan, and R. Zelis, Clinical experience with amrinone in patients with advanced congestive heart failure, J.Clin. Pharmacol., 22:433-440 (1982).

55. A. A. Alousi and J. Edelson, Amrinone, in; "Pharmacological and Biochemical Properties of Drug Substances," 3:120-147 (1981).

56. J. Bayliss, M. Norell, R. Canepa-Anson, S. R. Reuben, P. A. Poole-Wilson, and G. C. Sutton, Acute haemodynamic comparison of amrinone and pirbuterol in chronic heart failure. Additional effects of isosorbide dinitrate, Brit.Heart J., 49:214-221 (1983).

57. P. T. Wilsmurst and M. M. Webb-Peploe, Side effects of amrinone therapy, Brit.Heart J., 49:447-451 (1983).

58. J. Thormann, W. Kramer, M. Schlepper, and M. Gottwick, AR-L
     115BS in the treatment of heart failure, Eur.Heart J.,
     3(suppl.D):87–95 (1982).
59. G. Haeusler, Ro 13-6438, a new original positive inotropic
     agent, World Conf.Clin.Pharmacol.Ther. (London), abstract
     0057 (1980).
60. S. Gasic, Preliminary experience with i.v. and oral Ro 13-6438
     in healthy volunteers. Roche Clinical Unit, Univ. of Vienna.
     Rocher files (unpublished data) (1980).
61. W. Kiowski, F. Burkart, M. Pfisterer, and Z. Y. Lu, Die positive
     inotrope wirkung von Ro13-6438, einer nicht-digitalis-
     ähnlichen/nicht sympathomimetisch wirkenden substanz bei
     patienten mit herzinsuffizienz. 51 Annual Meeting of the
     Swiss Soc. of Internal Medicine, 5-7th May 1983, abstract
     No.8.
62. O. A. Smiseth, Inosine infusion in dogs with acute ischemic left
     ventricular failure: favourable effects on myocardial
     performance and metabolism, Cardiovasc.Res., 17:192–199
     (1983).

CHAPTER IV:

DISCUSSION

RAINERI

I agree with you, Dr Guazzi, that Verapamil and Nifedipine give different results but I think that we have to stress that these two drugs are dose dependent in their effect, so I think that we have to be precise as to what the dose was in your experiments. I was impressed by your results regarding heart rate. Even if there were no statistical differences in heart rate, there was an increase of the heart rate with Verapamil and a decrease with Nifedipine. But we know these two drugs are of different trends: in fact we know that Verapamil acts on the AV node and so it has a tendency to lower the heart rate instead of Nifedipine because of catecholamines intervention.

GUAZZI

Especially as far as heart rate is concerned which is the observation of the reduction in heart rate with Nifedipine and the tendency to an increase with Verapamil is in the same way against our classical knowledge of the effect of these drugs on heart rate.

I know that Verapamil has a tendency to reduce heart rate especially through a reduction on the transmission, on the sympathetic synaptic transmission, and through a depression on the sinus node, and also we know well that following the vasodilation produced by Nifedipine there is a reflexed increase in heart rate. And I would interpret the discrepancy we have observed with my results in this way: these were patients with a baseline three-vessel dysfunction and probably elevated heart rate in the base line was a kind of mechanism of compensation.

And if the function of the heart was improved by Nifedipine we can assume that the heart rate was reduced in consequence of the withdrawal of a compensatory mechanism. And if the heart rate was increased with Verapamil it might be interpreted as a kind of compensation to the depressive effect of the drug itself.

As far as dosage is concerned I think I said during my presentation that Nifedipine was given at the dose of 20 milligrams and Verapamil at the dosage of 160 milligrams.

LEACHMAN

I was curious, from Doctor Guazzi's elegant observations, if you had another group of patients that had been treated and who were not hypertensive, that had congestive heart failure of some myocardial type, to know whether or not same general trend of effect might be true in the weakened myocardium rather than just in patients with excessive afterload due to hypertensive disease.

GUAZZI

I did not have enough time to present these data which we published a few years ago in the American Journal of Medicine concerning the treatment of acute decompensation in congestive cardiomypathy and also in patients with rheumatic valve disease, aortic insufficency and mitral regurgitation. We found that better results were observed in patients with rheumatic heart disease. I refer to acute studies in patients with acute left ventricular decompensation and we found an obvious improvement in the hemodynamic conditions of the patients and also a reduction in the regurgitation volume in patients with mitral and aortic regurgitation. So I would suggest that Nifedipine might be useful for treating acute conditions related to this valve disease, but I would be very careful if using this drug in patients with mitral stenosis or aortic stenosis.

And also - if I may - I would be very careful if using Nifedipine in hypertrophic cardiomyopathy, contrary to what is being claimed at the present time.

LEACHMAN

A question for Dr Cocco.

All the early studies that I was aware of pointed to the inotropic effect of Amrinone as an explanation of its therapeutic benefit. Several colleagues of mine who have been associated with the group of St. Thomas Hospital in London, have come to the conclusion that perhaps the major therapeutic benefit is a vasodilating effect rather than an inotropic effect.

And I wonder whether you have any insight or information about this.

COCCO

Thank you for the opportunity of discussing this point. All the drugs we are studying, and Amrinone is no exception, they are mainly vasodilating drugs with some inotropic efficacy. Indeed, just pure inotropic drugs are non existent.

Now, double blind studies between Amrinone and β_2-agonists show the same hemodynamics results, which means it is hemodynamically a vasodilator with some inotropic effects as well.

QUESTION

I would like to ask Dr Cocco if he has experience with Ibopamina or if he knows something about it.

COCCO

No, sorry, I have no direct experience.

CAPONNETTO

With regard to Ibopamina, I would like to say I have some experience because I was once involved in a trial on the drug. It is a dopamine that can be administered orally. In acute experiments we saw - and not only ourselves, but also many others - an increase in cardiac output, a reduction in pulmonary wedged pressure, and a decrease in pulmonary resistance, when this drug was administered to patients with heart failure.

We decided that the resulting hemodynamic improvement in these patients also depended on the initial degree of heart failure.

This is the case as regards acute administration.

In chronic studies, in patients with heart failure, that do not respond to digitalis, and especially in congestive cardiomyopathy, we were able to observe that Ibopamina undoubtedly has a positive inotropic action, and in many cases even manages to solve the therapeutic problems of these heart failure patients.

GUAZZI

Dr Cocco, I put this question just for my personal clarifi-
cation. When you talk about the use of α-receptors in the treatment
of heart failure you refer to drugs specifically acting at the level
of the heart, drugs which also have some peripheral effect.

COCCO

All the drugs we got for experimentation, and I mentioned very
clearly that these are experimental drugs for clinical staging, are
specific. Up to now - to anticipate one second question you might
ask, and I think you would - they both, α_1 and α_2 are agonists, are
effective. So we think there must be a peripheral and myocardial
effect.

GUAZZI

So if I may continue, if it is so, how can we manage with the
peripheral effects of these drugs, not only the effect on arterioles,
but also referring to their effect on the renin system and the consequent vasoconstriction, because we know very well that an activation of α-receptors reduces the sodium content in the tubules and
activates the macula densa and activates the renin system and consequently vasoconstriction, so I think we are getting into a vicious
circle.

COCCO

You are right! It will be difficult to evaluate these drugs
clinically but I didn't mention that these drugs are also active as
positive inotropic drugs in the isolated heart, and that is the
point.

Indeed, I studied, together with Dr Opie in Cape Town, these
drugs in baboons. They had a lot of baboons there to kill when we
went there, and there are some 3000 animals going around and they are
quite dangerous being 95 to 100 pounds heavy and they can be very
nasty. Now, some of these animals get killed or captured. We tested
the heart of these animals which is quite like the heart of man, and
isolated the heart, cooled it and put it in a temperature of 20 to 25
degrees, which is something surgeons know. These are the data: it is
not reacting very well to beta agonist, it is not reacting very well
to glucagon, it is not reacting well to many things. When you use
α-agonist of an experimental type, and I told you no specific, and
probably there are no specifics, there is a dramatic increase up to
40% increase of contractile force. If you use it in anesthetized

animals, you get a tremendous increase in blood pressure, you get several complex reactions which make the interpretation of the isolated data quite difficult. But what struck us was that the heart rate was barely if at all increasing!

GAASCH

Dr Guazzi mentioned earlier that he had some questions in his mind about the use of Nifedipine and hypertrophic cardiomyopathies. I wonder if he could perhaps expand on that for a moment, Nifedipine as well as Verapamil. What is the mixture that makes the use of those agents in hypertrophic cardiomyopathies critical.

GUAZZI

The potent vasodilating effect of Nifedipine and also the reflex sympathetic activation on the heart induced by these drugs is not beneficial at all and the clinical harm supports this view: not beneficial at all in patients with hypertrophic cardiomoypathy with obstruction and also without obstruction. I believe that drugs like β-blockers, of course, and in the calcium channel blocking group, Verapamil, which has a negative inotropic effect and which does not induce reflex tachicardia is much better suited than Nifedipine for treatment of hypertrophic cardiomyopathy.

GAASCH

I was just going to comment on the use of various calcium blockers in hypertrophic cardiomyopathy.

I believe there is only one publication in the United States showing hemodynamic improvement in diastolic parameters, which is the Peter Bent Brigham Hospital study, and I don't think that any of us today are using Nifedipine as a primary agent of choice by any means in IHSS either obstructive or non obstructive.

However, we are using Verapamil, as the first choice of drug, except in patients who - as Dr Leachman referred to, have a very high pulmonary wedge pressure, and I think the data from the N.I.H., which have been published both in Circulation and in the American Medical Journal of Cardiology, clearly demonstrate that about 10% of these patients have a predominant vasodilator response to Verapamil, and in fact can worsen in response to it, or if they have very high wedge pressure, they may go into pulmonary edema when you give Verapamil to them. So with those two caveats we go ahead and use Verapamil. I found in my own practice that the major advantage of it over Propanolol is that if you do get some improvement in diastolic function, there are symptoms of shortness of breath that are considerably improved over what you see with Propanolol.

And I also think that if they tolerate some of the other side effects of Verapamil, the major one being constipation, they stay on the drug and are happier with it than with Propanolol. That doesn't mean to say that Propanolol is not a good drug for IHSS, in fact, it is quite good.

Then I would like to mention another agent that has recently been published as being effective, and that's Disopyramide or Norpace, which is a very potent negative inotropic agent. There is one publication from the United States showing a very remarkable improvement in the outflow gradient in patients with IHSS who are put on that agent.

LEACHMAN

The other part of the comment: you are really very exact according to our experience. We believed that Verapamil may be a better drug for people but we still have to emphasize that the patient is his best own control, and one of the reasons why they tolerate Verapamil better than Propanolol, particularly the large dose of Propanolol, is the effective change in cardiac output with exercise. As the patient taking Verapamil can still increase it with exercise whereas many of these people are so dependent on heart rate to increase cardiac output that if you block the response with Propanolol they feel terrible, and they depend on increasing AV oxygen difference as a compensatory mechanism for their trouble.

GAASCH

I think that's absolutely right. Verapamil has of course a negative chronotropic effect that can be adverse in the patients that have a relatively stiff ventricle. And just as you said, they do better, for example, at a heart rate of 80 than they do at 60 or 70.

In some patients who have not been tolerant to Propanolol, consequently, and where we wanted to use Verapamil, sometimes we put in AV sequential pacemaker, allowing maintenance of the atrial contribution to diastolic filling, controlling their rate and then gave them Verapamil and pehaps we found the best symptomatic response with that combination than almost any conservative therapeutic measures we have taken in IHSS.

RAINERI

This section is over with this final intervention.

I am sure that the aim of the course has been achieved. During this week we have had the opportunity to deepen our knowledge in a field of cardiology which is the most difficult one. We accepted the challenge and, in my opinion, it is possible to conclude saying that everybody has contributed to get the results obtained.

I should like to take this opportunity of thanking the co-directors of the Course who have played an important role in this meeting.

I should like also to thank all the members of the Faculty for finding time to come to Erice and giving us the contribution of their knowledge and experience. I think that the participants have to be thanked too, for showing interest in this Course.

LECTURERS AND INVITED SPEAKERS

P. Assennato
Cattedra di Fisiopatologia
  Cardiovascolare
Università di Palermo
Italy

L. Bonandi
Cattedra di Cardiologia
Università di Brescia
Italy

P. P. Campa
Cattedra di Cardiologia
Università de L'Aquila
Italy

S. Caponnetto
Cattedra Malattie dell'Apparato
  Cardiovascolare
Università di Genova
Italy

A. Cherchi
Cattedra di Cardiologia
Università di Cagliari
Italy

A. Chiddo
Cattedra Malattie Cardiovascolari
Università di Bari
Italy

G. Cocco
Dept. of Internal Medicine
Klinik Barmelweid
CH-5017 Barmelweid
Switzerland

H. Denolin
Hôspital Universitaire
  Saint-Pierre
Bruxelles
Belgium

S. Dalla Volta
Cattedra di Cardiologia
Università di Padova
Italy

J. S. Forrester
Cedars-Sinai Medical Center
University of California
USA

W. H. Gaasch
VA Medical Centre
Boston
USA

A. Galassi
Divisione di Cardiologia
Ospedale Garbaldi
Catania
Italy

E. Geraci
Divisione di Cardiologia
Ospedale Cervello
Palermo
Italy

M. Guazzi
Cattedra di Cardiologia
Università di Milano
Italy

J. J. Kellermann
The Chaim-Sheba Medical Center
Tel-Hashomer
Israel

R. D. Leachman
Texas Heart Institute
Houston
USA

M. Mariani
Cattedra di Cardiologia
Università di Pisa
Italy

G. Mattioli
Cattedra di Cardiologia
Università di Modena
Italy

M. Niederberger
Dept. of Cardiology
University Hospital
Vienna
Austria

K. L. Peterson
University Hospital
San Diego
USA

A. Raineri
Cattedra di Fisiopatologia
  Cardiovascolare
Università di Palermo
Italy

A. Reale
Cattedra di Cardiologia
Università di Roma
Italy

P. Rizzon
Cattedra Malattie Cardiovascolari
Università di Bari
Italy

M. Rothlin
University Hospital
Zurich
Switzerland

O. Visioli
Cattedra di Cardiologia
Università di Brescia
Italy

A. M. Weissler
Rose Medical Center
Denver
Colorado
USA

P. Zardini
Cattedra di Cardiologia
Università di Verona
Italy